Environmental Chemistry

D0082346

Many controversial environmental issues revolve around complex scientific arguments which can be better understood with at least a minimal knowledge of the chemical reactions and processes going on in the world around us. *Environmental Chemistry* offers an introduction to chemical principles and concepts and applies them to relevant environmental situations and issues.

Environmental Chemistry first considers some basic chemical concepts, including the structure of the atom, the elements, isotopes, radioactive decay, electronic configurations, chemical reactivity and bonding, the mole as a unit and chemical solution concentration and pH. It then examines such topics as:

- planet Earth and the origin of our environment – the formation of elements and Earth's atmosphere, hydrosphere and lithosphere;
- the Earth as a finite resource – renewable and non-renewable resources;
- risk and hazards – risk assessment and management and hazard identification;
- energy, entropy and rates of reaction – an introduction to chemical reactions occurring in the environment;
- an introduction to the lithosphere and its erosion and pollution;
- the chemistry of the atmosphere and its pollution;
- the properties of natural waters and their pollution;
- organic chemicals and their environmental effects;
- energy production.

Environmental Chemistry makes the subject accessible to those with little or no previous knowledge of chemistry. It is highly illustrated with global case studies, figures and tables and contains end of chapter summaries, discussion questions and annotated guides for further reading.

John Wright is Principal Lecturer and Head of Programme for Geography and Applied Environmental Science in the School of Education and Theology at York St John, College of the University of Leeds, UK.

Routledge Introductions to Environment Series
Published and Forthcoming Titles

Titles under Series Editors:
Rita Gardner and A.M. Mannion

Environmental Science texts

Atmospheric Processes and Systems
Natural Environmental Change
Biodiversity and Conservation
Ecosystems
Environmental Biology
Using Statistics to Understand the
   Environment
Coastal Systems
Environmental Physics
Environmental Chemistry

Titles under Series Editor:
David Pepper

Environment and Society texts

Environment and Philosophy
Environment and Social Theory
Energy, Society and Environment, 2nd edition
Environment and Tourism
Gender and Environment
Environment and Business
Environment and Politics, 2nd edition
Environment and Law
Environment and Society

Forthcoming:
Environmental Policy (July 2003)
Environmental Values (September 2003)
Representing the Environment (October 2003)
Environment and the City (January 2004)
Environment and Sustainable Development
   (December 2004)

Routledge Introductions to Environment

Environmental Chemistry

John Wright

Routledge
Taylor & Francis Group

LONDON AND NEW YORK

Routledge Introductions to Environment

First published 2003
by Routledge
11 New Fetter Lane, London EC4P 4EE

Simultaneously published in the USA and Canada
by Routledge
29 West 35th Street, New York, NY 10001

Routledge is an imprint of the Taylor & Francis Group

© 2003 John Wright

Typeset in Times and Franklin Gothic by
Florence Production Ltd, Stoodleigh, Devon
Printed and bound in Malta by
Gutenberg Press

British Library Cataloguing in Publication Data
A catalogue record for this book is available from the British
Library

Library of Congress Cataloging in Publication Data
Wright, John
Environmental chemistry / John Wright.
     p. cm. – (Routledge introductions to environment series)
   Includes bibliographical references and index.
   1. Environmental chemistry.   I. Title.   II. Series.
   TD193.W75 2003
   540–dc21   2002014941

ISBN 0–415–22600–7 (hbk)
ISBN 0–415–22601–5 (pbk)

This book is dedicated to my family,
Mary, Matthew and Beth, for all their patience and support,
and to my mother who was so ill during
its final stages of preparation.

Contents

Series editors' preface
Environmental Science titles

The last few years have witnessed tremendous changes in the syllabi of environmentally-related courses at Advanced Level and in tertiary education. Moreover, there have been major alterations in the way degree and diploma courses are organised in colleges and universities. Syllabus changes reflect the increasing interest in environmental issues, their significance in a political context and their increasing relevance in everyday life. Consequently, the 'environment' has become a focus not only in courses traditionally concerned with geography, environmental science and ecology but also in agriculture, economics, politics, law, sociology, chemistry, physics, biology and philosophy. Simultaneously, changes in course organisation have occurred in order to facilitate both generalisation and specialisation; increasing flexibility within and between institutions is encouraging diversification and especially the facilitation of teaching via modularisation. The latter involves the compartmentalisation of information which is presented in short, concentrated courses that, on the one hand, are self-contained but, on the other hand, are related to prerequisite parallel and/or advanced modules.

These innovations in curricula and their organisation have caused teachers, academics and publishers to reappraise the style and content of published works. Whilst many traditionally-styled texts dealing with a well-defined discipline, e.g. physical geography or ecology, remain apposite there is a mounting demand for short, concise and specifically-focused texts suitable for modular degree/diploma courses. In order to accommodate these needs Routledge have devised the Environment Series which comprises Environmental Science and Environmental Studies. The former broadly encompasses subject matter which pertains to the nature and operation of the environment and the latter concerns the human dimension as a dominant force within, and a recipient of, environmental processes and change. Although this distinction is made, it is purely arbitrary and is made for practical rather than theoretical purposes; it does not deny the holistic nature of the environment and its all-pervading significance. Indeed, every effort has been made by authors to refer to such interrelationships and to provide information to expedite further study.

This series is intended to fire the enthusiasm of students and their teachers/lecturers. Each text is well illustrated and numerous case studies are provided to underpin general theory. Further reading is also furnished to assist those who wish to reinforce and extend their studies. The authors, editors and publishers have made every effort to provide a series of exciting and innovative texts that will not only offer invaluable learning resources and supply a teaching manual but also act as a source of inspiration.

A.M. Mannion and Rita Gardner
1997

Series International Advisory Board

Australasia: Dr P. Curson and Dr P. Mitchell, Macquarie University

North America: Professor L. Lewis, Clark University; Professor L. Rubinoff, Trent University

Europe: Professor P. Glasbergen, University of Utrecht; Professor von Dam-Mieras, Open University, The Netherlands

Note on the text

Bold is used in the text to denote words defined in the Glossary. It is also used to denote key terms.

Preface

Many students who have a keen interest in the environment and want to study it do not always have the same degree of interest in chemistry. They do, though, need to be aware that answers to a range of environmental questions cannot be provided unless some key areas in basic chemistry are understood. This book tries to link the learning of this basic chemistry to its application in the explanation of, and the solving of, environmental problems and tries to bridge the gap between the less and more advanced books on environmental chemistry. It should be understandable by readers who have a basic knowledge of chemistry. The chemistry underpins many examples of environmental problems and concerns such as radon in the environment, the erosion of the stonework of York Minster, nuclear accidents, the asbestos time-bomb, Itai–Itai disease and so on. Some chapters have been included, such as risk assessment and the origin and development of Earth, which are not normally found in environmental chemistry textbooks.

This book should be useful to A-level students, first-year undergraduates or to anyone else who has a limited background in chemistry. It starts at GCSE chemistry or combined science level, and quickly proceeds to about A-level standard and beyond in some areas of chemistry. Students who have to study environmental chemistry as part of some qualification in environmental science/studies/management should also find the contents useful.

Acknowledgements

I would like to thank all of the following for their help in producing this book:

- My wife Mary and son Matthew for taking nearly all of the photographs, and daughter Beth for checking some of the layout.
- Rob Gendler, a US physician and astronomer, for permission to reproduce Figure 4.1.
- Mr S. Mills, Superintendent of York Minster Stone Works, for permission to reproduce Figure 1.1.
- The US Geological Survey for permission to use their data.
- The National Atmospheric Emissions Inventory.
- NASA for permission to reproduce Figure 11.1.
- The editors and referees for their very useful comments and suggestions.

1 Some basic chemical concepts

- ◉ **Phases of matter and their interrelationships**
- ◉ **The structure of the atom and the properties of the proton, neutron and electron**
- ◉ **The stability of the nucleus, radioactivity and the properties of emitted particles**
- ◉ **The mole and its use**
- ◉ **Electronic structures of atoms**
- ◉ **Structure of the Periodic Table**
- ◉ **Elementary chemical bonding theory**
- ◉ **Water as a solvent, and Lowry–Brønsted acids and bases**
- ◉ **Oxidation and reduction. The use of oxidation numbers**

The erosion of York Minster, UK

York Minster is over 500 years old and requires very expensive maintenance to ensure its existence for future generations. It is composed of a wide variety of materials, which have been subjected to erosion. The external fabric of stone has been eroded both by natural weathering and by chemicals put into the atmosphere by man's actions (Figure 1.1). Winning the battle between the Minster and its environment lies not only in reducing atmospheric pollution but also in the replacement of damaged stone. This replacement is not a simple matter because, although there is fresh stone in abundance, the continual replacement of stone can undermine the integrity of a building by causing the structure to become unstable. At York Minster the stonemasons try, as far as possible, to incorporate all old stone into any repair work because there is the danger that if too much fabric is replaced there could be a serious loss of authenticity (Brimblecombe and Bowler 1990).

What chemicals then have caused the erosion of the stone?

The pollutants sulphur dioxide (SO_2), nitrogen oxides (NO_x) and ozone (O_3) are believed to be the main culprits. Although York Minster has suffered from anthropogenic effects, there is no method by which these effects can be readily separated from other contributions such as poor construction techniques, wrongly chosen materials, natural weathering and biological attack. The rate of destruction can be significantly increased during heavy rainfall if vast quantities of water containing pollutants percolate through the stone carrying away any reaction products in the run off (Baer and Snethlage 1997). The reaction products are mainly salts which are formed by the reaction of negatively charged anions (e.g. SO_4^{2-}) from pollutant gases or acids, with positively charged cations in the stone (e.g. Ca^{2+}).

Figure 1.1 Stone erosion York Minster 2000. *(above)* An eroded vestibule buttress on the Chapter House (north side of York Minster). *(below)* The same buttress after restoration.

Source: Reproduced with permission from Mr S. Mills.

York Minster is made of mainly two types of stone: a crystalline, granular dolomite, $MgCO_3.CaCO_3$, and a more porous, granular oolithic limestone made largely of calcite, $CaCO_3$. As construction materials, the dolomite stone is much more resistant to erosion than the limestone (Mills 2000), lasting up to four times longer.

In the presence of water, sulphur dioxide reacts with dolomite limestone as follows,

$$CaCO_3.MgCO_3(s) + 2SO_2(g) + O_2(g) \xrightarrow{\;H_2O\;} CaSO_4.2H_2O(s) +$$
$$MgSO_4.7H_2O(s) + CO_2(g)$$

Dry oolithic limestone reacts with sulphur dioxide and oxygen thus,

$$CaCO_3(s) + SO_2(g) + 1/2\,O_2(g) \longrightarrow CaSO_4(s) + CO_2(g)$$

Again, the presence of water leads to the formation of hydrated calcium sulphate.

The calcium sulphate (gypsum), produced in the above reactions is much more soluble in water than the stone from which it is derived and consequently more is lost as a result of solution in rainwater. Solid crystalline gypsum also has a more open structure, which leads to an increase in volume of about 100 per cent. The result is an increase in internal pressure that causes cracks to develop together with crumbling and bursting of the stone.

Whilst it is true that sulphur dioxide emissions in York have greatly declined since the 1950s, the stone decay continues. Why this is so remains unclear. It may be linked with ozone and nitrogen oxide concentrations at ground level caused by emissions from vehicles (Haneef *et al.* 1990). There is, for example, strong evidence to suggest a synegetic effect between sulphur dioxide and nitrogen oxides, which enhances stone corrosion (Haneef *et al.* 1990). Here sulphuric acid together with nitrogen oxide (nitrogen monoxide) are formed,

$$SO_2(g) + NO_2(g) + H_2O(l) = H_2SO_4(aq) + NO(g)$$

Nitrogen oxides (NO_x) are known to form nitric acid with water which will react with limestone to form the much more soluble salt calcium nitrate, $Ca(NO_3)_2$. However, studies made on York Minster have found no evidence that NO_x gases have had any direct effects on limestone decay (Cook and Gibbs 1996).

The effects of atmospheric pollutants on limestone decay are complex and there are a number of uncertainties concerning the reliance of one chemical on the presence of another in order for enhanced corrosion to take place. What is certain is that, since the Industrial Revolution and the corresponding increase in atmospheric pollution, the rate at which York Minster stone has eroded has substantially increased.

What is environmental chemistry?

It is clear from the opening section that chemistry is a discipline much involved in the study of human interaction with the environment. Chemistry is the study of the composition, structure and properties of materials and how they undergo chemical and physical changes. Environmental chemistry is the study of those changes that have had an effect on both living organisms and non-living matter in the environment.

Chemicals have a poor reputation! Some are known to be a source of pollution and many are hazardous if used incorrectly. However, it is important to realise that all forms of matter in our environment whether synthetic or natural are made of chemicals. Many materials in common use such as paper, cloth, plastics, metals, etc. have undergone some form of chemical treatment and change during their manufacture, and will probably undergo more change before they become waste.

There are many ways in which humans and other living organisms are exposed to chemicals such as detergents, paints, drugs, exhaust fumes, industrial effluents, pesticides, natural toxins in plants and animals, etc. in their everyday existence. When chemicals are a main source of pollution, then that pollution is usually caused by human error, lack of understanding and knowledge, greed, or by inefficient technology. Chemicals may well be the cause of a number of environmental problems but it is also the use of chemicals that often provides the answers to those problems. Many chemicals are dangerous but many are also beneficial! There is no doubt that, without chemicals and the chemical industry, human life would be far less enjoyable.

A revision of the elementary classification of matter

Matter can be classified by the state it is normally found in, i.e. as a solid, liquid or gas. These states are called the phases of matter.

The connections between the three main phases of matter can be established by examination of what happens to water when it is cooled and heated. At a temperature of $-10\,°C$ (Celsius) and 1 atmosphere pressure, pure water exists as ice (Figure 1.2, point A). If it is heated to a temperature of $0\,°C$ (A to B), at 1 atmosphere pressure, ice will start to melt and become liquid water. Its temperature will remain constant at $0\,°C$ until all the ice has melted (B to C). Thus water has a melting point of $0\,°C$ at 1 atmosphere pressure. If the pressure is kept constant and the heating continued until the

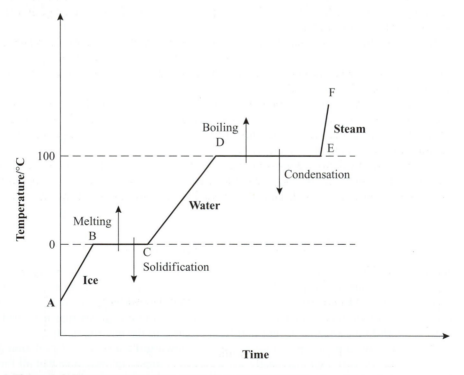

Figure 1.2 Temperature vs time graph for water heated at a constant rate.

temperature reaches 100 °C (C to D), the liquid water will start to boil and invisible gaseous water or steam is formed. Again, the temperature will remain constant until all of the liquid water has been turned into a gas (D to E). Water has a boiling point of 100 °C at 1 atmosphere pressure. Continued heating will only make the steam hotter (E to F).

If steam at 200 °C and 1 atmosphere pressure is cooled down (F to E), it will start to form a liquid at 100 °C. Its temperature will remain the same until all the steam has liquefied or condensed (E to D), and then it will cool down further (D to C) until solid ice starts to form. This will occur at 0 °C, the solidification or freezing point. The temperature of the liquid/solid mixture will remain at 0 °C until all of the water has solidified/undergone freezing (C to B). Cooling down to a temperature below 0 °C then involves no further phase change (B to A).

Evaporation is different from boiling. When liquid water is placed in an open container at room temperature, evaporation will occur from its surface until there is none left. The liquid changes into a gas which, because it is formed below the boiling point, is known as a vapour. When water is warmed, the rate of evaporation from its surface is increased. At its boiling point, liquid water is turned into bubbles of gaseous water inside liquid and not just at its surface. Water vapour therefore exists over liquid water at all times up to and including its boiling point. It is identical to steam except it is much cooler and, like steam, is invisible.

Sublimation occurs when matter changes from solid to gas or vice versa without an intermediate liquid phase being formed. For example, when solid iodine is gently warmed, it will change to a gas without the intermediate liquid phase being observed, gaseous iodine will also condense back to the solid phase without the liquid phase again being observed. Water under normal environmental conditions does not sublime.

Figure 1.3 shows the connections between the three main phases. Changes brought about by cooling and heating which cause phase changes without a change in composition are **physical changes**.

A piece of pure copper has a uniform composition and is therefore homogeneous. The smallest particle that is still identifiable as being copper is an **atom** of copper. Atoms cannot be sub-divided further by chemical means and so they are seen as the smallest building blocks of all materials.

During chemical changes, atoms are rearranged and recombined with each other to form different materials. Groups of atoms joined together by some form of chemical bonding are called **molecules**, e.g. dioxygen is composed of two oxygen atoms joined together to form the dioxygen molecule, O_2.

During a chemical change the total mass of the matter before reaction is the same as the total mass after the reaction is complete. Hence, none of the atoms taking part are destroyed or new ones created. This is expressed as the **Law of Conservation of Matter**,

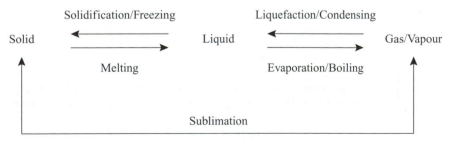

Figure 1.3 The phases of matter and their interrelationships.

i.e. during a chemical reaction matter is neither created nor destroyed. This law is the basis on which chemical equations are used, and calculations made concerning the amounts of matter involved as the starting materials (the reactants) and those which are the result of reaction (the products).

A substance like copper, which cannot be broken down by chemical means into anything simpler than itself, is called an **element**. Each element has a set of properties that fingerprint that element, such as its melting point, density, electrical conductivity, etc. The combination of two or more of these elements by chemical bonds leads to a wide variety of new substances called **compounds**, each with their own peculiar properties. For example, sodium, Na, is a metallic element that reacts vigorously with water to give another element hydrogen, H, as one of its products. Chlorine, Cl, is another element and is a toxic, choking gas. Individually, sodium and chlorine are dangerous chemical elements. When they combine, they form the compound sodium chloride, NaCl, which is used as common table salt. If a new substance is produced when elements react with each other, then a **chemical change** has occurred. A compound is composed of elements joined together in definite proportions. Thus formulae such as SO_2 for a molecule of sulphur dioxide, $CaSO_4$ for calcium sulphate and H_2O for water, are written to show the ratio of the atoms present in a compound. A pure compound like a pure element will be homogeneous in composition.

When two or more pure substances are mixed together, then a heterogeneous mixture usually results. The components of a mixture are not chemically combined, can be relatively easy to separate and retain their own individual properties. Mixtures are common, e.g. the atmosphere is a mixture of elements and compounds and seawater compounds mixed with compounds. The stone of York Minster, although composed of mainly one compound, is a mixture of several compounds. Matter is thus classified into elements, compounds and mixtures.

Matter can also be separated into **metals**, **semi-metals** and **non-metals**, which can be elements, mixtures or compounds. A metal can be defined as a material that conducts electricity well, e.g. copper, silver, iron, mercury, steel. A non-metal is a poor conductor (insulator) of electricity, e.g. oxygen, sulphur, iodine, calcium carbonate, whilst a semi-metal has intermediate electrical conductivity, e.g. silicon, germanium. Metals also conduct heat well, and usually have high melting points, high boiling points and high densities. They are malleable, ductile and, when freshly cut, show lustre.

The elementary structure of the atom

An atom can be viewed as being a sphere consisting mainly of empty space. A typical **atomic radius** is about 3.8×10^{-10} m, which corresponds to a volume of 2.3×10^{-28} m^3. An atom carries no overall electric charge and is therefore **electrically neutral**.

The **mass** of an atom is also unimaginably small, e.g. 3.8×10^{-26} kg for a sodium atom. Experiments show that the mass of an atom is concentrated in a central region, the **nucleus**. The nucleus of an atom has a typical radius of 6.8×10^{-15} m and hence a volume of 1.3×10^{-42} m^3. The nucleus is thus very small in volume compared with the total volume of the atom by a factor of about 10^{14}.

For most chemical purposes, the nucleus can be considered as being composed of two **fundamental particles** or **nucleons**, the **neutron (n)** and the **proton (p)**. Their charges and masses are listed in Table 1.1. The nucleus of an atom carries a number of positive charges equal to the number of its protons. The particle that ensures the electrical neutrality of an atom is the **electron (e)** which 'orbits' the nucleus, i.e. electrons are extra-nuclear particles. The electron has a charge equal in magnitude but opposite

Table 1.1 *The rest masses and electrical charges of the neutron, proton and electron*

Nucleon	Charge/ °C	Rest mass/ kg
Neutron	0	1.67493×10^{-27}
Proton	$+1.60210 \times 10^{-19}$	1.67262×10^{-27}
Electron	-1.60210×10^{-19}	9.10939×10^{-31}

in sign to that of the proton. The mass and electrical charge on the electron are also given in Table 1.1.

The masses of the neutron, proton and electron and their charges are cumbersome to use in many situations, so each mass is divided by the mass of the proton and each charge by the charge on the proton. This removes the need to know either the magnitudes or the units of the mass or charge of these particles and thus the **relative charges and relative masses** can be used, as in Table 1.2. Since the mass of the electron is so small compared to the nucleons, its mass can be taken to be zero for most chemical purposes.

A more accurate definition of an element is a substance made up of atoms which contain the same number of protons in their nuclei. For example, copper has 29 protons therefore any atom containing 29 protons is an atom of copper and nothing else. There are some 92 naturally occurring elements.

The number of protons in a particular atomic nucleus is the **atomic number (Z)** of that atom. For example, if $Z = 92$ for the element uranium (U), then it has 92 protons in its nucleus. An atom of uranium will also contain 92 electrons.

The total number of protons and neutrons is called the **mass number (A)** of the element. A particular atom of uranium has a mass number of 238. Hence, the total number of protons plus neutrons is 238. If the atom has 92 protons, then it must also contain 146 neutrons.

Hydrogen (H) is the simplest of all the elements with an atomic number, $Z = 1$, and a mass number, $A = 1$. Thus it has one proton in its nucleus, and one extra-nuclear electron. This atom is represented by,

$$^{1}_{1}\text{H}$$

where the superscript is the mass number and the subscript the atomic number. The uranium atom described earlier would be represented by,

$$^{\text{mass number } 238}_{\text{atomic number } 92}\text{U}$$

Two other types of atoms of hydrogen exist, one which contains one proton and one neutron, and the other one proton and two neutrons. The former is called **deuterium** and the latter **tritium**. Both of these atoms differ from ordinary hydrogen in having neutrons in their nuclei. Atoms of the same element that have different numbers of neutrons are called **isotopes**. Hydrogen thus has three naturally occurring isotopes (Table 1.3).

Table 1.2 *Relative masses and relative charges of the electron, neutron and proton*

	Relative charge	Relative mass
Proton	1	1
Neutron	0	1
Electron	−1	1/1,840

If an atom loses or gains an electron, it acquires a net electric charge. For example, the neutral oxygen atom, O, has eight protons and eight electrons. If it acquires two extra electrons, it is no longer a neutral atom but a negatively charged **ion** O^{2-}. A negative

Table 1.3 *Isotopes of hydrogen*

Hydrogen	^1_1H
Deuterium	^2_1H
Tritium	^3_1H

ion is known as an **anion**. The calcium atom, Ca, may lose two electrons. This would leave two positive charges on the atom because two protons in its nucleus would no longer be counter-balanced by two electrons. A Ca^{2+} ion would be formed. A positively charged ion is called a **cation**. Why different elements form different ions is explained on pp. 21–2.

Radioactivity

Radioactivity is the spontaneous disintegration of an energetically unstable atomic nucleus. It is characterised by the emission of various types of particles and electromagnetic radiation (Box 1.1). Chemical or physical changes do not have any effect on the type or amount of emissions from a radioactive material. The rate of disintegration depends upon the element that is present. Nuclear disintegration causes the formation of new elements. There are three natural radioactive decay series, which result in a number of radioactive materials occurring in the environment. The one for uranium-238 is shown in Figure 1.4. The overall risk associated with these series is that posed by the parent radioactive element plus its daughter elements. In particular, the uranium-238 series produces the only known natural radioactive atmospheric pollutant, radon gas.

The stability of a nucleus depends on its neutron to proton ratio, the most energetically stable nuclei having a ratio of 1:1. Figure 1.5 shows the neutron to proton ratio of all naturally occurring elements plotted against their atomic numbers. A nucleus may tend to break down in order to establish this ratio. This can occur in a number of ways.

If a nucleus has a high neutron to proton ratio, i.e. too many neutrons, then the number of neutrons can decrease via a neutron changing to a proton, with the emission of a **beta** particle,

$$^1_0n \longrightarrow ^1_1p + ^{\ 0}_{-1}\beta$$

e.g. $^{15}_6C \longrightarrow ^{15}_7N + ^{\ 0}_{-1}e$ (β-ray) $+ ^0_0\gamma$ (gamma ray)

If a nucleus has a low neutron to proton ratio, i.e. too many protons, then a proton can change to a neutron with the consequent emission of a **positron**,

$$^1_1p \longrightarrow ^1_0n + ^{\ 0}_{+1}\beta \text{ (positron)}$$

e.g. $^{11}_6C \longrightarrow ^{11}_5B + ^{\ 0}_{+1}\beta$

A nucleus with a large excess of neutrons results in **alpha** particle decay,

e.g. $^{238}_{92}U \longrightarrow ^{234}_{90}Th + ^4_2\alpha$ (alpha particle) $+ ^0_0\gamma$

Table 1.4 *The half-lives of some isotopes*

Isotope	Half-life
^8_4Be	2×10^{-16} second
$^{35}_{16}S$	88 days
$^{14}_6C$	5,730 years
$^{238}_{92}U$	4.51×10^9 years

The stability of a radioactive isotope is reflected in its half-life, $t_{1/2}$. This is the time taken for the number of atoms present at a particular time to decay to half that number. Depending upon the atom in question this can range from a fraction of a second to many thousands of years (Table 1.4).

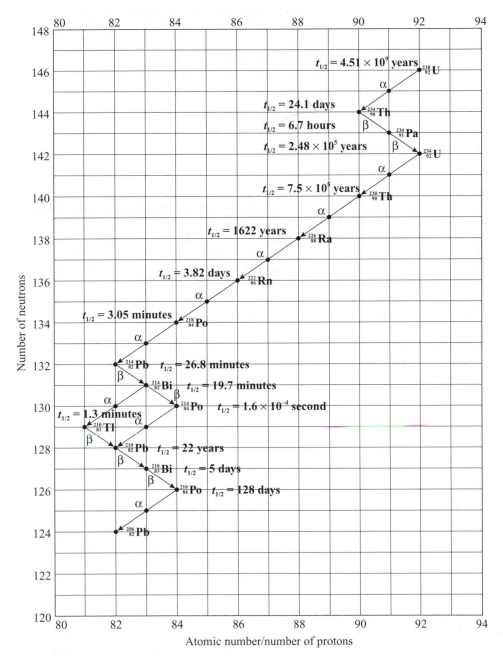

Figure 1.4 The natural decay series for U-238. (*Note*: The half-life of the parent and daughter products are listed.)

Box 1.1

The properties of particles emitted during radioactive decay

Alpha particles

The alpha particle or α-particle is composed of two neutrons and two protons. It therefore carries a double positive charge. This particle is represented by either of the following symbols since it is effectively a helium nucleus,

$$_2^4 \text{He or } _2^4 \alpha$$

The alpha particle has the following properties:

- the least penetrating of radioactive emissions
- a high velocity
- most strongly ionising when it interacts with matter, and therefore the most biologically damaging
- deviated by magnetic and electric fields
- stopped by human skin or a sheet of paper.

Alpha particles are also produced when helium atoms, $_2^4\text{He}$, are ionised,

$$\text{i.e. He} -2e^{-1} \longrightarrow \text{He}^{2+}$$

Beta particles

A beta particle or β-particle is a very fast electron emitted from the atomic nucleus. It is represented by,

$$_{-1}^0 \beta \text{ or } _{-1}^0 e$$

Beta particles have the following properties:

- more penetrating than alpha rays
- less ionising than alpha rays
- very fast, velocities are about half the speed of light
- markedly deviated by electric and magnetic fields
- stopped by a few millimetres of metal.

Positrons

A positron is similar to the beta particle but carries a single positive electric charge. The mass of the positron is taken to be zero in the same way that the electron's mass is zero. The symbol for the positron is,

$$_{+1}^0 \beta$$

Gamma rays

Gamma rays are produced by an energetically excited nucleus. These rays are represented by the symbol,

$$_0^0 \gamma$$

Gamma rays have the following properties:

- they are electromagnetic radiation and therefore travel at the speed of light
- not deflected by electric and magnetic fields
- stopped by centimetres of lead, i.e. are the most penetrating
- less ionising than any of the other radiations.

(*Note*: Particles such as neutrinos and antineutrinos are also emitted during a nuclear disintegration. These two particles have neither mass nor carry an electrical charge.)

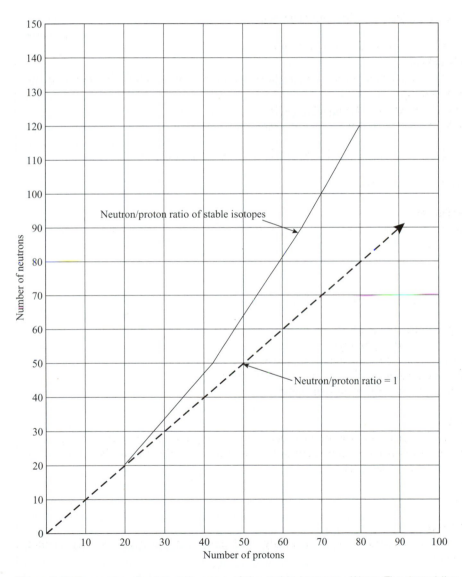

Figure 1.5 The neutron/proton ratio curve of the stable isotopes. (*Note*: The dotted line shows the position the isotopes would take if n/p = 1.)

Radon in the environment

Radon as a harmful natural pollutant in the UK was extensively studied in the 1980s. It was established that radon and its daughter products were a cause of lung cancer in human beings. Thus, in 1990, the National Radiation Protection Board gave advice on the level of radiation at which action was to be taken against radon (NRPB 1990). By 1996, some 250,000 homes in England had been assessed for the level of radon concentrations. Several areas were identified that may require remedial action: Devon, Cornwall, Northants, Somerset, Dorset, Lincolnshire, Oxfordshire and Shropshire. Parts of Wales and Scotland also became designated areas. Work carried out in the UK on the effects of radon has been further pursued in other countries such as Switzerland, Norway and the US.

In 1998 an American report (The National Academy of Sciences 1998) concluded that radon gas in homes is a cause of lung cancer in the general population. The number of cases of cancer, based on studies of miners other than coal miners, has been predicted to be between 3,000 and 32,000 per year. Such numbers indicate that there is a health risk to the general public, and that radon gas is second only to cigarette smoking in causing lung cancer. The report examined evidence for a link between lung cancer and people who both smoked and were exposed to radon. Although not conclusive, the report states that it is likely that most of the radon-related deaths amongst smokers would not have occurred if the victims had not smoked. Some kind of synergic mechanism may be at work. Simple protective measures such as sealing the floor, construction joints and cracks in walls, using appropriate ventilation and extractor fans, and the use of wallpaper, though not excluding radon altogether, would considerably reduce deaths caused by lung cancer.

Table 1.5 *The isotopes of radon*

Isotope	Half-life
$^{222}_{86}$Rn	3.82 days
$^{220}_{86}$Rn	55.6 seconds
$^{219}_{86}$Rn	3.96 seconds

Radon, one of the noble gases, is a naturally occurring radioactive gas. It is odourless, colourless and tasteless. The three isotopes of radon that exist in the environment are shown in Table 1.5. Since their half-lives are so short, then the only way that these isotopes can exist in the environment is if they are being continuously formed. The most abundant isotope, radon-222, is a product of the uranium-238 decay series, whereas radon-220 and radon-219 are products of the thorium-232 and uranium-235 decay series, respectively.

Because of its longer half-life and thus relative greater stability, radon-222 is responsible for the vast majority of the annual radon radiation dose received by people. Indeed, over 50 per cent of the total radiation dose from natural sources is provided by exposure to radon-222 and radon-220.

Radon exists in the atmosphere, in the soil and in water. The main problem arises from the inhalation of the radioactive gas and its solid radioactive daughter products such as polonium-218, polonium-214, lead-214 and bismuth-214. These solids can become attached to aerosol-sized particles in the air and can, together with radon gas, be inhaled. Unfortunately, radon-222, polonium-218 and poloniuim-214 all emit α-particles as they decay. It has been well established that these particles are the most biologically damaging and increase the likelihood of cellular damage, which gives rise to cancers, genetic damage and accelerated ageing. Normally, radon concentrations are not at a level that can cause such effects. Those areas in the UK, and elsewhere in the world, which have excessively high concentrations of radon gas are associated with the rock-forming granites that naturally contain relatively high levels of uranium, some ironstone deposits and some sedimentary rock deposits. Indeed, minute quantities of uranium are found in all earth and building materials so radon is found inside all buildings. Normally, radon is not a problem in most houses and workplaces. Unfortunately, it is possible to be living in a house that has a high radon concentration in its atmosphere because of the type of rock structure it has been built over, particularly if that rock is porous or is heavily fractured. The atmospheric pressure inside a building is slightly lower than that outside because of higher internal temperatures and the effects of wind. Radon can collect in the basement of houses and underneath floorboards from where it can, because of the small difference in pressure, penetrate the main body of the house. Outside, radon is readily dispersed in the atmosphere where it ceases to be a problem.

Relative atomic mass and relative molecular mass

Relative atomic mass (A_r), was formerly known as 'atomic weight'. This latter name is still in use but should be avoided. The mass of the carbon isotope $^{12}_6C$ atom determined by experiment is 1.99×10^{-26} kg. This isotope contains 12 nuclear particles (6 protons and 6 neutrons). Since the nucleus is where the mass of the atom is centred, it has been decided that the mass of the $^{12}_6C$ atom represents the mass of 12 units of atomic mass. Using $^{12}_6C$ as our standard mass, the mass of one unit of atomic mass can be determined by dividing the atomic mass of the atom by 12,

$$\frac{\text{atomic mass of } ^{12}_6C}{12} = \frac{1.99 \times 10^{-26}}{12} \text{ kg} = 1.66 \times 10^{-27} \text{ kg} = 1.66 \times 10^{-24} \text{ g}$$

Thus the mass of an atom of $^{12}_6C$ has been used to define the **unit of atomic mass**.

The mass of a sodium atom is 3.815×10^{-26} kg. It is now possible to determine how many atomic mass units (based on the $^{12}_6C$ calculation) this mass contains, i.e.

$$\frac{\text{atomic mass of the sodium atom}}{\text{the unit of atomic mass}} = \frac{3.815 \times 10^{-26} \text{ kg}}{1.66 \times 10^{-27} \text{ kg}} = 22.98$$

This is the same as:

$$\frac{\text{atomic mass of sodium atom}}{(\text{atomic mass of } ^{12}_6C \text{ atom})/12} = \frac{\text{atomic mass of Na atom} \times 12}{\text{atomic mass of } ^{12}_6C \text{ atom}} = 22.98$$

or the sodium atom has a mass of '22.98 times that of the unit of atomic mass'.

The atomic mass of the sodium atom has been expressed **relative** to the *unit* of atomic mass based on the mass of the $^{12}_6C$ atom.

Thus, the **relative atomic mass** of any element is given by,

$$A_r = \frac{\text{atomic mass of that element} \times 12.00}{\text{the atomic mass of } ^{12}_6C \text{ atom}}$$

The relative atomic mass of any atom is **unitless**!

The relative atomic mass gives an indication of the *total* number of neutrons and protons there are in a particular atom, e.g. relative atomic mass of sodium atom =

22.98 = 23. Hence, the mass number of sodium is equal to the relative atomic mass rounded up to the nearest integer. Relative atomic masses are much simpler numerical values than actual atomic masses to handle in calculations. The relative atomic mass of a particular isotope of an element is some multiple of the atomic mass unit. However, when tables of the relative atomic masses of the elements are consulted, it will be noticed that some are not close to an integer value. What is being quoted is a weighted average value for the naturally occurring atoms of the elements that take into account the presence of isotopes. The further away the relative atomic mass is from an integer indicates that the element is a mixture of its isotopes and is not composed of a single type of atom. For example, in a sample of natural chlorine atoms, 75.8 per cent of the atoms are $^{35}_{17}Cl$ and 24.2 per cent $^{37}_{17}Cl$. The relative atomic mass of $^{35}_{17}Cl$ is 34.97 and that of $^{37}_{17}Cl$ 36.97, both very close to integer values.

The *average* relative atomic mass is thus,

$$\left(\frac{75.8}{100} \times 34.97\right) + \left(\frac{24.2}{100} \times 36.97\right) = 35.5$$

the figure normally found in tabulated data.

Relative molecular mass (M_r), formerly known as 'molecular weight', is the sum of all the relative atomic masses of each of the atoms that make up a molecule or an ion. Using the relative atomic masses of 22.9898 for sodium atoms, 32.064 for sulphur atoms and 15.9994 for oxygen atoms, the relative molecular mass of sodium sulphate, Na_2SO_4, is given by $(2 \times 22.9898) + (1 \times 32.064) + (4 \times 15.9994) = 142.0412$. Notice again the answer has no units.

The extra-nuclear electrons

If gaseous dihydrogen (H_2) at low pressure is subjected to electrical discharge, the molecules are broken down into energetically excited atoms (H). These atoms emit visible radiation which, when passed through a prism, show a discrete set of wavelengths (Figure 1.6) – an atomic spectrum is observed.

The emitted radiation is not a continuous spectrum as in the case of visible light from the Sun. When gaseous hydrogen atoms emit (or absorb) radiation, they are able to do so only at certain fixed, or 'allowed', wavelengths. These wavelengths are characteristic of the element hydrogen, irrespective of its source. Such discrete wavelengths indicate

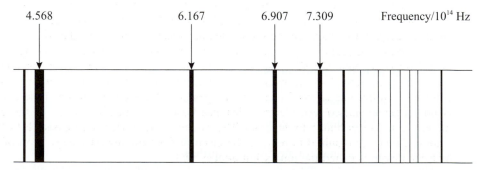

Figure 1.6 Emission spectrum of hydrogen in the visible and near ultraviolet regions.

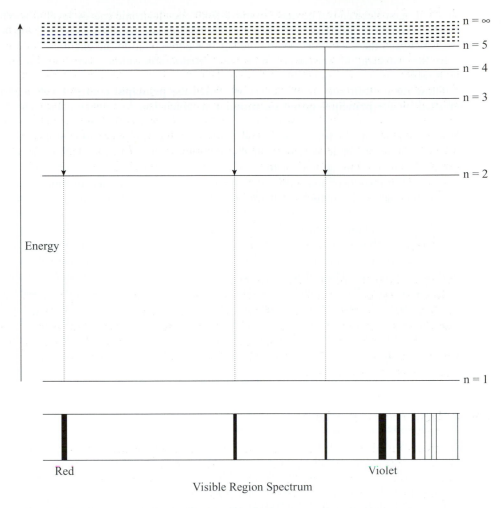

Figure 1.7 Energy level diagram for atomic hydrogen.

that the amounts of energy represented by them are themselves discrete in nature. Energy that is discrete is said to be 'quantised'. The existence of these wavelengths is interpreted as being due to the movement of the hydrogen atom's single extra-nuclear electron between different allowed main electronic energy levels (Figure 1.7). These observations are also found to be true of all of the other elements. However, the more complex the atom is, then the more complex the spectrum that is produced.

The atomic spectrum of hydrogen also shows that the main electronic energy levels of a hydrogen atom are not equally spaced and get closer and closer together at higher energies until they eventually converge. The point at which the energy levels converge and become a broad band of energies is called the **continuum**. Here energy is no longer quantised. An electron that has been energetically excited to the continuum is no longer bound to the atom. The hydrogen atom becomes ionised (H^+). The amount of energy required to remove the electron from the lowest energy level of the hydrogen atom is called the **ionisation energy (E_i)**,

$$H(g) \longrightarrow H^+(g) + e^-(g)$$

The hydrogen electron can be made to move from one electronic energy level to another by the emission or absorption of energy of the appropriate wavelength. It is the *differences* between these electronic energy levels that correspond to the energies associated with the wavelengths measured from the emission/absorption atomic spectrum of hydrogen.

These main electronic energy levels are called the **principal energy levels** and are labelled by the **principal quantum number n** which has an integer value. The first seven principal energy levels are labelled 1, 2, 3, 4, 5, 6 and 7. The first main energy level is thus the n = 1 level, the second the n = 2 level and so on. In the case of the hydrogen atom, its single electron normally occupies the n = 1 level, which is the lowest energy level called the ground state.

The determination of ionisation energies of successive electrons in more complex atoms, together with photoelectron spectroscopy, confirm that electrons are contained in main energy levels. They also show that, with the exception of the first, these main energy levels contain sub-energy levels, i.e. shells of electronic energies exist containing sub-shells. As above, each main shell/energy level is labelled with an n value. However, the sub-levels are identified by a **second quantum number *l***, which takes the values from zero to (n−1). Thus, if n = 1, then $l = (1−1) = 0$; if n = 2, then $l = 0$ and 1; if n = 3, then $l = 0$, $l = 1$ and $l = 2$. The sub-energy level with the lowest energy, i.e. $l = 0$ is symbolised by **s**, and successive higher ones by **p** ($l = 1$), **d** ($l = 2$) and **f** ($l = 3$). These letters indicate the existence of different **types** of sub-energy levels. So, for the first main energy shell there is only one type of energy level, and this is depicted by the letter **s**. Hence, the first main energy shell is fully labelled as **1s.**

Each shell has a maximum limit to the number of electrons it can accommodate. The first shell is complete with two electrons, the second with eight, with successive shells holding larger numbers.

In the second main energy shell, two sub-levels exist. The first is labelled, as before, the **s sub-shell** and the second (of higher energy) the **p sub-shell**. Hence, if the main energy level is depicted by n = 2, then the two sub-energy levels are depicted as **2s** and **2p**. Again, only two electrons can be placed in the 2s sub-shell. In the p sub-shell a maximum of six electrons can be accommodated, thus making a total of eight electrons that can be placed in the second principal energy level.

In the third principal energy level, i.e. n = 3, there are three sub-energy levels. The first is labelled the **3s**, the second the **3p** and the third the **3d**. As before, a maximum of two electrons is associated with the 3s, six with the 3p but ten electrons can be accommodated in the 3d. This would enable a maximum of 18 electrons to be accommodated in the third principal energy level.

In the fourth energy level, i.e. n = 4, there are four sub-shells. The first is labelled **4s**, the second **4p**, the third **4d** and the fourth **4f**. Again, two electrons can be placed in the 4s, six in the 4p, ten in the 4d and fourteen in the 4f. Hence a maximum of 32 electrons can be placed in the fourth principal energy level.

There are further main or principal energy levels/electron shells to be considered but, for the needs of environmental chemistry, the description can be terminated at the fourth one.

A beam of hydrogen atoms, when subjected to a non-uniform magnetic field, is split into two deflected beams. This suggests that the atom behaves like a magnet caused by the electron in the hydrogen atom being able to spin in one of two directions, clockwise and anti-clockwise. These directions are indicated by a third quantum number called the **magnetic spin quantum number**. These numbers are labelled $m_s = +1/2$ and $m_s = −1/2$. Thus, although like electrical charges repel, the first energy shell (1s) will accommodate a maximum of two electrons with opposite spins.

A p sub-shell can accommodate six electrons. Because only two electrons of opposite spin can exist closely together, then there must be 3 'compartments' or orbitals in this sub-shell. These orbitals are of equal energy and are therefore said to be degenerate. Evidence for the existence of such degenerate orbitals is shown when atomic spectra are produced inside a strong magnetic field. The degeneracy of these levels is raised and found to split into separate energy levels. A final quantum number is assigned to these initially degenerate levels called the **magnetic quantum number m_l**, having the values $-l$ to $+l$.

Every electron in an atom can therefore be characterised by four quantum numbers, n, l, m_l and m_s.

A description of how the electrons in an atom are arranged is called its **electronic configuration**. The atomic number of a particular element indicates how many electrons it contains. First, assuming the atom is in its ground state, the electrons are fed into its innermost/lowest energy sub-shells. Within each sub-shell, the electron spin is maximised, i.e. the number of unpaired electrons is a maximum (Hund's Rule).

So far the following principal energy levels and sub-levels have been identified,

1s
2s 2p
3s 3p 3d
4s 4p 4d 4f

The *order* in which these energy levels are filled by electrons is not necessarily consecutive. The electrons of the first 18 elements are added in a regular manner, each shell being filled to its own limit before a new shell is stated. At the nineteenth element, the outermost electron starts a new shell before the previous shell has been completely filled. The actual order is,

Hence, iron, Fe, has 26 electrons and therefore its electronic configuration is written,

$1s^2 2s^2 2p^6 3s^2 3p^6 4s^2 3d^6$

As will be seen on pp. 19–21, the electronic configuration of an element is related to its position in the Periodic Table, and to its chemical activity.

The Periodic Table

The Periodic Table of the elements is a listing of the elements in order of increasing atomic number from left to right and from top to bottom (Figure 1.9). The horizontal rows of elements are called **periods** and the vertical columns of related elements are

Box 1.2

Another look at the electron

If an attempt is made to determine the position and momentum of an electron then, because it is so small, the method used will change either the electron's position or its momentum. It is not practically possible to determine simultaneously an electron's position and momentum (Heisenberg's Uncertainty Principle). It is only possible to determine the probability of finding an electron of a particular energy at a point in space about the nucleus. This gives rise to the concept of indistinct and overlapping 'electron probability clouds' to approximate the position of an electron. The density of such a cloud is not uniform but where the cloud is dense there is a high probability of finding the electron and vice versa. There is a larger probability of finding the electron of a hydrogen atom closer to the nucleus than at greater distances. The application of quantum mechanics identifies the same quantum numbers n, l and m_l as seen above. An electron is identified by quantum numbers that take into account its distance from the nucleus (n and l), its angular distribution (shape) relative to the nucleus (l and m_l) and its energy (n only). The volume of space in which there is a 95 per cent chance of finding the electron is called an **atomic orbital**. The quantum number l describes the shape of the orbital occupied by an electron and, as before, takes the values 0, 1, 2, 3 . . . etc. When $l = 0$ the orbital is spherical and describes as an s-orbital, when $l = 1$ the orbital has a dumb-bell shape and is termed a p-orbital and when $l = 2$ it is referred to as a d-orbital. The shapes of some of these three-dimensional orbitals are shown in Figure 1.8.

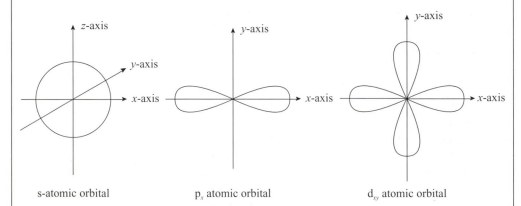

| s-atomic orbital | p$_x$ atomic orbital | d$_{xy}$ atomic orbital |

Figure 1.8 Some atomic orbitals.

Because of the shape of the ns-orbital, an electron that contributes to one is more likely to be found very close to the nucleus than an np-orbital. Similarly an np-electron is more likely to be found nearer the nucleus than an nd. Hence the energy of an electron depends on the orbital it is found in and on the effective nuclear charge it experiences. The effective nuclear charge depends upon the atomic number, the charge on the ion and the number and type of inner and therefore screening electrons.

called **groups**. This table is organised on the repeated pattern to be found in the electronic configurations of the elements.

The electronic configurations of the elements hydrogen and helium are:

First short period

Name	Atomic number	Electronic configuration
Hydrogen H	1	$1s^1$
Helium He	2	$1s^2$ (full)

These two elements form the first short period and involve the filling of the 1s shell.

The second and third short periods are composed of the following elements:

Second short period

Name	Symbol	Atomic number	Electronic configuration
Lithium	Li	3	$1s^2 2s^1$
Beryllium	Be	4	$1s^2 2s^2$
Boron	B	5	$1s^2 2s^2 2p^1$
Carbon	C	6	$1s^2 2s^2 2p^2$
Nitrogen	N	7	$1s^2 2s^2 2p^3$
Oxygen	O	8	$1s^2 2s^2 2p^4$
Fluorine	F	9	$1s^2 2s^2 2p^5$
Neon	Ne	10	$1s^2 2s^2 2p^6$ (full)

Third short period

Name	Symbol	Atomic number	Electronic configuration
Sodium	Na	11	$1s^2 2s^2 2p^6 3s^1$
Magnesium	Mg	12	$1s^2 2s^2 2p^6 3s2$
Aluminium	Al	13	$1s^2 2s^2 2p^6 3s^2 3p^1$
Silicon	Si	14	$1s^2 2s^2 2p^6 3s^2 3p^2$
Phosphorus	P	15	$1s^2 2s^2 2p^6 3s^2 3p^3$
Sulphur	S	16	$1s^2 2s^2 2p^6 3s^2 3p^4$
Chlorine	Cl	17	$1s^2 2s^2 2p^6 3s^2 3p^5$
Argon	Ar	18	$1s^2 2s^2 2p^6 3s^2 3p^6$ (full)

The above elements can be grouped into two rows:

Li	Be	B	C	N	O	F	Ne
Na	Mg	Al	Si	P	S	Cl	Ar

For the elements that are starting to form a column, e.g. Li and Na, the outer electronic configurations are seen to be similar, and of the form ns^1 where n is the principal quantum number. They both have a single electron in their outer shell. Similarly, C and Si electronic configurations are of the form $ns^2 np^2$, both elements having four electrons

Group 1	Group 2	Group 3	Group 4	Group 5	Group 6	Group 7	Group 8	Group 9	Group 10	Group 11	Group 12	Group 13	Group 14	Group 15	Group 16	Group 17	Group 18
1 **H** 1.0079																	2 **He** 4.0026
3 **Li** 6.941	4 **Be** 9.01218											5 **B** 10.81	6 **C** 12.011	7 **N** 14.0067	8 **O** 15.9994	9 **F** 18.9984	10 **Ne** 20.179
11 **Na** 22.9898	12 **Mg** 24.305											13 **Al** 26.9815	14 **Si** 28.086	15 **P** 30.9738	16 **S** 32.06	17 **Cl** 35.453	18 **Ar** 39.948
19 **K** 39.098	20 **Ca** 40.08	21 **Sc** 44.9559	22 **Ti** 47.90	23 **V** 50.944	24 **Cr** 51.996	25 **Mn** 54.9380	26 **Fe** 55.847	27 **Co** 58.9332	28 **Ni** 58.71	29 **Cu** 63.546	30 **Zn** 65.38	31 **Ga** 1.0079	32 **Ge** 72.59	33 **As** 74.9261	34 **Se** 78.96	35 **Br** 79.904	36 **Kr** 83.80
37 **Rb** 85.467	38 **Sr** 87.62	39 **Y** 89.909	40 **Zr** 91.22	41 **Nb** 92.9064	42 **Mo** 95.94	43 **Tc** 98.9062	44 **Ru** 101.07	45 **Rh** 102.9055	46 **Pd** 106.4	47 **Ag** 107.868	48 **Cd** 112.40	49 **In** 114.82	50 **Sn** 118.69	51 **Sb** 121.75	52 **Te** 127.60	53 **I** 126.9045	54 **Xe** 131.30
55 **Cs** 132.9054	56 **Ba** 137.34	57 **La*** 138.9155	72 **Hf** 178.49	73 **Ta** 180.9479	74 **W** 183.85	75 **Re** 186.2	76 **Os** 190.2	77 **Ir** 192.22	78 **Pt** 195.09	79 **Au** 196.9665	80 **Hg** 200.59	81 **Tl** 204.37	82 **Pb** 207.2	83 **Bi** 208.9804	84 **Po** 209*	85 **At** 210*	86 **Rn** 222*
87 **Fr** 223	88 **Ra** 226.0254	89 **Ac**** 227*															

Lanthanides *	58 **Ce** 140.12	59 **Pr** 140.9077	60 **Nd** 144.24	61 **Pm** 145*	62 **Sm** 150.4	63 **Eu** 150.4	64 **Gd** 157.25	65 **Tb** 158.9254	66 **Dy** 162.50	67 **Ho** 164.9304	68 **Er** 167.26	69 **Tm** 168.9342	70 **Yb** 173.04	71 **Lu** 174.97
Actinides **	90 **Th** 232.038	91 **Pa** 231.0359	92 **U** 238.029	93 **Np** 237.0482	94 **Pu** 244*	95 **Am** 243*	96 **Cm** 247*	97 **Bk** 247*	98 **Cf** 251*	99 **Es** 254*	100 **Fm** 258*	101 **Md** 258*	102 **No** 255*	103 **Lr** 256*

Figure 1.9 Periodic Table of the elements.

in their outer shell. A table is starting to form in which atoms having similar outer electronic configurations are occurring at intervals. By grouping these elements into columns, a Periodic Table is being constructed that represents periodicity in terms of electronic configurations. The elements so far described are referred to as the typical elements, forming a block of elements on the left-hand side of the Periodic Table termed the s-block elements and one on the right called the p-block elements. Note that helium, neon and argon have 'complete' electronic configurations.

Periods 4 and 5 each have 18 elements. The first two in each period (K, Ca and Rb, Sr) have outer electronic configurations ns^1 and ns^2, respectively. The next shell to be filled is the $(n-1)d$ resulting in two sets of elements known as the **transition elements** because they bridge the gap between the groups of typical elements. The filling of the np shell then completes the periods.

Period 6 is composed of 32 elements. Again, the first two (Cs, Ba) have an outer electronic configuration of ns^1 and ns^2. The next element (La) begins the filling of the $(n-1)d$ shell but, because the 4f and 5d are so close energetically, the following element (Ce) has the outer electronic configuration of $[Xe]6s^2 4f^1 5d^1$, i.e. the $(n-2)f$ sub-shell is being filled. Thereafter, for the next 13 elements it is the 4f that fills and not the 5d. This group of elements are collectively known as the **lanthanides**. A third series of 10 transition elements, beginning with hafnium (Hf), then continue the filling of the $(n-1)d$ sub-shell. Typical elements then complete the period with the filling of the np sub-shell.

Finally, Period 7 starts with two elements (Fr, Ra), which complete the s-block, with outer electronic configurations of ns^1 and ns^2, respectively. The next element actinium (Ac) begins the filling of the $(n-1)d$ sub-shell, i.e. $[Rn]7s^2 6d^1$. Again, because the 6d and the 5f sub-shells are energetically very close, the next sub-shell to be filled is the $(n-2)f$ which is completed by 14 elements known as the **actinides**. After the actinides the $(n-1)d$ sub-shell begins to be filled at element 103, thus starting a new series of transition elements. *Most* of the actinides, and the newly discovered elements not shown in Figure 1.9, are probably only of academic interest to the environmental chemist.

In summary, the Periodic Table consists of two blocks of typical elements. The first block is situated on the left-hand side of the table, and these are characterised by having one and two electrons in their outer ns-electron shells. The second block of elements is situated on the right-hand side of the table, and these are characterised by the gradual filling of the np-electron shells. Between these two main blocks is a third block characterised by the filling of the $(n-1)d$ sub-energy level. These are referred to as the transition elements. A fourth block of elements (the lanthanides and actinides) precedes the transition elements and these are characterised by the filling of the $(n-2)f$ sub-shell. These are referred to as the inner transition elements.

The position of an element in the Periodic Table enables a description of the electronic configuration of that element to be written down, thus enabling some of its chemical characteristics to be explained. If some properties of an element are known in a group, then it is possible to predict the possible properties of other elements in the group.

An introduction to chemical bonding

Ionic bonding

Those elements on the far right of the Periodic Table that have all of their electronic shells complete, helium, neon, argon, krypton, xenon and radon, are particularly chemically stable. They are known as the noble gases and exist as monatomic gases. These elements do not react with normal laboratory reagents or with chemicals found in the environment. The link between their electronic configuration and their chemical inertness is used in the elementary theories of chemical bonding.

These theories are based on the assumption that, because the electronic configurations of the noble gases are so stable, many elements react in order to attain the nearest noble gas configuration. There are two basic ways that elements can achieve this – by electron transfer from one atom to another, or by the sharing of pairs of electrons.

Sodium has the electronic configuration $1s^2 2s^2 2p^6 3s^1$ or $[Ne]3s^1$. It has one electron above the nearest noble gas configuration. In order to achieve the stability associated with this configuration, the sodium atom will readily lose its single electron to become a positively charged sodium cation. Metals like sodium that readily release one or more electrons are called **electropositive elements**,

$$Na^x - e^- \longrightarrow Na^{+1}$$
$$\text{atom} \qquad\qquad \text{cation}$$
$$[Ne]3s^1 \qquad\qquad [Ne]$$

In the case of chlorine, its electronic configuration is $1s^2 2s^2 2p^6 3s^2 3p^5$ or $[Ne]3s^2 3p^5$. To achieve the nearest noble gas electronic configuration, it is energetically much easier to gain a single electron to achieve the argon configuration than lose seven to attain that of neon. Hence, chlorine will readily accept a single electron to become a negatively charged anion. Elements like chlorine that readily accept one or more electrons are referred to as the **electronegative elements**,

$$\overset{xx}{\underset{xx}{_x\text{Cl}^x}} + e^- \longrightarrow Cl^-$$
$$\text{atom} \qquad\qquad \text{anion}$$
$$[Ne]\,3s^2 3p^5 \qquad [Ar]$$

Unlike electrostatic charges attract each other. Hence, one sodium cation is balanced by one chlorine anion, and the formula for sodium chloride is NaCl, or Na^+Cl^- if the nature of the bonding is to be indicated. Compounds in which elements are joined as a result of the gain and loss of electrons forming oppositely charged ions are called **ionic compounds** and the bonding **ionic bonding**. Ionic bonding is the result of the combination of an electropositive element with an electronegative element.

Calcium has the electronic configuration $[Ar]4s^2$ and oxygen $[He]2s^2 2p^4$. When calcium joins with oxygen, it will form an ionic compound called calcium oxide. This time two electrons are lost by the calcium atom and two are gained by the oxygen atom,

$$Ca - 2e^- \longrightarrow Ca^{2+}$$
$$\text{atom} \qquad\qquad \text{cation}$$
$$[Ar]4s^2 \qquad\qquad [Ar]$$

$$\overset{xx}{\underset{xx}{_x\text{O}}} + 2e^- \longrightarrow O^{2-}$$
$$\text{atom} \qquad\qquad \text{anion}$$
$$[He]2s^2 2p^4 \qquad [Ne]$$

The chemical formula for calcium oxide is thus CaO, or $Ca^{2+}O^{2-}$.

Table 1.6 *Some common cations*

Hydrogen, H^+	Magnesium, Mg^{2+}	Aluminium, Al^{3+}
Sodium, Na^+	Calcium, Ca^{2+}	Iron(III), Fe^{3+}
Potassium, K^+	Strontium, Sr^{2+}	
Copper(I), Cu^+	Barium, Ba^{2+}	
Mercury(I), Hg_2^{2+}	Zinc, Zn^{2+}	
	Cadmium, Cd^{2+}	
	Iron(II), Fe^{2+}	
Ammonium, NH_4^+	Mercury(II), Hg^{2+}	
	Copper(II), Cu^{2+}	
	Lead(II), Pb^{2+}	

Table 1.7 *Some common anions*

Fluoride, F^-	Oxide, O^{2-}	Phosphate, PO_4^{3-}
Chloride, Cl^-	Sulphide, S^{2-}	
Bromide, Br^-	Sulphate, SO_4^{2-}	
Iodide, I^-	Sulphite, SO_3^{2-}	
Hydroxide, OH^-	Carbonate, CO_3^{2-}	
Hydrogen Sulphide, HS^-	Chromate, CrO_4^{2-}	
Nitrate, NO_3^-	Dichromate, $Cr_2O_7^{2-}$	
Nitrite, NO_2^-	Hydrogen Phosphate, HPO_4^{2-}	
Hydrogen Sulphate, HSO_4^-		
Hydrogencarbonate, HCO_3^-		
Permanganate, MnO_4^-		
Dihydrogen Phosphate, $H_2PO_4^-$		

If calcium were to react with chlorine, then two chlorine atoms would be required to accept the two electrons donated,

$$Ca - 2e^{-1} \longrightarrow Ca^{2+}$$

$${}_x^x\!Cl^x_{xx}{}^{xx} + e^- \longrightarrow Cl^-$$

$${}_x^x\!Cl^x_{xx}{}^{xx} + e^- \longrightarrow Cl^-$$

Hence, the chemical formula for calcium chloride is $CaCl_2$ or $Ca^{2+}(Cl^{-1})_2$.

Note that, in all of the above examples, the individual charges on the ions result in an ionic compound that is overall electrically neutral.

Tables 1.6 and 1.7 list the most common simple and complex cations and anions.

Covalent bonding

The element chlorine exists as the gaseous diatomic molecule Cl_2 or dichlorine. Examination of the chlorine atom's electronic configuration $[Ne]3s^23p^5$ shows that another way the nearest noble gas electronic structure can be attained is when two chlorine atoms share a pair of electrons. One way of showing this is to concentrate just on the outer seven electrons, i.e. on those involved in bonding. The underlying electronic configuration can be assumed. The two chlorine atoms can be represented as,

$$\overset{\text{xx}}{\underset{\text{xx}}{\text{x}\overset{}{\text{Cl}}\text{x}}} \quad \text{and} \quad \overset{\text{oo}}{\underset{\text{oo}}{\text{o}\overset{}{\text{Cl}}\text{o}}}$$

where each dot and cross represents one electron on each atom. If the two atoms share a pair of electrons, then both appear to have a complete noble gas configuration, i.e.

$$\overset{\text{xx}\;\text{oo}}{\underset{\text{xx}\;\text{oo}}{\text{x}\text{Cl}\overset{}{\text{x}}\overset{}{\text{o}}\text{Cl}\text{o}}}$$

Such a description is called a Lewis Structure. The pair of electrons joining the two atoms is called a **covalent bond**. A covalent bond is often represented by a single straight line between the two joined atoms, so the above dot and cross diagram becomes Cl—Cl.

Water, H_2O, is a covalently bonded compound. Oxygen has six electrons in its outer shell, hydrogen has 1. To obtain the [Ne] and [He] electronic configurations, respectively, then oxygen will share two of its electrons with two separate hydrogen atoms,

$$\text{H}^\text{x} + \overset{\text{oo}}{\underset{\text{oo}}{\text{o}\text{O}\text{o}}} + {}^\text{x}\text{H} \qquad \overset{\text{oo}}{\underset{\text{oo}}{\text{H}\overset{}{\text{x}}\text{O}\overset{}{\text{x}}\text{H}}} \qquad \text{or}$$

The existence of oxygen as the molecule dioxygen, O_2, may be explained by assuming that the oxygen atoms achieve the stable noble gas configuration by sharing two pairs of electrons,

$$\overset{\text{xx}}{\underset{\text{xx}}{\text{x}\text{O}}} + \overset{\text{oo}}{\underset{\text{oo}}{\text{O}\text{o}}} \qquad \overset{\text{xx}\;\text{oo}}{\underset{\text{xx}\;\text{oo}}{\text{x}\text{O}\overset{}{\text{x}}\overset{}{\text{o}}\text{O}\text{o}}} \qquad \text{or} \quad \text{O}{=}\text{O}$$

The two covalent bonds form a **double bond**.

Carbon dioxide, CO_2, becomes,

$$\overset{\text{xx}}{\underset{\text{xx}}{\text{x}\text{O}}} \quad \overset{\bullet}{\underset{\bullet}{\bullet\text{C}\bullet}} \quad \overset{\text{oo}}{\underset{\text{oo}}{\text{O}\text{o}}} \qquad \overset{\text{xx}}{\underset{\text{xx}}{\text{x}\text{O}\text{x}}}{:}\text{C}{:}\overset{\text{oo}}{\underset{\text{oo}}{\text{o}\text{O}\text{o}}} \qquad \text{or} \quad \text{O}{=}\text{C}{=}\text{O}$$

Ionic bonding and covalent bonding can be found in the same compound. For example, sodium hydroxide is found to have the formula NaOH. If the OH group of atoms is treated as if it is covalent, then it can be written as,

$$\text{H}^\text{x} + \overset{\text{oo}}{\underset{\text{oo}}{\text{o}\text{O}\text{o}}} \qquad \text{or} \qquad \overset{\text{oo}}{\underset{\text{oo}}{\text{H}\overset{}{\text{x}}\text{O}\text{o}}}$$

The OH group still only has seven electrons associated with the oxygen atom at any time. An extra electron is added via the sodium atom to complete the required eight,

i.e.

$$\overset{\text{oo}}{\underset{\text{oo}}{\text{H}\overset{}{\text{x}}\text{o}\text{O}\text{o}\bullet}}{}^{-1}$$

Note that we do not know where this single electron will go on the oxygen atom, so we place the negative sign outside a set of brackets in this explanation. The hydroxide group can thus be written $(O—H)^{-1}$ but it is normally represented by OH^-. Sodium hydroxide is NaOH or more precisely Na^+OH^-.

Extended covalent bonding

The covalently bonded elements and compounds covered so far are composed of discrete molecules. However, some substances form **extended covalent** structures, e.g. carbon in diamond and silicon dioxide (SiO_2) in silica.

Carbon can form four single covalent bonds, which are arranged in a regular tetrahedron. This gives rise to the large three-dimensional crystalline structure known as diamond (Figure 1.10).

Because of this arrangement of carbon atoms, and the fact that the C—C single bond is so strong, diamond is the hardest known substance and has a high melting point. The formula of silicon dioxide, SiO_2, at a first glance, resembles that of carbon dioxide, CO_2. One example of silicon dioxide found in nature is quartz. This is not a triatomic gas, but a very hard, high melting point solid. In this substance, silicon is joined to four oxygen atoms arranged in a tetrahedron via strong Si—O single covalent bonds. Each oxygen atom is shared with another silicon atom. Thus both atoms achieve the nearest noble gas configuration. Because of the type of bonding, the properties of ionic and covalent compounds differ. In the case of ionic compounds, the ions act like charged spheres. Thus they exert their attractive forces equally in all directions. This gives rise to strongly bonded, three-dimensional networks of ions, which causes the ionic compounds to be hard, crystalline solids with high melting points and boiling points. In the molten state ions in ionic compounds can move under the influence of a potential difference and therefore conduct electricity. When a solid ionic compound is placed in contact with water, the ions can interact with the water molecules to such an extent that they can enter solution and become dissociated. In many ionic solids the interactions between the constituent ions are so great, and the ionic structure so stable, that

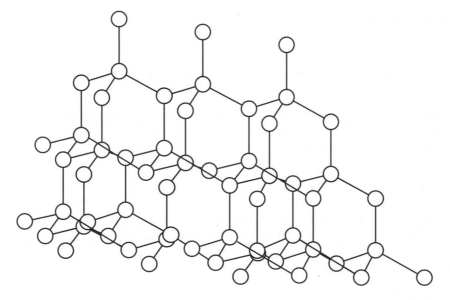

Figure 1.10 The diamond structure.

the interactions between water molecules and the ions are not strong enough to overcome the ionic interactions, and therefore dissolving does not take place. When solutions containing ions chemically react with each other then, because of the intimate nature of the mixing and the three-dimensional nature of ionic forces of attraction, reaction is observed to be almost instantaneous and precipitates are often seen to form.

Covalent substances tend not to dissolve in water. A large number of them have a low melting point and boiling point. In the molten state ions are not formed and therefore they do not conduct electricity. Their structure often consists of discrete molecules, or three-dimensional extended covalent crystalline structures. The chemical reactions of covalent substances also tend to be slower, e.g. boiling over a long period of time may be necessary in order for bond breaking and making to occur.

Metallic bonding

In metals and metallic alloys, the electropositive nature of the atoms means that they give up the electrons in their outer shells to form positive ions. The electrons form a 'sea' of negative charge in which the positive ions are embedded. Each positive ion is attracted to the 'sea' of electrons, and vice versa. One electron does not belong to one particular ion. Since electrostatic attraction is exerted in all directions, metals form gigantic structures and give rise to the properties described previously. The strength of the bonding is related to the number of electrons that can be released. For example, sodium atoms release their single electron to give Na^+ ions in a sea of electrons. The bonding in this case is weak because only one electron per atom is released; therefore, sodium has a low melting point but conducts electricity well under the influence of a potential difference.

The structures of metals can be described in terms of the close packing of their atoms, treated as if they are identical spheres. The spheres are packed in such ways as to occupy the smallest volume of space whilst leaving the minimum of empty space. Figure 1.11 shows the two main ways of stacking the spheres in a close packing arrangement. In Figure 1.11(a), the atoms of Layer 2 (B) are stacked in the dips formed by the atoms of Layer 1 (A). Layer 3 (C) is stacked in the dips created by the second layer so that they are not directly over the atoms of Layer 1 (Fig 1.11(b)). This will give a repeating structure of the form ABC, ABC, ABC, etc. Alternatively, Layer 3 can be placed in the dips formed by the second layer immediately over the atoms in Layer 1 (Fig 1.11(c)). The repeating structure in this case will be AB, AB, AB, etc.

The ABC type structure is called the cubic close-packed structure (ccp). Each atom in each layer has three neighbours in the layer below, six in its own layer and three in the layer above. Such an atom is said to have a **co-ordination number** of twelve. It is possible to identify the smallest portion of a crystal, called the unit cell, which will, if repeated, construct the crystal. In the case of the ccp structure the unit cell is depicted in Figure 1.12 and is called the face-centred cubic unit cell. Note the use of dots to represent atoms in order to make the structure type more clear.

The AB structure is called the hexagonal close-packed structure (hcp). Again, since each atom in a layer has three neighbours in the layer below it, six in its own layer and three in the one above, its co-ordination number is twelve. The unit cell of this structure is shown in Figure 1.13.

Some metals take up a much more open structure than the ones just described. The commonest one is the body-centred cubic (bcc) structure (Fig 1.11(d)). In this structure each atom has eight neighbours and therefore its co-ordination number is eight. The unit cell is shown in Figure 1.14.

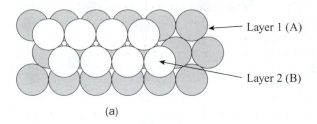

(a)

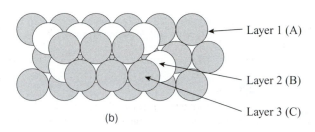

(b)

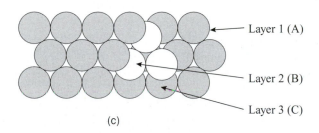

(c)

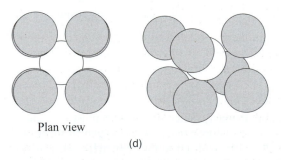

Plan view

(d)

Figure 1.11 The close packing of spheres and metal structures. (a) Layer 1 spheres packed to maximise use of space. Layer 2 lies in the dips created by the first layer. (b) As above but with Layer 3 placed in the dips of Layer 2. (c) Layer 3 placed immediately above Layer 1. (d) The body-centred cubic structures.

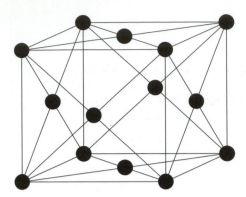

Figure 1.12 Face-centred cubic unit cell. Example: aluminium and copper.

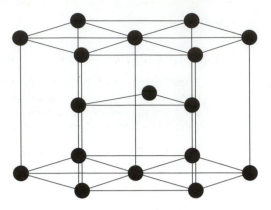

Figure 1.13 Hexagonal close-packed unit cell. Example: magnesium and zinc.

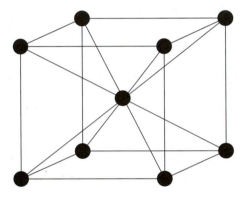

Figure 1.14 Body-centred cubic unit cell. Example: iron and sodium.

Oxidation, reduction and oxidation numbers

Oxidation and **reduction** are chemical reactions that occur frequently in the environment. Oxidation may be viewed as the addition of oxygen to a species and reduction its removal. However, these definitions are too limited and so the electronic theory of oxidation and reduction is used. Here oxidation is the *loss* of one or more electrons and reduction the *gain* of one or more electrons. Chemical species that are able to gain electrons are called oxidising agents and those that can lose electrons are reducing agents. For example,

$$Cu - 2e^- \rightleftharpoons Cu^{2+} \text{ oxidation}$$

$$1/2\ O_2 + 2e^- \rightleftharpoons O^{2-} \text{ reduction}$$

Oxidation and reduction occur simultaneously, since one species must give electrons to a second accepting species. Thus, when copper reacts with oxygen to form copper(II) oxide, copper is the reducing agent and is oxidised to the copper(II) ion, whilst oxygen is the oxidising agent and is reduced to the oxide ion.

The term **oxidation number** is conveniently used to describe the state of a particular atom or ion. Thus manganese(II) describes the Mn^{2+} ion whilst manganate(VII) the MnO_4^{-1} ion. The rules for the use of oxidation numbers are shown in Box 1.3.

These rules, and the concept that all compounds can be treated *as if* they are ionic, can be applied to compounds irrespective of their bonding.

Box 1.3

Oxidation number rules

1 Where the *atom* of an element is involved, then its oxidation number is taken to be zero, i.e. it has not gained or lost electrons. Thus the oxidation numbers of the atoms of iron Fe, chlorine Cl, oxygen O, fluorine F, copper Cu or any other atom is zero. Copper metal would be referred to as copper(0).

2 The oxidation number of a monatomic ion is equal to the charge on that ion. For example Cu^{2+} is termed copper(II), Fe^{3+} iron(III).

3 The oxidation number of fluorine in any compound, including interhalogen compounds, is always -1 (minus one).

4 The oxidation number of oxygen is -2 except when it is bonded to fluorine, or in peroxides (O.N.$= -1$) where there is an O—O single bond.

5 The oxidation numbers of metal ions are always positive.

6 The oxidation number of hydrogen bonded to a metal is -1 and where bonded to a non-metal $+1$.

7 The sum of the oxidation numbers of each atom in an ion or compound is equal to the charge on the ion or compound.

Example 1: Carbon dioxide, CO_2, covalently bonded
Oxidation number of oxygen is -2, there are two oxygen atoms and so the overall oxidation number is -4. Since there is no overall charge on carbon dioxide, then the oxidation number of carbon must be $+4$.

Example 2: The manganate(VII) ion, MnO_4^-
The oxidation number of oxygen is -2. There are 4 of them and hence the overall oxidation number is -8. Since the manganate(VII) ion carries a -1 charge, then the manganese atom must have an oxidation number of $+7$. Hence the name manganate(VII). (This does not mean that the manganese atom carries a charge of $+7$.)

Example 3: The chromate(VI), CrO_4^{2-}, and dichromate(VI), $Cr_2O_7^{2-}$, ions
The chromate(VI) ion -4 oxygen atoms, hence an overall oxidation number of -8. This ion carries a charge of -2 and therefore the chromium atom is in an oxidation state of $+6$. In the dichromate(VI) ion there are a total of seven oxygen atoms and thus the total oxidation number would be -14. The two chromium atoms are linked by an oxygen atom. The dichromate(VI) ion carries an overall electronic charge of -2. Hence each chromium atom carries a formal charge of $+6$.

  If zinc(0) powder is added to an aqueous solution of blue copper(II) sulphate(VI) contained in a test-tube, then the following spontaneous reaction occurs,

$$Zn(s) + Cu^{2+}(aq) \rightleftharpoons Zn^{2+}(aq) + Cu(s)$$

Thus zinc(0) is oxidised to zinc(II) and goes into solution, whilst copper(II) is reduced to dark brown copper(0), which is deposited: and the blue solution loses its colour.

The reaction can be written,

$$Zn(s) - 2e^- \rightleftharpoons Zn^{2+}(aq) \text{ oxidation}$$

$$Cu^{2+}(aq) + 2e^- \rightleftharpoons Cu(s) \text{ reduction}$$

$$Zn(s) + Cu^{2+}(aq) \rightleftharpoons Zn^{2+} + Cu(s) \text{ overall reaction}$$

This is an example of a **redox reaction**.

A large number of redox reactions also involve the transfer of other material as well as electrons, for example the half reaction manganate(VII) in acid solution will undergo reduction to manganese(II) when an appropriate reducing agent is present,

$$MnO_4^-(aq) + 8H^+(aq) + 5e^- \rightleftharpoons Mn^{2+}(aq) + 4H_2O(l)$$
manganate(VII)   manganese(II)

Electronegativity

The elements all have a different ability for attracting or repelling electrons when forming chemical bonds. **Electronegativity** is a way of expressing the power of an element to attract electrons to itself when in chemical combination. Fluorine exerts the greatest attraction for electrons and is therefore the most electronegative element, whilst caesium is the least electronegative. The electronegativity of an element thus has a marked effect on the 'purity' of an ionic or covalent bond, and hence on the properties of a compound.

Water contains angular molecules in which hydrogen and oxygen are joined by single covalent bonds. Hydrogen has a lower electronegativity than oxygen, so these atoms do not equally share the electron pair making up a bond. The bond therefore becomes uneven with respect to the electron distribution. This results in a small residual or partial positive charge ($\delta+$) being left on the hydrogen atom, and a partial negative charge ($\delta-$) on the oxygen atom,

i.e. H—O becomes $^{\delta+}H$—$^{\delta-}O$

so that the water molecule is written,

$$^{-\delta-\delta}O$$
$$/ \qquad \backslash$$
$$^{+\delta}H \qquad ^{+\delta}H$$

Hence, the water molecule has a negative end and a positive end. It is this unequal sharing of electrons which causes water to have somewhat unique properties. Remember, this is not ionic bonding but an uneven distribution of the bonding electrons.

The water molecule is said to be **polar** in nature, and its covalent bonds **polarised**. Solvents, such as water, which possess polarised covalent bonds are referred to as **polar solvents**, whereas those that do not, such as hexane, are referred to as **non-polar** solvents. Water can dissolve a large number of ionic substances because of the interaction between the appropriate polarised end of the water molecule and an ion. For example, the negative end of water will be attracted to the Na^+ ion and the positive ends to the Cl^{-1} ion in sodium chloride. The interactions are such that the ions are dissociated and become distributed throughout the water.

The interactions that exist between the different ends of the water molecules themselves lead to a weak form of bonding called **hydrogen bonding**. It is the existence of this bonding which leads to pure water being a liquid at room temperatures and pressures instead of being a gas. The weakness of the hydrogen bonding is indicated in the ease with which water is evaporated.

The mole

The word 'mole' as used in chemistry is derived from a Latin word meaning 'massive heap'! The **mole (mol)** is the unit used for expressing large numbers of atoms, ions or molecules, and as a means of comparing those numbers.

The definition of a mole is the number of atoms occurring in exactly 12.0000 g of the $^{12}_{6}C$ isotope. Since an atom of $^{12}_{6}C$ contains six protons and six neutrons, the mass of one atom of $^{12}_{6}C$ can be taken as exactly 12 atomic mass units, i.e. $12 \times 1.66 \times 10^{-24}$ g or 1.99×10^{-23} g.

Therefore, the number of atoms in 12.0000 g of $^{12}_{6}C$ must be given by:

$$\frac{\text{Mass of the sample}}{\text{Mass of one } ^{12}_{6}C \text{ atom}} = \frac{12.0000 \text{ g}}{1.99 \times 10^{-23} \text{ g}} = 6.02(4) \times 10^{23} \text{ atoms}$$

This number is called the **Avogadro number (N_A)**. Thus, a mole of any substance contains the Avogadro number of particles of that substance, i.e. 6.02×10^{23} particles. This can be worked backwards!

The mass of a $^{12}_{6}C$ atom = 12 atomic mass units = 1.99×10^{-23} g. Hence, the Avogadro number of atoms will have a mass of 1.99×10^{-23} g $\times 6.02(4) \times 10^{23} = 12.00$ g, which corresponds to the original 12.00 g of $^{12}_{6}C$ containing a mole of atoms!

What would be the result if this approach was applied to any element, e.g. sodium? The mass of a sodium atom is 3.815×10^{-23}g. Hence, the mass of one mole is $3.815 \times 10^{-23} \times 6.02(4) \times 10^{23}$ g of sodium atoms = 22.98 g. So a mole of atoms of any element is simply its relative atomic mass expressed in grams. The **molar mass (M)** of any element is therefore the mass of one mole of its atoms in grams. This contains one mole of atoms and would have a mass corresponding to its relative atomic mass in grams. Its units are g mol^{-1}.

What is the molar mass of naturally occurring cadmium (Cd)? The relative atomic mass of Cd = 112.4. Hence, the average mass of the cadmium atom is:

$$112.4 \times 1.66 \times 10^{-24} \text{ g} = 1.866 \times 10^{-22} \text{ g}$$

Therefore N_A atoms will have a mass of 1.866×10^{-22} g $\times 6.02(4) \times 10^{23}$ g = 112.4 g, or you can just write down that one mole of cadmium atoms have a mass of 112.4 g; its molar mass is thus 112.4 g mol^{-1}.

Remember a mole of *anything* contains 6.02×10^{23} particles.

Example 1
One mole Cl *atoms* have a molar mass of 35.5 g mol^{-1}.
One mole Cl$_2$ *molecules* have a molar mass of 71.0 g mol^{-1}.
One mole Cl^{-1} *ions* have a molar mass of 35.5 g mol^{-1}.
Hence, the species being dealt with must be clearly identified.

Example 2
The molar mass of anhydrous sodium sulphate, Na_2SO_4, is its relative molecular mass expressed in g per mole, i.e.142 g mol^{-1}. However, Na_2SO_4 can dissociate into two Na$^+$

ions and an SO_4^{2-} ion. Hence, one mole of Na_2SO_4 molecules will give 2 moles of Na^+ ions and 1 mole of SO_4^{2-} ions when it dissociates.

Calculating numbers of moles

Ethanol has the molecular formula C_2H_6O (or C_2H_5OH) and therefore a molar mass of 46.069 g mol^{-1}. In 92.138 g of ethanol there would be 2 moles, and in 23.035 g, 0.5 mole of ethanol. Generally, the number of moles n of any substance is given by,

$$n = \frac{\text{mass of substance/g}}{\text{molar mass/g mol}^{-1}}$$

When handling moles, the amounts of substance are being compared.

For example, $2H_2(g)$ + $O_2(g)$ = $2H_2O(l)$

2 moles	1 mole	2 moles	numbers of moles
$2 \times N_A$	N_A	$2N_A$	numbers of particles
4 g	32 g	36 g	mass of chemical

The mole and solutions

There are many occasions when samples taken from the environment have to be chemically analysed. This often involves the making up of solutions involving distilled or deionised water as the solvent. A solution is an intimate homogeneous mixture of two or more substances, e.g. salt-water. In salt-water, water is the material that does the dissolving and is called the **solvent**. Salt is the substance that is dissolved and is thus called the **solute**.

Volumetric analysis involves the accurate measurement of volumes of liquids and of masses of solids and liquids. In such an analysis it is possible to produce a result with an error that does not exceed 0.2 per cent.

A volumetric analysis is carried out by first making up a solution of a particular material of known concentration called a **standard solution**, e.g. hydrochloric acid. The volume of the standard solution needed to chemically react exactly with a known volume of another solution is then determined. The equation for the chemical reaction must also be known. The course of the reaction is followed by some means, for example by an indicator that changes colour when the reaction is complete. An aqueous solution of potassium hydroxide of unknown concentration containing the indicator phenolphthalein will be pink. Eventually, on the addition of a known standard aqueous solution of hydrochloric acid, the solution just becomes colourless, and the acid will be in *slight* excess. From these results it is possible to calculate the unknown concentration. The process of adding a standard solution to another in order to determine its concentration is called **titration.**

In volumetric analysis, the units of concentration normally used are 'moles per cubic decimetre' (or moles per litre), abbreviated to mol dm^{-3} (mol l^{-1}). A term that is still in common use is 'molar solution', e.g. a 0.2 molar solution should be referred to as a 0.2 mol dm^{-3} solution.

What then is meant by a 1.000 mol dm^{-3} solution of sulphuric acid? Sulphuric acid (H_2SO_4) has a relative molecular mass of 98.213 (to three decimal places) and therefore one mole of it will have a mass of 98.213 g, i.e. its molar mass is 98.213 g mol^{-1}. If this mass of sulphuric acid is placed in a 1 dm^3 volumetric flask and its volume *made up* to 1,000 cm^3 using distilled water, then we have a 1.000 mol dm^{-3} solution of sulphuric acid. Similarly, if 24.553 g of sulphuric acid are placed in a 250 cm^3 volumetric flask and made up to 250 cm^3, then again we have a 1.000 mol dm^{-3} solution.

Often solutions of such 'high' strength are not required. Typically, standard solutions of 0.100 mol dm^{-3} and 0.010 mol dm^{-3} are more common. This would mean in the case of sulphuric acid a mass of 2.455 g made up to 250 cm^3 in the former case and 0.246 g made up to 250 cm^3 in the latter case. Such solutions also ensure a considerable saving in the *amount* of material used and are therefore less expensive.

Solutions and parts per million (ppm)

A convenient unit when dealing with both dilute aqueous solutions and gaseous mixtures is the 'part per million', abbreviated to ppm. It is convenient because it reduces cumbersome numbers to simple ones, is universally recognised, and is a characteristic of any volume taken from the solution or gas mixture in question. If, for example, 1,000 cm^3 of a sample of water was found to contain 50 ppm of lead in the form of Pb^{2+}, then 1 cm^3 or 0.001 cm^3 or 1,000 m^3 of the same solution, assuming uniform mixing, would also contain 50 ppm of lead.

Table 1.8 *Some volume equivalents*

Volume	Equivalent volume
1 dm^3	1 × 10^3 cm^3
1 m^3	1 × 10^6 cm^3
1 m^3	1 × 10^3 dm^3
1 dm^3	1 × 10^{-3} m^3
1 × 10^3 cm^3	1 × 10^{-3} m^3

From Table 1.8, if 1 g of lead as the Pb^{2+} ion was found in 1 m^3 of water then, since this is equivalent to 1×10^6 cm^3, the concentration can be written 1 g in 1×10^6 cm^3 or as 1 part per million. If water contained a concentration of 1.68×10^{-3} g dm^{-3} of Pb^{2+}, then it can also be written as 1.68 mg dm^{-3} (1×10^{-3} g is 1 mg). In 1 m^3 of water there would be 1.68 mg $\times 1 \times 10^3$ or 1.68 g of lead present. Thus this would be reported as having a concentration of 1.68 ppm. A water sample containing 1.68×10^{-3} g dm^{-3} also has a concentration of 1.68 ppm.

Where concentrations are smaller than can be reasonably expressed in ppm, then 'parts per billion' or ppb is used. One part per billion is equivalent to 1 g in 1×10^9 cm^3 or 1 g in 1×10^3 m^3. An aqueous solution containing 1.68 ppb of lead would have 1.68 g in 1×10^9 cm^3. Alternatively, 1.68×10^{-6} g in 1 dm^3 or 1.68×10^{-3} g in 1 m^3 will also be equivalent to 1.68 ppb.

A similar approach is adopted for the contamination of volumes of air, i.e. replace the word 'air' for 'water' in the above example, and perhaps ozone for lead. For the atmosphere, ppm is expressed with respect to volume measurements. Hence, 1×10^{-6} m^3 m^{-3} is 1 ppm volume for volume (v/v) and is equivalent to 1 cm^3 m^{-3}. Ppb (v/v) is 1×10^{-9} m^3 m^{-3} or 1×10^{-3} cm^3 m^{-3}.

For solids, mg kg^{-1} is equivalent to ppm, and μg kg^{-1} to ppb.

Water, acidity, alkalinity and pH

Perhaps the most useful definitions of an **acid** and a **base** to the environmental chemist are those that are based upon the Lowry–Brønsted theory. Here, an acid is a substance that can donate one or more protons, whilst a base is defined as a species that accepts one or more protons. These definitions apply in any solvent and to the gas phase.

A molecule of hydrogen chloride, upon dissolving in water, can donate a proton to a water molecule, and is thus a Lowry–Brønsted acid,

$$HCl(g) + H_2O(l) \longrightarrow H_3O^+(aq) + Cl^-(aq)$$

The H_3O^+ ion is called the **hydronium ion**. The formula for this ion is really an over-simplification, but it is sufficient for our needs.

Ammonia, NH_3, is an example of a Lowry–Brønsted base, since it can accept a proton from an acid,

$$NH_3(aq) + HCl(aq) \longrightarrow NH_4^+(aq) + Cl^-(aq)$$

Water can act as both a Lowry–Brønsted acid or base,

$$acid \quad NH_3(aq) + H_2O(l) \longrightarrow NH_4^+(aq) + OH^-(aq)$$

$$base \quad H_2S(aq) + H_2O(l) \longrightarrow H_3O^+(aq) + HS^-(aq)$$

Solutions in which the concentration of hydronium ions is large are said to be acidic, and those containing a large concentration of hydroxide ions alkaline.

A large number of acid–base reactions occur in water in the environment. Water can act both as a Lowry–Brønsted acid and a base, and in the absence of any solute it forms hydronium ions and hydroxide ions, i.e. **auto-ionisation** (sometimes referred to as **auto-protolysis**) occurs,

$$H_2O(l) + H_2O(l) \longrightarrow H_3O^+(aq) + OH^-(aq)$$

Equal numbers of hydroxide and hydronium ions are produced but their concentrations are very low. Water is therefore described as being neutral in the context of acidity and alkalinity. One way of expressing the strength of an acid in water is to use the **pH scale**. This is a measure of the concentration of hydronium ions present. The number produced is temperature dependent and therefore the pH should be quoted at 25 °C. At 25 °C, the concentration of the $H_3O^+(aq)$ ions are 1.00×10^{-7} mol dm^{-3}. To a good approximation, pH can be defined as the negative of the log to the base 10 of the magnitude of the hydrogen ion concentration,

i.e. $pH = -\log$ (concentration of H_3O^+/mol dm^{-3})

The reason for the taking of logs is that the pH scale is broad and only applies to dilute aqueous solutions. Therefore, the small numbers involved are made more manageable. The negative is taken of the log value in order to make the pH value a positive number. Thus, for water its pH is given by,

$$pH = -\log (1.00 \times 10^{-7}) = -(-7.00) = 7.00$$

Water has a pH value of 7.00 at 25 °C. This value of pH is used to define neutrality in aqueous solutions at this temperature.

The pH of a solution of hydrochloric acid in which the concentration of hydronium ions is 1.0×10^{-2} mol dm^{-3} is 2.00. The pH of a solution which contains 1×10^{-9} mol dm^{-3} is 9.00, and described as being alkaline. In this latter case, the hydronium ion concentration is lower than in the case of pure water and the hydroxide concentration is larger. A more detailed explanation of pH will be found in a later chapter. At this stage it is enough to appreciate that pH = 7.00 is neutral, any pH value less than 7.00 is acidic and any pH value greater than 7.00 is alkaline.

The importance of knowing the pH of a particular solution is reflected in its origin or use. Human blood has a pH which normally ranges between pH = 7.2 and 7.6 – if the pH falls below the minimal value by 0.4 or above the maximum value by 0.4, then death can result. In the UK, water is considered to be corrosive if the pH is less than 3.00 or greater than 12.5.

Summary

- The mass number of an atom is the sum of the protons and neutrons present in its nucleus. An element is defined by the number of protons in its nucleus, known as the atomic number. Isotopes have the same number of protons but differ in the number of neutrons.
- Radioactivity is due to the disintegration of an unstable nucleus. The stability of a nucleus depends on its neutron to proton ratio. The three main types of emissions are α-, β- and γ-radiation.
- The relative atomic mass of an element is unitless and is defined by,

$$A_r = \frac{\text{atomic mass of an element} \times 12.00}{\text{atomic mass of } {}^{12}_{6}\text{C atom}}$$

- There are three main types of chemical bonding that give rise to different chemical and physical properties. Ionic bonding is formed by ions achieving the noble gas configuration by electron transfer. Covalent bonding is formed by atoms achieving the noble gas configuration by sharing pairs of electrons. Metallic bonding is formed by positive ions with a noble gas configuration being held together by a 'sea' of electrons.
- Electronegativity is a measure of an atom's ability to attract electrons to itself in a chemical bond. Electronegativity strongly influences the degree of ionic bonding and covalent bonding found in a molecule.
- Two non-metallic elements with high or similar electronegativities form covalent substances. Lewis-type structures can be used to model such materials.
- A metallic element of low electronegativity and a non-metallic element of high electronegativity will form ionic compounds. These materials form three-dimensional arrangements of extended crystalline ionic structures.
- Elements with low electronegativities combine to form alloys.
- Extended covalent bonding occurs when particularly energetically stable three-dimensional structures can be formed between atoms or molecules.
- Metals have extended three-dimensional structures. The internal shape of these structures depends upon how the ions are packed together.
- Oxidation is the loss of one or more electrons, and reduction the gain of one or more electrons. Chemical species that are able to gain electrons are called oxidising agents and those that lose electrons are reducing agents.
- A mole of defined particles of any substance is the Avogadro number of those particles, i.e. 6.02×10^{23}. The molar mass is the mass of the Avogadro number of particles, and is equal to the relative atomic or relative molecular mass of that substance expressed in grams.
- The concentration of solutions can be expressed in mol dm^{-3} or ppm.
- The Lowry–Brønsted definition of an acid is any substance that is a proton donor, and that of a base is any substance that is a proton acceptor.
- The level of acidity of an aqueous solution can be expressed as its pH or,

$$\text{pH} = -\log (\text{concentration of } H_3O^+/ \text{ mol dm}^{-3})$$

Questions

1　Write down the electronic configurations of the elements with atomic numbers 32 and 56. What Group of the Periodic Table is each of these elements in?

2　Calcium carbonate is an ionic compound whilst octane is a covalent compound. What properties would you expect each compound to have?

　　Confirm your predictions by reviewing the properties of these compounds in an appropriate textbook.

3　How many moles of each substance do the following represent?

　(a)　1.256 g of iodine (I_2).

　(b)　123.2 g of sulphur in the form of the S_8 molecule.

　(c)　27.898 g of the metal mercury.

　In each case how many particles would be present?

4　Beer is a mixture that many chemistry students resort to! If it is assumed that a beer contains 4.4 per cent by mass of ethanol (C_2H_5OH), calculate how many moles of ethanol there are in a pint of this beer. You may assume that one pint is equivalent to 0.576 dm^3.

5　(a)　Calculate the mass (to 3 decimal places) of sodium chloride, NaCl, required to make 1 dm^3 of a standard aqueous solution of concentration 0.100 mol dm^{-3}.

　(b)　Calculate the mass (to 4 significant figures) of sodium carbonate, Na_2CO_3, required to make a 250 cm^3 aqueous solution of 0.500 mol dm^{-3} concentration.

　(c)　Calculate the mass of potassium hydroxide (KOH) required to make 100 cm^3 of 0.250 mol dm^{-3} aqueous solution. Quote the number of significant figures you have chosen.

　(d)　Calculate the mass of silver nitrate(V) ($AgNO_3$) to make a solution of concentration 0.150 mol dm^{-3}.

6　(a)　What mass of lead(II) nitrate(V), $Pb(NO_3)_2$, would be needed to make a solution of concentration 25 ppm with respect to Pb^{2+}(aq)?

　(b)　A sample of air was found to contain 190 ppb of nitrogen(IV) oxide, NO_2. How many grams of this compound does 1 m^3 of air contain?

　(c)　Back to beer again! A pint of beer was found to contain 1.27 mg of lead as Pb^{2+}(aq). Express this concentration in μg dm^{-3}. By how much does this concentration differ from the upper limit of 50 μg dm^{-3} for the lead concentrations in drinking water given in the EC Directive 80/778/EEC?

7　The pH of a solution was determined using a pH meter. The first reading showed a pH of 3.40. A second reading of the same solution was taken using the same pH meter which showed a pH of 3.60. Is this a significant difference?

8　(a)　What is the pH of a solution containing 1.237×10^{-3} mol dm^{-3} of hydronium ions?

　(b)　The pH of orange juice is found to be 5.4. What is the concentration of the hydronium ions?

　(c)　The EC Directive on water quality for human consumption requires that the maximum permissible level of pH is 9.5. What would be the hydronium ion concentration of drinking water at this pH?

9　(a)　Determine the oxidation state of the central atom in each of the following chemical species, and name the species.

　　(i) NO_2^-　(ii) NO_3^-　(iii) SO_2　(iv) SO_3　(v) SO_4^{2-}　(vi) SO_3^{2-}

(b) In the following reactions explain which species are reduced and which are oxidised.

(i) $2Fe^{3+} + S^{2-} = 2Fe^{2+} + S$

(ii) $Zn + 2HCl = ZnCl_2 + H_2$

(iii) $MnO_2 + 4H^+ + 2Cl^- = Mn^{2+} + 2H_2O + Cl_2$

(iv) $10I^- + 2\,MnO_4^- + 16\,H^+ = 5I_2 + 2\,Mn^{2+} + 8H_2O$

References

Baer, N.S. and Snethlage, R. (1997) *Saving Our Architectural Heritage*. John Wiley, Chichester.

Brimblecombe, P. and Bowler, C. (1990) *Pollution History of York and Beverley*, Vol. 1. University of East Anglia, Norwich.

Cook, R.U. and Gibbs, G.B. (1996) *Crumbling Heritage: Studies on Stone Weathering in Polluted Atmospheres*. National Power, Wetherby.

Haneef, S.J., Hepburn, S.J., Hutchinson, B.J., Thomson, G.E. and Wood, G.C. (1990) Laboratory exposure systems to simulate atmospheric degradation of building stone under dry and wet conditions. *Atmospheric Environment*, **24a**, 2585–92.

Mills, S. (2000) *Superintendent York Minster Stone Works*: private communication.

The National Academy of Sciences (1998) *Biological Effects of Ionising Radiation VI Report: The Health Effects of Exposure to Indoor Radon*. Environmental Protection Agency.

NRPB (1990) *Limitation of Human Exposure to Radon in Homes*. Docs NRPB 1, No. 1, Her Majesty's Stationery Office, London.

Further reading

Atkins, P.W. and Beran, J.A. (1992) *General Chemistry*, 2nd edn. Scientific American Books, W.H. Freeman, New York.
An easily read and understood basic chemistry textbook which is well illustrated. It starts at a fairly low level in all topics in chemistry and gradually builds up the reader's knowledge and understanding. Somewhere between A-level chemistry and first-year undergraduate level.

Earnshaw, A. and Greenwood, N.N. (1997) *Chemistry of the Elements*, 2nd edn. Butterworth–Heinemann, Oxford.
One of the standard university textbooks in its area. It contains a vast amount of information, and an excellent set of references. It not only covers the more traditional areas of inorganic chemistry but covers bio-inorganic chemistry, industrial chemicals, organo-metallic chemistry, etc.

Glugstone, M. and Flemming, R. (2000) *Advanced Chemistry*. Oxford University Press, Oxford.
A very well illustrated A-level textbook! The topics are well organised and the text easy to read and understand. As a preliminary read it is a difficult book to better. Contains numerous worked examples and additional problems.

Lewis, R. and Evans, W. (2001) *Chemistry*, 2nd edn. Palgrave, Basingstoke.
An easy to read textbook that covers a range of core topics. The minimal amount of maths and science background knowledge is required before you start! A very student-friendly text.

NRPB (1996) *Radon Affected Areas: England*, Docs NRPB 7, No. 2. Her Majesty's Stationery Office, London.
Everything you want to know concerning the occurrence of radon in England. The following texts are very similar in content.

NRPB (1998) *Radon Affected Areas: Wales*, Docs NRPB 9, No. 3. Her Majesty's Stationery Office, London.

NRPB (1999) *Radon Affected Areas: N. Ireland*, Docs NRPB 10, No. 4. Her Majesty's Stationery Office, London.

Williams, I. (2001) *Environmental Chemistry*. John Wiley, Chichester.
This assumes a basic knowledge of chemistry. It covers many key topics concerning the environment, such as the structure and composition of the Earth, mineral extraction, fossil fuels, etc.

② More advanced chemical concepts: energy, entropy and rates of reaction

- Thermodynamics is about if, and how far, a chemical reaction proceeds
- Reaction kinetics is the study of how fast chemical reactions proceed
- The First Law of Thermodynamics, internal energy and enthalpy changes
- The Second Law of Thermodynamics as a directional law
- Entropy and its meaning
- The significance of the Gibbs' Free Energy change
- Chemical equilibrium and the use of equilibrium constants
- Rates of reaction and catalysis
- The dependency of reaction rate on concentration – the establishment and interpretation of rate laws
- The dependency of rate of reaction on temperature
- Rate equations and the mechanism of reactions – the simple reaction and the multi-step reaction
- Kinetic vs thermodynamic control of chemical reactions

Introduction

There are two general questions concerning chemical reactions, which are of major importance. First, will the reaction proceed and if so how far will it go? The answer to this question enables both the viability of a chemical reaction and the amount of yield of a desired (or undesired!) product(s) from the reaction to be predicted. The second question is how fast will the reaction go? It may well be that a reaction can occur and give an excellent yield but it may do so at such a slow rate that it is not viable because of how long it takes. Answers to the first question involve the study of **thermodynamics** and the movement of energy in the form of **heat** and **work**. The answers to the second lie in the field of **reaction kinetics** which investigates the factors that affect the speed of a reaction.

There are many chemical reactions of environmental importance for which explanations of why, how far and how fast they occur are needed. For example, iron is a very useful metal but is subject to rusting. One equation representing rusting is,

$$2Fe(s) + 3/2O_2(g) = Fe_2O_3(s)$$

for which,

$$\Delta H^0_{f,298\,K} = -824.2 \text{ kJ mol}^{-1}$$
$$\Delta S^0_{f,298\,K} = -274.9 \text{ JK}^{-1} \text{ mol}^{-1}$$
$$\Delta G^0_{f,298\,K} = -742.2 \text{ kJ mol}^{-1}$$

What do these symbols and numerical values mean?

Nitric oxide (NO) is produced by high-flying jet aircraft. This chemical reacts with ozone,

$$NO(g) + O_3(g) \longrightarrow O_2(g) + NO_2(g)$$

and is then regenerated,

$$NO_2(g) + O(g) \longrightarrow NO(g) + O_2(g)$$

Overall, the nitric oxide has not been destroyed. It can be regarded as a catalyst for the destruction of ozone. What does the term catalyst mean?

Carbon dioxide dissolves in, and reacts with, water to produce an acidic solution,

$$CO_2(g) + H_2O(l) \rightleftharpoons H^+(aq) + HCO_3^-(aq)$$

This acidic solution can then take part in the weathering of limestone,

$$CaCO_3(s) + H^+(aq) \rightleftharpoons Ca^{2+}(aq) + HCO_3^-(aq)$$

What is meant by the symbol '$\rightleftharpoons$'?

This chapter attempts to offer answers to these and other questions.

The First Law of Thermodynamics

In chemistry the transfer of heat (q) into or out of a system, and the work (w) that is done by the system on its surroundings or vice versa are of considerable importance.

One of the fundamental laws of nature is the First Law of Thermodynamics, which states that the total amount of energy in an isolated system is the same before and after a change takes place. Alternatively, energy can neither be destroyed nor created, but it can change its form (Box 2.1).

If the system is *not* isolated, then the total energy of the system plus its surroundings must remain constant. For example, in an electric light bulb the electrical energy used is converted into light, thermal/heat and sound energy (Figure 2.1). By the First Law, the energy going into the bulb must equal what comes out, or $E = L + H + S$.

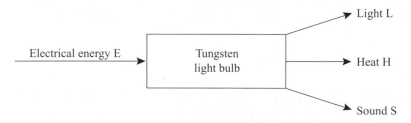

Figure 2.1 Energy types and the tungsten light bulb.

Box 2.1

Some terminology

In the investigation of any change it is important to clearly differentiate between the materials of interest and their **surroundings** (the rest of the Universe!). Any chemicals being investigated are referred to as the **system**, e.g. the contents of a test tube. The system is usually separated from the surroundings by a boundary wall, e.g. the thin glass wall of the test tube. An **isolated system** is insulated from its surroundings so that no energy can enter or leave it. What is added to, or subtracted from, the system either in terms of energy or matter can now be readily identified.

Thermodynamics concerns energy and its movement. If you were to push a light vehicle with wheels (the system), then it will move. An applied **force** has caused movement. If you continued to push the vehicle, then you would soon feel tired because of the work done on the vehicle. You have lost energy to effect the change. Since you are part of the surroundings, then it has also been changed. Assuming there are no frictional forces involved, then the system will gain the energy lost by the surroundings. There has been a transfer of energy from the surroundings to the system and **work** has been done on the system by the surroundings. Thus a definition of energy is the ability to do work. Something that contains a lot of energy has the potential to do a lot of work.

There are various types of energy, e.g electricity, sound, light and matter. Energy which is stored is called potential energy, e.g. chemical energy stored in a battery or in petrol, whilst kinetic energy is energy associated with movement.

If a force (F) is applied continuously to a body, then, providing there are no other forces acting on it, the body will accelerate (or decelerate) for as long as the force is applied. The unit of force, the **Newton (N)**, is that amount of force required to give a mass (m) of 1 kg an acceleration (a) of 1 m s^{-2}. The relationship between F, m and a is given by,

$F = ma$

Work (w) is defined as the transfer of mechanical energy to/from a system to somewhere else. It is energy in transit and has the units of energy. In the case of the vehicle, the work done on it is given by the distance (d) through which the force acts multiplied by the force applied,

$w = Fd$

Work has units of N m or **Joule** (J). The amount of energy transferred when 1 N moves through a distance of 1 m is 1 J.

Heat (q) is the transfer of energy as a consequence of a temperature difference, and has the units of energy, i.e.

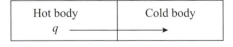

Hot body	Cold body
q ⟶	

higher temperature   lower temperature

The final temperature of the two bodies will be the same for both if their volumes do not change, i.e. some energy has not been lost to the surroundings as a result of the expanding second body pushing and therefore doing work against the atmosphere.

Internal energy changes and enthalpy changes

Heat transferred from the surroundings to a system is given a positive sign $(+q)$; if it is transferred from the system to the surroundings it carries a negative sign $(-q)$. Similarly, if work is done on a system by its surroundings, it carries a positive sign $(+w)$, whereas if work is done by the system on the surroundings it has a negative sign $(-w)$. Figure 2.2 shows heat being added and work being done on a system. The consequence of adding heat to, and doing work on, the system is that a change takes place because the energy of the system has increased. The energy transferred from the surroundings is now inside the system. Before this energy transfer, the system had a certain amount of internal energy 'U_1' and existed in a particular state/condition, State 1. This internal energy is the sum of all the different types of atomic/molecular energies it contains, e.g. translational, rotational, vibrational, electronic and nuclear energies. The addition of more energy thus changes the internal energy to U_2, or State 2. The change in internal energy will be given by,

Figure 2.2 Transferring heat q and doing work w on a system.

$$\Delta U = U_2 - U_1 = q + w$$

where 'Δ' stands for 'change in'.

Often in chemical reactions, heat changes are observed, together with work being done by liberated gases expanding against the atmosphere. The sign of w in such cases will be negative. For example, the addition of hydrochloric acid to granulated zinc in a test tube causes the transfer of heat to the surroundings and the escape of hydrogen gas to the atmosphere.

The heat change to/from a chemical system when measured at constant atmospheric pressure is called the enthalpy change, ΔH, of the reaction. The enthalpy change takes into account both the internal energy change and the work that is done by any expanding gas on the surroundings. The use of enthalpy change rather than change in internal energy is much more convenient in the laboratory or in the field because it is measured at constant surroundings pressure. If the sign of ΔH is *negative* for a chemical reaction, then it is **exothermic**, that is heat is evolved. If the sign is *positive*, then the reaction is **endothermic** and heat is taken in by the reaction.

Standard enthalpy changes

Because an enthalpy change value depends upon the prevailing atmospheric pressure and temperature, then it is convenient to use standard enthalpy changes, measured under standard conditions. Such a change is symbolised by $\Delta H^0_{298\ \text{K}}$. The superscript denotes an atmospheric pressure of $1 \times 10^5\ \text{N m}^{-2}$ and the subscript a temperature of 298 K (or more accurately 298.15 K). For example, the standard enthalpy change of formation

Table 2.1 *Some standard enthalpy changes and their definitions*

Name	Symbol	Definition	Example
Standard enthalpy change of atomisation	$\Delta H^0_{atom,298K}$	The heat change at constant pressure when 1 mole of gaseous atoms are formed from the element in its physical state under standard conditions	$1/2\,H_2(g) \rightarrow H(g)$ $\Delta H^0_{atom,298K}[1/2\,H_2(g)]$ $= +218\,kJ\,mol^{-1}$
Standard enthalpy change of combustion	$\Delta H^0_{comb,298K}$	The heat change at constant pressure when 1 mole of substance undergoes complete combustion under standard conditions	$C_{12}H_{22}O_{11}(s) + 12O_2(g)$ $\rightarrow 12CO_2 + 11H_2O(1)$ $\Delta H^0_{comb,298K}[C_{12}H_{22}O_{11}(s)]$ $= -5\,645\,kJ\,mol^{-1}$
Standard enthalpy change of solution	$\Delta H^0_{soln,298K}$	The heat change at constant pressure when 1 mole of substance in its standard state is dissolved in a solvent such that further dilution produces no further energy change	$NaCl(s) \rightarrow$ $Na^+(aq) + Cl^-(aq)$ $\Delta H^0_{soln,298K} = +3.9\,kJ\,mol^{-1}$

of a substance, $\Delta H^0_{f,298K}$, refers to the enthalpy change of formation of one mole of that substance from its elements under standard conditions. For an element in its standard state, $\Delta H^0_{f,298K} = 0.0\,kJ\,mol^{-1}$. However, this value does depend on the physical state, e.g. $\Delta H^0_{f,298K}[Hg(l)] = 0.0\,kJ\,mol^{-1}$ but $\Delta H^0_{f,298K}[Hg(g)] = +60.84\,kJ\,mol^{-1}$. The endothermic nature of the latter enthalpy change shows that energy is required to change liquid mercury at 298 K to gaseous mercury at the same temperature.

Some various $\Delta H^0_{298\,K}$ values are listed in Table 2.1.

Thus, for the rusting of iron the term $\Delta H^0_{f,298K} = -824.2\,kJ\,mol^{-1}$ is the standard enthalpy change of formation of $Fe_2O_3(s)$ and is exothermic.

The standard enthalpy change of chemical reactions may be determined indirectly using Hess's Law. This states that the total energy change accompanying a chemical reaction is independent of the route by which the reaction proceeds, and is another way of expressing the First Law of Thermodynamics. The use of Hess's Law can be illustrated by examining the determination of $\Delta H^0_{f,298K}$ for methane gas, CH_4,

$$C(s) + 2H_2(g) \longrightarrow CH_4(g)$$

This quantity cannot be directly determined in the laboratory so other reactions are used whose enthalpy changes are known. The values of $\Delta H^0_{f,298K}\{CO_2(g)\}$, $\Delta H^0_{f,298K}\{H_2O(l)\}$ and $\Delta H^0_{comb,298K}\{CH_4(g)\}$ can be used in the following way,

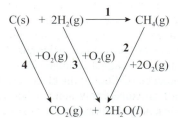

The formation of carbon dioxide and water can be achieved via steps 1 and 2 or by going through steps 3 and 4. According to Hess's Law, no matter what route is taken the overall enthalpy change will be the same.

Thus,

$$1 + 2 = 3 + 4$$

or

$$\Delta H^0_{f,298\,K}\{CH_4(g)\} + \Delta H^0_{comb,298\,K}\{CH_4(g)\} = \Delta H^0_{f,298\,K}\{CO_2(g)\}$$
$$+ 2\Delta H^0_{f,298\,K}\{H_2O(l)\}$$

where

$$\Delta H^0_{comb,298\,K}\{CH_4(g)\} = -890\ kJ\ mol^{-1}$$

$$\Delta H^0_{f,298\,K}\{CO_2(g)\} = -393.5\ kJ\ mol^{-1}$$

$$\Delta H^0_{f,298\,K}\{H_2O(l)\} = -285.9\ kJ\ mol^{-1}$$

Hence,

$$\Delta H^0_{f,298\,K}\{CH_4(g)\} = -393.5\ kJ\ mol^{-1} + (2 \times -285.9\ kJ\ mol^{-1})$$
$$- (-890\ kJ\ mol^{-1})$$
$$= -75.3\ kJ\ mol^{-1}$$

Thus, the enthalpy change of formation of methane is exothermic.

A similar cycle of reactions can be written involving just the enthalpy change of formation of reactants and products. Such cycles show that for any reaction,

$$\Delta H^0_{298\,K}(\text{reaction}) = \Sigma\Delta H^0_{f,298\,K}(\text{products}) - \Sigma\Delta H^0_{f,298\,K}(\text{reactants})$$

where 'Σ' means 'the sum of'.

The enthalpy change of reaction can thus be calculated by subtracting the sum of the enthalpy changes of formation of the reactants from the sum of the enthalpy changes of formation of products (Box 2.2).

The Second and Third Laws of Thermodynamics

The Second Law of Thermodynamics is a directional law – it tries to explain why a chemical change, or any change, proceeds in the way it does.

One way of deciding whether or not a change will take place is to calculate, using statistics, the probability of that change occurring. A high probability would indicate a likely change.

Consider an insulated and therefore isolated container consisting of two halves separated by a partition. In one half there is a gas, in the other a vacuum. If the partition was removed, then the gas would naturally and spontaneously expand until it fills all available space. There is no overall change in internal energy of the gas because the container is insulated. The energy of the gas is the same after the change. Why has the change taken place? Why can't the particles all collect in one half of the container?

Statistically it can be shown that there *is* a chance of finding all of the molecules in one half of the container. However, this chance is exceedingly small when compared with finding the particles evenly distributed between both sides. In the case of this gas,

Box 2.2

The use of standard enthalpy changes of formation in calculating the enthalpy change of a chemical reaction

Calculate the $\Delta H^0_{m,298\,K}$ for the reaction,

$$2N_2O_5(g) = 4NO_2(g) + O_2(g)$$

given that $\Delta H^0_{f,298\,K}$ for $N_2O_5(g)$ is $+11.3$ kJ mol^{-1} and $\Delta H^0_{f,298\,K}$ for $NO_2(g)$ is $+33.2$ kJ mol^{-1}.

The equation can be thought of as being composed of the following cycle:

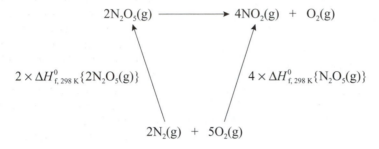

Hence,

$$\Delta H^0_{m,298\,K}\{\text{reaction}\} + 2 \times \Delta H^0_{f,298\,K}\{N_2O_5(g)\} = 4 \times \Delta H^0_{f,298\,K}\{NO_2(g)\}$$

$$\Delta H^0_{m,298\,K}\{\text{reaction}\} = 4 \times \Delta H^0_{f,298\,K}\{NO_2(g)\} - 2 \times \Delta H^0_{f,298\,K}\{N_2O_5(g)\}$$

$$= 4 \times (+33.2) - 2 \times (+11.3)$$

$$= +110.2 \text{ kJ mol}^{-1}$$

Note that the number of moles taking part in the stoichiometry of the reaction must be noted and that the standard enthalpy change of any element, in this case dioxygen, is zero.

chance determines how its particles are distributed with some of the distributions being very much rarer than others. The change observed can be viewed as being statistically by far the most likely one. This calculated probability of a particular state existing, and therefore of the change taking place, is referred to as its W-number.

Any isolated material at constant temperature (T_1) will have a fixed amount of internal energy. If this energy is considered to be quantised, then it is possible to statistically determine how the quanta of energy are distributed between the atoms/molecules of the material. If this distribution of energy is assumed as occurring by chance then it can be shown that the result will be an exponential distribution of the quanta (Figure 2.3).

If another isolated material is at a higher temperature (T_2), then it has a higher internal energy and there are more quanta of energy to distribute. It is also found that the distribution follows a less steep exponential curve. On placing both materials in contact, energy will flow, by chance, naturally and spontaneously from the hot to the cold

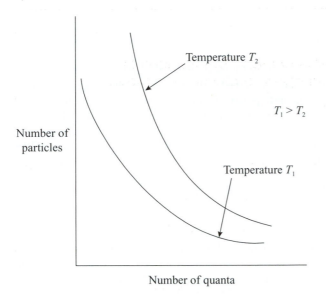

Figure 2.3 The distribution of energy quanta between numbers of particles.

system until both exponential distributions are equally steep. That is, heat flows from a hot body to a cold body until the temperatures are the same. No one has observed heat to flow naturally and spontaneously from a cold body to a hot body without the aid of some external force.

It is found that (i) the more quanta of energy there are to distribute between a fixed number of gas atoms, then the greater is the number of ways (W-number) of doing it, and (ii) this number of ways is increased in a multiplicative manner. As each additional quantum of energy is added that increases the overall number of quanta, then the number of ways of distributing the energy is multiplied by a definite factor 'r'. Hence, if an extra quantum of energy ε is added then this multiplies W by r. Adding another quantum of energy multiplies W by r again, i.e. each additional quantum of energy ε adds $\ln r$ to $\ln W$. As the temperature of a system declines, then the number of quanta available for distribution is less and r becomes larger. The absolute temperature (T) of a system can thus be written,

$$T \propto \varepsilon/(\ln r)$$

Hence, $kT = \varepsilon/(\ln r)$, where k = constant of proportionality, known as the Boltzmann constant.

If the initial W-number is W_1, then, if by adding one quantum it is increased to W_2, the change in W-number is given by,

$$W_1 \times r = W_2$$

Thus, $\ln W_2 - \ln W_2 = \ln r$

or $\quad \Delta \ln W = \ln r$

Since $\quad kT = \varepsilon/(\ln r)$, then $k \, \Delta \ln W = \varepsilon/T$

For many quanta of energy equivalent to a flow of heat 'q', then,

$$k \, \Delta \ln W = q/T$$

Thus q/T becomes a 'counter' of changes in the number of ways the energy can be distributed. This is called the **change in entropy (ΔS)** of the system, i.e. $\Delta S = k \, \Delta \ln W$

or $\quad S_2 - S_1 = k(\ln W_2 - \ln W_1)$

It is possible to calculate the W-number of a particular state, and this number can be related to the chance that the state can exist. The larger the W-number, the more likely is the state to occur. Hence, if a natural spontaneous change takes place in an isolated system, it must be accompanied by an increase in W-number, i.e. a change takes place from a less likely state of lower W-number to a more likely one of higher W-number.

When a maximum W-number is reached, no further change takes place and **equilibrium** is said to be achieved. Thus,

$$\Delta \ln W = \underset{\text{state 2}}{(\ln W_2} - \underset{\text{state 1}}{\ln W_1)}$$

The value of $\Delta \ln W$ is thus positive when W_2 is greater than W_1.

In an isolated system a spontaneous change will take place when ΔS is positive. The Second Law of Thermodynamics therefore states that, in an isolated system, a change will occur if there is an accompanying increase in the entropy of the system.

The change in entropy is given by,

$$\Delta S = q/T$$

The units of ΔS are J K^{-1} mol^{-1}.

At absolute zero of temperature (0 K) all material motions cease and matter is in a state of lowest energy. The number of ways of achieving this state is one, i.e. W-number $= 1$. Thus, $S_0 = k \ln W = 0$. At absolute zero, if matter is all assumed to be in the form of perfect crystals, then their entropy will be zero. This latter statement is known as the Third Law of Thermodynamics.

Entropy values

Entropy values have been tabulated for the elements and a wide range of their compounds. They are quoted at 298.15 K at a pressure of 1 atmosphere. The symbol used is $S^0_{298\,K}$ and not $\Delta S^0_{298\,K}$ because they are measured relative to absolute zero when $S_{0\,K} = 0$.

The order of magnitude of $S^0_{298\,K}$ depends upon the 'freedom' of the particles of a system. Since gaseous particles are able to move much more freely, are less ordered and have more ways available for distribution of energy than more tightly bound solid particles, gases have higher entropy values. For example,

$$S^0_{298\,K} \; \text{Li(s)} \quad = \quad 28.0 \; \text{J K}^{-1} \, \text{mol}^{-1}$$

$$S^0_{298\,K} \; H_2O(l) = \quad 69.9 \; \text{J K}^{-1} \, \text{mol}^{-1}$$

$$S^0_{298\,K} \; F_2(g) \quad = \quad 203.0 \; \text{J K}^{-1} \, \text{mol}^{-1}$$

To calculate the entropy change accompanying a chemical reaction, then, the following formula can be used,

$$\Delta S^0_{298\,K} = \Sigma(S^0_{298\,K})_{\text{products}} - \Sigma(S^0_{298\,K})_{\text{reactants}}$$

Tabulated data are inserted appropriately:

e.g. $\qquad\qquad\qquad\qquad$ $H_2(g) + Cl_2(g) \longrightarrow 2HCl(g)$

$S^0_{298\,K}/J\,K^{-1}\,mol^{-1}$ $\qquad$ 130.6 $\qquad$ 223 $\qquad\qquad$ 186.7

Hence, $\Delta S^0_{298\,K} = (2 \times 186.7) - (130.6 + 223) = 19.8\ J\,K^{-1}\,mol^{-1}$

The reaction between hydrogen and chlorine is accompanied by an increase in entropy. It would appear that this reaction is thermodynamically favourable because $\Delta S^0_{298\,K}$ has a positive value.

Gibbs' Free Energy

Does the kind of calculation described above give a clear enough indication of how thermodynamically favourable or otherwise a chemical reaction is? In order to answer this question, consider again the reaction,

$$2Fe(s) + 3/2O_2(g) \;=\; Fe_2O_3(s)$$

$$\Delta H^0_{f,298\,K} \;=\; -824.2\ kJ\,mol^{-1}$$

$$\Delta S^0_{298\,K} \;=\; -274.9\ J\,K^{-1}\,mol^{-1}$$

The value of $\Delta S^0_{298\,K}$ is negative. According to previous discussion, for an isolated reaction to proceed $\Delta S^0_{298\,K}$ must be positive. It would appear that the Second Law has been violated. What if the above is *not* an isolated system/reaction? Most reactions taking place in the environment are not isolated reactions – they occur in conjunction with their surroundings. The negative enthalpy change indicates that energy is being transferred from the system to the surroundings. The entropy of the surroundings will be changed as well as that of the system. The total entropy change is now given by,

$$\Delta S^0_{298\,K}(total) \;=\; \Delta S^0_{298\,K}(system) + \Delta S^0_{298\,K}(surroundings)$$

$$=\; \Delta S^0_{298\,K}(system) + q/T(surroundings)$$

Now, $q(surroundings) = -\Delta H^0_{298\,K}(system)$ because 'q' has the opposite sign of $\Delta H^0_{298\,K}(system)$ since the surroundings have acquired energy equal in magnitude to that lost by the system.
   Therefore,

$$\Delta S^0_{298\,K}(total) = \Delta S^0_{298\,K}(system) - \Delta H^0_{298\,K}/T(system)$$

So, for the above reaction,

$$\Delta S^0_{298\,K}(total) \;=\; \left[-274.9 - \left(\frac{-824.2 \times 10^3}{298.15} \right) \right] J\,K^{-1}\,mol^{-1}$$

$$=\; +3{,}039.3\ J\,K^{-1}\,mol^{-1}$$

As a result of the transfer of heat of 824.2 kJ to the surroundings the *total* entropy change is +3,039.3 $J\,K^{-1}\,mol^{-1}$. The Second Law has *not* been violated.

A much more useful definition of the Second Law is that, if the system and the surroundings are the same as the Universe, then the entropy of the Universe is always increasing.

In the above example, all of the available energy has been used to warm the surroundings. All that is needed is to transfer just enough energy to the surroundings to ensure that the total entropy change is not negative, i.e. to compensate for -274.9 J K^{-1} mol^{-1} the entropy of the surroundings must increase by a minimum of $+274.9$ J K^{-1} mol^{-1}, so that $\Delta S^0_{298\,K}(\text{total}) = 0.0$ J K^{-1} mol^{-1}.

Hence, if $\quad \Delta S^0_{298\,K}(\text{surroundings}) = +q/T(\text{surroundings})$

$$\text{then} \quad +q(\text{surroundings}) = T\,\Delta S^0_{298\,K}(\text{surroundings})$$

$$= 298.15\ \text{K} \times +274.9\ \text{J K}^{-1}\ \text{mol}^{-1}$$

$$= +81.96\ \text{kJ mol}^{-1}$$

The surroundings need only gain 81.96 kJ mol^{-1} of energy.

The rest $(824.2 - 81.96) = +742.2$ kJ mol^{-1}, could be made usefully available.

$$\text{If} \quad \Delta S^0_{298\,K}(\text{total}) = \Delta S^0_{298\,K}(\text{system}) - \Delta H^0_{298\,K}(\text{system})$$

$$\text{then} \quad T\,\Delta S^0_{298\,K}(\text{total}) = T\,\Delta S^0_{298\,K}(\text{system}) - \Delta H^0_{298\,K}(\text{system})$$

$$\text{Hence,} \quad -T\,\Delta S^0_{298\,K}(\text{total}) = -T\,\Delta S^0_{298\,K}(\text{system}) + \Delta H^0_{298\,K}(\text{system})$$

$$\text{Let} \quad -T\,\Delta S^0_{298\,K}(\text{total}) = \Delta G^0_{298\,K}$$

$$\text{then} \quad \Delta G^0_{298\,K} = \Delta H^0_{298\,K}(\text{system}) - T\,\Delta S^0_{298\,K}(\text{system})$$

This latter equation can now be used to determine the thermodynamic feasibility of the reaction (or any other reaction!),

$$\text{i.e.} \quad \Delta G^0_{298\,K} = \left(-824.2 - \frac{298.15 \times -274.9}{10^3}\right)\ \text{kJ mol}^{-1}$$

$$= -742.2\ \text{kJ mol}^{-1}$$

It is the value of $\Delta G^0_{298\,K}$ that defines the criterion for a reaction to be thermodynamically favourable, i.e. $\Delta G^0_{298\,K}$ must have a negative value. $\Delta G^0_{298\,K}$ is called the **Gibbs' Free Energy change**.

Chemical equilibrium

Equilibrium may be defined as the point when no further change is seen to take place in a system, i.e. $\Delta S = 0$, and the change is complete.

At high temperatures, hydrogen and iodine will react to produce hydrogen iodide,

$$H_2(g) + I_2(g) \longrightarrow 2HI(g)$$

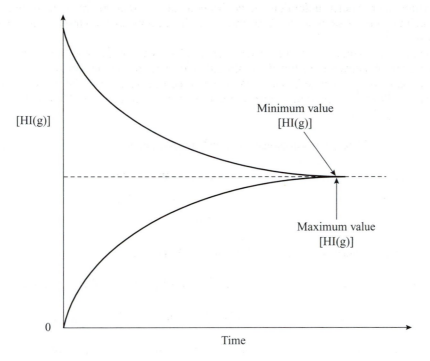

Figure 2.4 A graph of concentration of HI(g) vs time.

If this reaction is attempted in an enclosed container, eventually it will be observed that all chemical species will be present and the concentration of the hydrogen iodide will reach a maximum value. Similarly, if hydrogen iodide is decomposed under the same conditions as before,

$$2HI(g) \longrightarrow H_2(g) + I_2(g)$$

then again all species will be present and the concentration of HI will reach a minimum value. The maximum and minimum values will be identical and define the position of chemical equilibrium (Figure 2.4).

The two reactions can be represented by,

$$H_2(g) + I_2(g) \rightleftharpoons 2HI(g)$$

It is found experimentally that, if the concentrations of the three chemical species are measured at equilibrium,

$$\frac{[HI(g)]^2_{eq}}{[H_2(g)]_{eq}[I_2(g)]_{eq}} = \text{a constant}$$

The constant is called the **experimental equilibrium constant (K_c)**. It is a characteristic constant for a chemical reaction for the conditions under which equilibrium has been established.

For the general reaction,

$$aA + bB \rightleftharpoons cC + dD$$

where A, B, etc. are the chemical species and a, b, etc. the stoichiometric coefficients,

$$K_c = \frac{[C]_{eq}^c [D]_{eq}^d}{[A]_{eq}^a [B]_{eq}^b}$$

This equation represents the **Law of Chemical Equilibrium**. Note that the concentrations are always measured at equilibrium.

Applying the equilibrium law to the reaction,

$$2SO_2(g) + O_2(g) \rightleftharpoons 2SO_3(g)$$

$$K_c = \frac{[SO_3(g)]_{eq}^2}{[SO_2(g)]_{eq}^2 [O_2(g)]_{eq}}$$

This will have units of $[(mol\ dm^{-3})^2]/[(mol\ dm^{-3})^2(mol\ dm^{-3})]$ or $dm^3\ mol^{-1}$.

Different stoichiometries will produce different expressions for this equilibrium constant and therefore the units are important.

With gases it is often more convenient to work in terms of their partial pressures, i.e. the pressure an individual gas would exert if it occupied the container by itself. In such cases, K_c would be replaced by K_p and concentration [ ] by partial pressures measured at equilibrium, hence, for the above chemical reaction,

$$K_p = \frac{\{p(SO_3)\}^2}{\{p(SO_2)\}^2 p(O_2)}$$

If, in this expression for K_p, the unit of pressure used is the atmosphere, then the units now will be atm^{-1}.

The relationship between K_c and K_p is given by,

$$K_p = K_c(RT)^{\Delta n}$$

where
$\Delta n = $ {number of moles of gaseous products} $-$ {number of moles of gaseous reactants}
$R = $ universal gas constant
$T = $ absolute temperature

Hence, for the SO_2 reaction, $K_p = K_c(RT)^{2-3} = K_c(RT)^{-1}$

or $K_p = K_c/RT$

When writing an experimental equilibrium constant for an equilibrium involving more than one phase, then the concentrations of all pure liquids and solids are constants. For example, the thermal decomposition of calcium carbonate at constant temperature produces calcium oxide and carbon dioxide,

$$CaCO_3(s) \rightleftharpoons CaO(s) + CO_2(g)$$

for which

$$K = \frac{[CaO(s)]_{eq}[CO_2(g)]_{eq}}{[CaCO_3(s)]_{eq}}$$

For solids the concentration is constant, hence,

$$\frac{K[CaCO_3(s)]_{eq}}{[CaO(s)]_{eq}} = [CO_2(g)]_{eq} = K_c$$

Gibbs' Free Energy and equilibrium

It can be shown that $\Delta G^0 = -RT \ln K^0$ where K^0 is termed the thermodynamic equilibrium constant. Now K^0 is numerically the same as K_c or K_p *but* is unitless. For example,

$$K_c = \frac{\{[SO_3]/mol\ dm^{-3}\}^2}{\{[SO_2]/\ mol\ dm^{-3}\}^2\{[O_2]/\ mol\ dm^{-3}\}}$$

Now, $\quad\quad\quad\quad\quad \Delta G^0 = \Delta H^0 - T\Delta S^0$

Therefore, $\quad -RT \ln K^0 = \Delta H^0 - T\Delta S^0$

or $\quad\quad\quad\quad\quad \ln K^0 = [-\Delta H^0/RT] + [T\Delta S^0/RT]$

Therefore, $\quad\quad \ln K^0 = [-\Delta H^0/RT] + [\Delta S^0/T]$

K^0 is thus defined under standard conditions and is termed the **standard** or **thermodynamic equilibrium constant**. If ΔS^0, ΔH^0 and T are known, then the value of K^0 can be determined.

For the reaction,

$$2SO_2(g) + O_2(g) \rightleftharpoons 2SO_3(g)$$

the entropy change and enthalpy change for 1 mole of product should be calculated by,

$$\Delta S^0_{298\,K} = -94.0\ J\ K^{-1}\ mol^{-1}$$

$$\Delta H^0_{298\,K} = -98.9\ kJ\ mol^{-1}$$

Therefore, $\quad \ln K^0 = \dfrac{-94.0}{8.314} - \dfrac{-98.9 \times 1000}{8.314 \times 298.15}$

$$= 28.59$$

$$K^0 = 2.6 \times 10^{12}$$

The value of K^0 indicates the position of an equilibrium, i.e. a large positive value indicates that the numerator in the expression for K^0 is much larger than the denominator.

That is, the equilibrium lies well over to the right/products side of the chemical equation. A large negative value indicates that the position of the equilibrium lies well over to the left/reactants side with very little product(s) being present at equilibrium. In the above example, the value of K^0 indicates that the production of sulphur trioxide from sulphur dioxide and oxygen under standard conditions is highly thermodynamically favourable.

What happens if the prevailing conditions change or a product is removed from an equilibrium? One useful 'rule of thumb' is called le Chatelier's Principle. This states that an equilibrium will adjust in order to annul the effect of any change that has been forced upon it. In the case of the sulphur dioxide reaction when equilibrium is achieved, it is accompanied by a reduction in pressure because fewer gas molecules result. If the pressure were to be increased by the application of an external force, then the reaction would move from left to right to annul this pressure increase. There would thus be an increase in amount of product in the reaction mixture. The reaction is also exothermic and therefore gives out heat. If the temperature was increased by adding more heat, then the equilibrium would again adjust to reduce the temperature. Thus, the reaction would move from right to left and less product would result. If sulphur trioxide were to be removed from the equilibrium, then it would again move from left to right to fill the 'hole' created.

Rates of reaction

For the reaction,

$$H_2(g) + 1/2\ O_2(g) \longrightarrow H_2O(l)$$

$$\Delta H^0_{298\,K} = -285.8\ \text{kJ mol}^{-1}$$

$$\Delta S^0_{298\,K} = -163.4\ \text{J K}^{-1}\ \text{mol}^{-1}$$

$$\Delta G^0_{298\,K} = -273.1\ \text{kJ mol}^{-1}$$

The reaction is thus thermodynamically favourable. However, if a stoichiometric mixture of hydrogen and oxygen is allowed to stand at 298 K in a container, the reaction does not proceed naturally and spontaneously. Upon applying a lighted spill the reaction proceeds violently, suggesting that there is some kind of barrier to be overcome. The lit match provides the initial energy to start the reaction. Reactions like this are said to be **kinetically controlled**.

Figure 2.5 shows the reaction profile for the above reaction, where the reaction co-ordinate represents the extent of reaction, or how far the reaction has proceeded. The energy barrier that a reaction has to overcome is termed the **activation energy (E_a)**. It is this energy barrier that prevents the reaction occurring.

Simple laboratory experiments show that the rate of a chemical reaction depends on a number of factors. Generally the rate will be increased by:

- increasing the temperature;
- increasing the concentration of the reactants – the relationship between the rate and concentrations of chemical species involved is the rate equation or rate law;
- increasing the state of division of reactants – in the case of solids, the smaller the particles the greater the surface area and the faster the reaction;
- the presence of a **catalyst** – this is a specific substance which remains chemically unchanged at the end of the reaction.

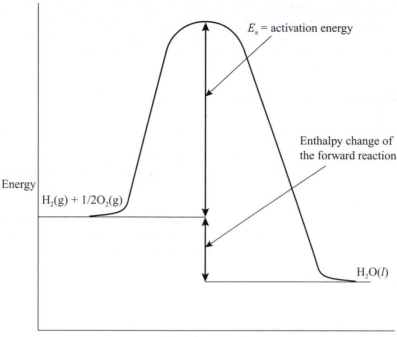

Figure 2.5 Reaction profile for $H_2(g) + 1/2 O_2(g) = H_2O(l)$.

Catalysis

A catalyst enhances the rate of a chemical reaction. Alternatively, such a substance may slow down a chemical reaction. In the latter case it is referred to as an **inhibitor**. If a catalyst is in the same phase (e.g. in solution) as the reactants, then it is taking part in homogeneous catalysis. If it is in a different phase to the reactants, e.g. gaseous reactants using a solid catalyst, it is taking part in heterogeneous catalysis.

Chemical reactions proceed by one or more steps called the reaction pathway or mechanism. The stoichiometry of a reaction represents the overall reaction and does not indicate the underlying steps. Catalysts act by providing an alternative, less energetically demanding, reaction pathway, i.e. a new **mechanism** exists. A larger number of steps are involved than in the case of the uncatalysed reaction. The lower activation energy involved means that the reaction will proceed more rapidly. In all catalysed reactions, the position of an equilibrium is unchanged and so the amount of required product(s) produced is the same. Both the rate of the forward reaction and reverse reaction are speeded up to the same extent. However, equilibrium is achieved faster. A catalyst is nearly always quite specific to the reaction it catalyses. How selective a catalyst is in reducing the number of alternative reactions is crucial in its selection.

The transition elements are components of many catalysts used in nature and in industry. These metals are good catalysts for a number of reasons. First, they form a range of stable oxidation states, e.g. iron(II), Fe^{2+}, and iron(III), Fe^{3+}. Second, transition metal ions can form a wide range of complex molecules/ions with a wide range of other ions and neutral molecules, including organic molecules. Third, they can adopt a range of co-ordination numbers, i.e. the number of molecules and/or ions they can form bonds with, and therefore molecular shapes. Fourth, how a transition metal complex will react is influenced by the ions or molecules it is attached to – hence a transition metal complex can be designed to do a particular chemical job.

Reaction rate and concentration

There are two ways in which the dependency of the rate of reaction on concentration can be expressed/determined. The first involves measuring how the concentration of each individual reactant varies with time, at a constant temperature. Graphs of concentration (or some property that is proportional to concentration) vs time, called **reaction profiles**, can then be plotted (Figure 2.6). The chemical species being folowsed must be clearly identified. From the graph of concentration vs time, the rate of reaction 'J' at various concentrations can be calculated from tangents drawn to these reaction profiles. The rate J can then be written as the change in concentration (d[A]) of a chosen species A with respect to change in time (dt), i.e. for the rate of disappearance of A:

$$\text{rate} \ = \ J \ = \ -\text{d}[A]/\text{d}t$$

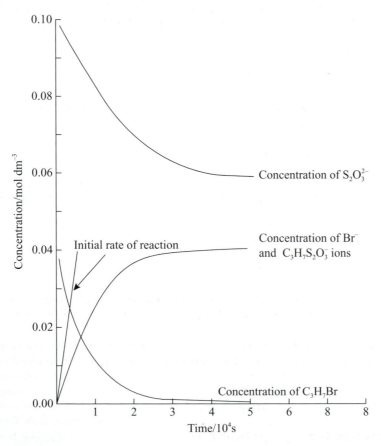

Figure 2.6 The reaction profile at 310 K for the chemical reaction between thiosulphate ion and 1-bromopropane.

Note that, if the rate of disappearance of a reactant is being followed, then the rate will have a negative sign, and for the rate of appearance of a product a positive sign.

For the general reaction,

$$aA + bB \rightleftharpoons cC + dD$$

the equation showing the relationship between **rate** and **concentration** takes the form,

$$\text{rate} = k_r[A]^\alpha[B]^\beta$$

In this expression, k_r is the rate constant which is a characteristic of the reaction being investigated. A and B are the reacting species. The exponents α and β are the **partial order** of reaction with respect to A and B, respectively. The **overall order** is the sum of α and β.

If the reaction were first order in A and first order in B, then the data would fit the equation,

$$\text{rate} = k_r[A][B]$$

The overall order is second (i.e. $\alpha + \beta = 2$).

To illustrate this further, if the reaction has a partial order of zero with respect to B but first order with respect to A, then the observed data would fit the equation,

$$\text{rate} = J = -d[A]/dt = k_r[A]^1[B]^0$$
$$= k_r[A]$$

and the overall order would be first. Note that the units of J are mol dm^{-3} s^{-1}.

The sign of J is very important. Consider, for example, the reaction,

$$2N_2O_5(g) = 4NO_2(g) + O_2(g)$$

From the stoichiometry of this equation, N_2O_5 would react/disappear at twice the rate that O_2 would be formed, and NO_2 would be formed at four times the rate of that of dioxygen. Hence,

$$J = -\frac{1}{2}\frac{d[N_2O_5]}{dt} = \frac{1}{4}\frac{d[NO_2]}{dt} = \frac{d[O_2]}{dt}$$

Equations involving rate of reaction are called **differential rate equations**. Details of how the differential rate law is determined experimentally are given in Boxes 2.3 and 2.4.

The value of a rate constant is independent of the concentrations of the reactants or products but the units depended upon the form of the rate law. For example, for the reaction,

$$S_2O_3^{2-}(aq) + C_3H_7Br(1) = C_3H_7S_2O_3^-(aq) + Br^-(aq)$$

the rate law has been found by experiment to be,

$$J = k_r[S_2O_3^{2-}][C_3H_7Br]$$

Box 2.3

The determination of the rate law – initial rate method

1   The stoichiometry of the chemical reaction is first established,

e.g. $2N_2O_5(g) = 4NO_2(g) + O_2(g)$

2   The expression for the rate of reaction is written,

$$J = \frac{-1}{2}\frac{d[N_2O_5]}{dt} = \frac{1}{4}\frac{d[NO_2]}{dt} = \frac{d[O_2]}{dt}$$

3   Assume $J = k_r[N_2O_5]^\alpha$ and taking logs to the base 10,

$$\log J = \log k_r + \alpha \log [N_2O_5]$$

4   How the concentration of N_2O_5 varies with time as the reaction progresses is determined, and the reaction profile plotted.

5   Values of J at various concentrations along the reaction profile are calculated.

6   The **initial rate method** is used when the products are able to decompose or interfere in some way with the progress of the reaction. In this case a tangent is drawn at $t = 0$ to determine the initial rate J_0. Several experiments are done to determine the initial rate at different starting concentrations $[N_2O_5]_0$.

7   If the initial rate is used, then we write,

$$\log J_0 = \log k_r + \alpha \log [N_2O_5]_0$$

A plot of $\log J_0$ vs $\log [N_2O_5]_0$ will give a straight line graph of slope equal to 'α'. The rate constant can then be calculated.

The units of k_r are thus given by,

$$k_r = J/[S_2O_3^{2-}][C_3H_7Br]$$
$$= \text{mol dm}^{-3}\text{ s}^{-1}/\text{mol dm}^{-3} \times \text{mol dm}^{-3}$$
$$= \text{dm}^3\text{ mol}^{-1}\text{ s}^{-1}$$

Note that, according to the latter rate law, the reaction has an overall order of two but has partial orders of one with respect to both the reactants, thiosulphate ($S_2O_3^{2-}$) ion and 1-bromopropane (C_3H_7Br). Figure 2.6 shows a reaction profile at 310 K for this reaction. Note that at about 5×10^4 seconds into the reaction, the concentrations of the products and reactants are approximately constant. Hence the reaction is kinetically complete. If the concentrations of the products and the reactants are summed at any time, then it will be seen that this is a constant value, and is equal to the sum of the

Box 2.4

The determination of the rate law – the isolation technique

When more than one species is involved in the rate equation, then the initial rate technique is coupled with the **isolation technique**. This technique is used to prevent complex solutions to finding the rate law. When two or more reactants are involved, the isolation technique requires that all but one reactant is present in large excess. Thus, the change in concentration of those reactants in large excess can be ignored – their concentrations can be considered to stay constant. Consider the reaction,

$$S_2O_3^{2-}(aq) + C_3H_7Br(l) = C_3H_7\,S_2O_3^{2-}(aq) + Br^-(aq)$$

The rate law can be written,

$$J = k_r[S_2O_3^{2-}]^\alpha [C_3H_7Br]^\beta$$

When the concentration of the thiosulphate ion ($S_2O_3^{2-}$) is very much greater than that of the 1-bromopropane (C_3H_7Br), then the former's concentration will not differ significantly from its concentration at time zero! Thus the rate law can now be written as,

$$J = k_r[S_2O_3^{2-}]_0^\alpha [C_3H_7Br]^\beta$$

Since $K_r[S_2O_3^{2-}]_0^\alpha$ is constant (k_a), then J becomes,

$$J = k_a[C_3H_7Br]^\alpha$$

The new constant k_a is called the pseudo rate constant. A plot of $\log J$ vs $\log [C_3H_7Br]$ will give a straight line of slope 'β'. Thus the partial order of the reaction with respect to 1-bromopropane can be found.

   The above process is then repeated for thiosulphate ion, i.e. the concentration of the 1-bromopropane is made in large excess and 'α' found. This procedure can be carried out for any number of reactants.

initial concentrations of reactants. Thus, this chemical reaction shows time-independent stoichiometry, i.e. the chemical equation as written above applies at any time and the reaction is a simple one.

   The second way is to see if the concentration vs time data fit an equation derived on the basis of an assumed order. For example, if the partial order of reaction is assumed to be first in 'A' and zero order in 'B', the data would fit the equation,

$$\ln [A]_0 - \ln [A] = ak_r t$$

where $[A]_0$ is the concentration of 'A' at time zero. A plot of $\ln [A]$ vs time (t) would give a straight line of slope equal to ak_r.

This latter equation is the mathematically integrated form of the rate equation,

$$J = k_r[\text{A}]^1[\text{B}]^0$$

If the reaction was first order in A and first order in B, then the appropriate equation would be,

$$\ln\{[\text{A}]/[\text{B}]\} = \{[\text{A}]_0 - [\text{B}]_0\}k_r t + \ln\{[\text{A}]_0/[\text{B}]_0\}$$

In this case, if the data fitted the above equation, then a plot of $\ln\{[\text{A}]/[\text{B}]\}$ vs t would yield a straight line. This latter equation is the 'mathematically integrated' form of the rate equation,

$$J = k_r[\text{A}][\text{B}]$$

Equations of the type used in these two examples are called **integrated rate equations**. Such equations are particularly useful when more than 50 per cent of the reaction has been completed.

A third way of determining the rate equation uses the half-life approach. The half-life ($t_{1/2}$) is the time taken for the concentration of a reactant to fall to half its original value, i.e. $[\text{A}]_0$ to $[\text{A}]_0/2$. Thus from a reaction profile it is possible to determine a number of half-lives merely by selecting a concentration at any time and finding the difference between this figure and the time taken to reach half that concentration. The half-life depends upon the order of reaction and the initial concentrations of reactants, i.e.

$$t_{1/2} \text{ proportional to } 1/[\text{A}]_0^{n-1}$$

where n = the partial order with respect to A.

Thus, a plot of $t_{1/2}$ vs $1/[\text{A}]_0$ gives a straight line for a second-order ($n = 2$) reaction in A, since $(n-1) = 1$.

So far, elementary reactions have been considered, i.e. one that occurs in a single step process and passes through a single activated complex. Many chemical reactions undergo more than one elementary step and are referred to as **composite reactions**. The steps constitute a **composite mechanism**. How is a composite reaction detected?

The first piece of evidence lies in the examination of the concentrations of reactants and products as the reaction progresses. If the stoichiometric equation is followed throughout a chemical reaction, then there is a simple relationship between the stoichiometric coefficients and the extent of reaction. For example, suppose a reaction occurs according to the equation,

$$2\text{A} + \text{B} = \text{C} + \text{D}$$

If the initial concentration of A was 1.0 mol dm^{-3} and after a certain time its concentration was found to be 0.5 mol dm^{-3}, and of B, C and D all 0.25 mol dm^{-3}, then a simple reaction could be suspected. If this ratio of 2:1:1:1 of the concentrations of A, B, C and D, respectively, was sustained throughout the reaction, then the reaction is simple and has time-independent stoichiometry. If time-*dependent* stoichiometry was found, i.e. a changing ratio of concentrations was detected, then the reaction is a composite one. In such reactions the stoichiometric equation applies only at the beginning of a reaction and at its end.

The second piece of evidence lies in the detection of intermediates. If a reaction is composed of a number of steps, then during its course it may be possible to detect short-lived intermediate chemical species.

Third, if the stoichiometry includes more than three reactants, then it is a composite reaction because the likelihood of three or more chemical species all meeting at the same time with the right energy to react would be remote.

Finally, if the rate law is complex (with fractional orders, negative orders), then a composite reaction is involved. There is no relationship between the stoichiometric equation and the experimentally determined rate equation. For example, the rate law for the reaction,

$$H_2(g) + Br_2(g) = 2HBr(g)$$

is given by,

$$J = \frac{k_1[H_2][Br_2]^{1/2}}{[Br_2] + k_2[HBr]}$$

where k_1 and k_2 are two different constants.

The form a rate law takes, and what the values of the rate constants are, can only be established by experiment. The more complex the relationship between rate and the concentrations of species involved, the more complex is the 'story' that underpins the stoichiometry of the reaction.

Rate equations and the mechanism of a reaction

Establishing the order of a reaction with regard to each of the chemical species that affect the rate is crucial from the point of trying to elucidate how a chemical reaction may be proceeding. If this is understood at the molecular level, then it can lead to the control of that reaction. The number of steps involved are a matter of speculation, but they must be chemically viable. The **reaction pathway** is the series of steps that the reaction is thought to undertake.

A mechanism lists the species colliding in each step, and includes details of any so-called **activated complexes**. These complexes are suggested entities of a very transient nature that are formed when the necessary activation energy is acquired. Such complexes have never been isolated and are usually placed on the top of the reaction profile diagram. For example, the single step reaction between diiodine and dihydrogen can be represented by,

$$H_2(g) + I_2(g) \longrightarrow H_2I_2^{\ddagger} \xrightarrow{\text{very fast step}} 2HI(g)$$

Here, the two reactant molecules have formed the activated complex $H_2I_2^{\ddagger}$ so the step is called a bimolecular step. Note that this reaction is a simple one in that it is first order with respect to diiodine and first order with respect to dihydrogen. If only one molecule is involved in a step in a mechanism, it is referred to as a unimolecular step, and if three are involved a termolecular step. Three-molecule collisions are very rare and there is no need to consider collisions involving more than three molecules.

The slowest step in a mechanism is the one that controls the overall rate of reaction and is called the **rate-determining step**. For each step the molecularity and the order of reaction are the same, e.g. a bimolecular step follows second-order kinetics.

How then is the link made between the stoichiometry of a reaction, its rate law and a mechanism? Consider the reaction between nitrogen(II) oxide and dioxygen,

$$2NO(g) + O_2(g) = 2NO_2(g)$$

The experimental rate equation for this reaction is,

$$d[NO_2]/dt = k_r[NO]^2[O_2]$$

The stoichiometry would suggest that the reaction is termolecular but these are rare. Investigations have also shown that the short-lived species N_2O_2 is present in the reaction mixture. A possible explanation for these observations is via a two-step composite mechanism,

Step 1 $NO(g) + NO(g) \rightleftharpoons N_2O_2(g)$

Step 2 $N_2O_2(g) + O_2(g) = 2NO_2(g)$

Step 1 is the pre-equilibrium step for which an equilibrium constant can be written,

$$K_c = \frac{[N_2O_2(g)]}{[NO(g)]^2}$$

Since N_2O_4 is a short-lived intermediate, we can apply a very useful rule to its formation known as the **Steady State Approximation**. The concentration of this intermediate never becomes significant compared with those of the reactants and product. The assumption that the rate of formation of the intermediate is equal to zero can be made, i.e. that its concentration does not significantly change with time. Thus the rate equation for the intermediate will be equal to zero.

The rate of establishment of the pre-equilibrium step is very much faster than Step 2. Hence the slowest step and therefore the rate-determining step is Step 2. This step is bimolecular, and so the rate law will be,

$$J = k_2[N_2O_2][O_2]$$

Rearranging the above expression for K_c,

$$[N_2O_2(g)] = K_c[NO(g)]^2$$

and substituting into the rate law gives,

$$J = k_2 K_c[NO(g)]^2[O_2]$$

Hence, the theoretically derived rate equation is the same as the experimentally determined one where $k_r = k_2 K_c$. Note also that the addition of the steps in the proposed mechanism produces the overall stoichiometric equation.

It is also interesting to note that the rate for this reaction decreases with increasing temperature (see p. 65). It is an exothermic reaction so that, as the temperature increases, the pre-equilibrium will shift to the left-hand side. Therefore K_c will decrease in value and so therefore will $k_2 K_c$. Again, the mechanism explains experimental observation.

Rate of reaction and temperature

The requirement for reaction between two chemical species is that they must collide in the correct orientation with respect to each other and have the right amount of energy. The rate will depend upon the number of collisions, but how exactly? In a gaseous mixture, for example, there will be many collisions occurring every second, but not all will lead to chemical reaction. In order to react, chemicals must have the minimum amount of energy called the **activation energy (E_a)**. By increasing the temperature of a reaction, the number of molecules possessing this minimal amount of energy will increase and thus so will the rate.

For the first-order rate equation,

$$\text{rate} = k_r[A]_0$$

if the initial concentration of A is kept constant, it will be found that as the temperature increases so will the rate. Since $[A]_0$ is constant, then it must be k_r that increases. The rate constant is temperature dependent. The relationship between k_r and the temperature of a reaction is given by,

$$k_r = A\exp(-E_a/RT)$$

where A = pre-exponential factor
E_a = activation energy
R = universal gas constant
T = absolute temperature

This equation is called the **Arrhenius Equation**. Alternatively, it may be more usefully written,

$$\ln k_r = \ln A - E_a/RT$$

A plot of $\ln k_r$ vs $1/T$ will give a straight line of slope $(-E_a/R)$, from which the activation energy can be determined (Figure 2.7).

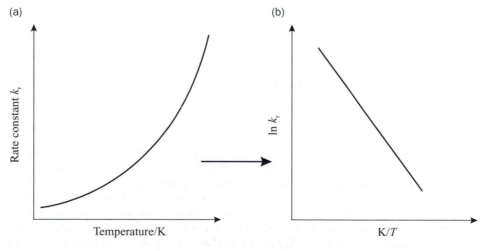

(a)　　　　　　　　　　　　　　　　　(b)

Rate constant k_r ... Temperature/K

$\ln k_r$ ... K/T

Figure 2.7 (a) The shape of the k_r vs temperature graph. (b) The Arrhenius plot.

Rate of reaction and thermodynamics

Returning to the reaction,

$$2SO_2(g) + O_2(g) \rightleftharpoons 2SO_3(g)$$

which is used industrially to produce sulphuric acid as well as occurring in a polluted atmosphere. Because it is an exothermic reaction, by le Chatelier's Principle, lowering the temperature will cause the equilibrium to move to the right and more sulphur trioxide will be produced. However, if the temperature is lowered too much, the rate of reaction becomes too slow. Hence, in industry a catalyst of vanadium(V) oxide is used to enable a compromise to be reached between rate and thermodynamic viability. The reaction is also run with a slight excess of oxygen at a temperature of 450 °C to give a yield of the order of 95 per cent. An increase in pressure would also favour a higher yield. Since such a good yield is produced just by temperature control, the added expense of higher pressures can be avoided.

Remember a particular reaction may be thermodynamically possible (i.e. ΔG^0 is negative) but kinetically too slow to be of importance! Some reactions are so thermodynamically viable with a low energy of activation that negative catalysts, known as inhibitors, have to be added to prevent fast and violent reactions.

Consider the following reactions of methanol, CH_3OH, with dioxygen:

$$CH_3OH(l) + 3/2O_2(g) = CO_2(g) + 2H_2O(l)$$
$$\Delta G^0_{298\,K} = -702.5 \text{ kJ mol}^{-1}$$

$$CH_3OH(l) + 1/2\,O_2(g) = HCHO(g) + H_2O(l)$$
$$\Delta G^0_{298\,K} = -169.2 \text{ kJ mol}^{-1}$$

Both reactions are thermodynamically favourable. The former though has a large activation energy. Hence, to obtain the required product of methanal in the second reaction, conditions are controlled so that this reaction proceeds faster than the first, i.e. an appropriate catalyst is used.

Summary

- The First Law of Thermodynamics – energy can neither be created nor destroyed.
- The standard enthalpy change of reaction $\Delta H^0_{298\,K}$ is the heat change accompanying a chemical reaction at constant pressure.
- The Second Law of Thermodynamics – in an isolated system a change will be possible if there is an increase in entropy S of that system, i.e. ΔS^0(reaction) is positive.
- The entropy change of a reaction is a measure of how likely the reaction is to occur.
- A chemical reaction will be thermodynamically favourable if the Gibbs' Free Energy change is negative.
- The experimental equilibrium constant K_c is a characteristic of a chemical reaction for a given set of conditions under which equilibrium was established.
- The value of standard equilibrium constant K^0 indicates the position of an equilibrium and therefore the ratio of the concentrations of reactants to products at equilibrium.
- Le Chatelier's Principle is used to qualitatively predict the outcome of changing variables on the position of an equilibrium.
- Catalysts affect the rate of a chemical reaction, but are not themselves consumed.

- The rate of a chemical reaction may depend upon the concentration of its reactants. This dependency is determined by experiment and expressed in the form of a rate law.
- The overall order of a chemical reaction is the sum of the partial orders to which the reactants' concentrations have been raised in the differential rate law.
- Some reactions are simple and composed of only one step but many are multi-step or composite reactions. The mechanism of a reaction is the proposed steps that explain the way the chemical reaction has proceeded. The rate of a chemical reaction depends upon its temperature.
- A reaction may be thermodynamically favourable but kinetically too slow.

Questions

1   The standard enthalpy change for the following reaction is $-93.4\,\text{kJ mol}^{-1}$ of ammonia, and the standard Gibbs' Free Energy change $-33.6\,\text{kJ mol}^{-1}$ at 298 K.

$$N_2(g) + 3H_2(g) \rightleftharpoons 2NH_3(g)$$

Estimate the Gibbs' Free Energy change at (a) 500 K and (b) 1,000 K.
Is the reaction spontaneous at a room temperature of 20 °C?
Is the formation of ammonia thermodynamically favoured or disfavoured by a rise in temperature? What would be the effect of increasing the pressure on the reaction?

2   Calculate the standard thermodynamic equilibrium constant for the reaction,

$$N_2O_4(g) \rightleftharpoons 2NO_2(g)$$

for which $\Delta H^0_{298\,K}(\text{reaction}) = +58.0\,\text{kJ mol}^{-1}$ and
$\Delta S^0_{298\,K}(\text{reaction}) = +176.6\,\text{JK}^{-1}\,\text{mol}^{-1}$

3   Calculate the value and state the units of K_p for the reaction,

$$2NO(g) = N_2(g) + O_2(g)$$
$$\Delta G^0(\text{reaction}) = -87\,\text{kJ mol}^{-1}$$

The reaction will not occur spontaneously. Why?

4   The decomposition of nitrosyl chloride, NOCl, proceeds at 500 K according to the stoichiometric equation,

$$2NOCl(g) = 2NO(g) + Cl_2(g)$$

The rate law is,

$$-d[NOCl]/dt = k_r[NOCl]^2$$

Given sufficient kinetic data, what graph would you plot to determine the rate constant for this reaction? Quote the units of the rate constant.

5   The rate of a reaction was found to depend only on the concentration of reactant A and reactant B. When the concentration of A was doubled and that of B kept constant, the rate was found to quadruple. When the concentration of B was doubled and that of A kept constant, the rate was found to remain unchanged.
   What is the partial order of this reaction with respect to reactant A and the partial order with respect to reactant B? What is the overall order of reaction?

6   The reaction between dichlorine and trichloromethane (CHCl$_3$) is believed to occur via the following mechanism,

   Step 1 $Cl_2(g) \rightleftharpoons 2Cl(g)$ (both forward and reverse reactions are fast)
   Step 2 $CHCl_3(g) + Cl(g) \longrightarrow CCl_3(g) + HCl(g)$ (slow)
   Step 3 $CCl_3(g) + Cl(g) \longrightarrow CCl_4(g)$ (fast)

What is the overall chemical equation for this mechanism?
Identify any reaction intermediates.
Deduce the rate law for this reaction.

7 The decomposition of dinitrogen pentoxide, N_2O_5, follows first-order reaction kinetics between 0 °C and 65 °C. Using the data in Table 2.2, determine the activation energy of the reaction.

Table 2.2

Temperature/ °C	Rate constant/ k_r/s^{-1}
0	7.9×10^{-7}
25	3.5×10^{-5}
35	1.4×10^{-4}
45	5.0×10^{-4}
55	1.5×10^{-4}
65	4.9×10^{-3}

Further reading

Atkins, P.W. (2001) *The Elements of Physical Chemistry*, 3rd edn. Oxford University Press, Oxford.
A less intense and mathematical version of his standard university textbook on physical chemistry. It provides many worked examples. However, you do need previous knowledge of physical chemical topics to really cope even with this book!

Atkins, P.W. and Beran, J.A. (1992) *General Chemistry*, 2nd edn. Scientific American Books, W.H. Freeman, New York.
See Chapter 1.

The Open University (1996) S342 Physical Chemistry, Principles of Chemical Change, Block 1: *Scope and Limitations of the Thermodynamic Approach*; Block 2: *An Introduction to Chemical Kinetics*. Oxford University, Milton Keynes.
This book forms an introduction to a Level 3 physical chemistry course. Although a knowledge and understanding of basic chemical principles and mathematics are needed, the content is excellent. The book is well illustrated throughout and there are many self-assessment questions complete with answers.

The Open University (1997) S205 The Molecular World, Book 4: *The Measure of Chemical Change: Thermodynamics and the Reactions of Metals*; Book 5: *Chemical Kinetics and Mechanisms*. Open University, Milton Keynes.
An integrated approach to physical, inorganic and organic chemistry is adopted in these books. A basic understanding of chemistry is needed but they are well written. Short case studies are included.

Price, G.J. (1998) *Thermodynamics of Chemical Processes*. Oxford Chemistry Primers, Oxford Science Publications, Oxford.
A very readable first-year undergraduate primer. It clearly explains the ideas behind enthalpy, internal energy and entropy changes leading to an understanding of the Gibbs' Free Energy change. It is well structured and laid out. Some understanding of maths is required but this is kept to a minimum.

Smith, E.B. (1990) *Basic Chemical Thermodynamics*, 4th edn. Oxford Chemistry Series, Oxford University Press, Oxford.
Although a little long in the tooth, a very understandable approach to equilibrium thermodynamics. The maths required may be a little difficult in places since a basic understanding of integration and differentiation is required.

3 An introduction to organic chemicals

- Bonding and molecular structure – the hybridisation of the carbon atom
- The hydrocarbons – alkanes and alkenes, cycloalkanes and cycloalkenes, aromatic hydrocarbons
- Representation of formulae and structures
- Homologous series
- Structural isomerism – skeletal, positional, functional
- Saturated and unsaturated compounds – their reactivities compared
- Geometric isomerism
- Optical isomerism and chirality. Living organisms and chirality
- Common functional groups
- Heterocyclic compounds
- Important types of reactions – oxidation and reduction, addition reactions, condensation reactions, hydrolysis, substitution reactions

Introduction

The life forms on this planet are based on the element carbon. In the nineteenth century it was thought that only living organisms could make carbon compounds, and so the study of carbon-based compounds became known as organic chemistry. Although many carbon compounds are made in the laboratory and by industrial processes, and therefore have nothing to do directly with living organisms, the term 'organic chemistry' is still synonymous with 'carbon compounds'. Normally, the oxides of carbon and the carbonates are classified as 'inanimate' inorganic chemicals. There are many millions of different organic chemicals known – their number far exceeds the number of inorganic compounds. The influence of these compounds in our lives has been great in that they form the basis of not only our own existence but of drugs, pesticides, food additives, plastics, fuels, etc.

Bonding and molecular structure

Carbon atoms can form bonds with other carbon atoms leading to long chains, branched chains and ring structures. This process is termed **catenation**. Single, double and triple bonds can also be formed between pairs of carbon atoms and with other appropriate atoms (see Table 3.1 for bond lengths and average bond enthalpies of some of these bonds).

The simplest compounds containing carbon and hydrogen only are called **hydrocarbons**. The simplest of these is methane gas, CH_4 (Figure 3.1).

Table 3.1 *The bond lengths and average bond enthalpies of some selected bonds*

Bond type	Bond length/ pm	Average bond enthalpy/ kJ mole^{-1}
C—H	109	+412
C—C	154	+348
C=C	134	+612
C≡C	120	+837
C—O	143	+360
O—H	96	+463

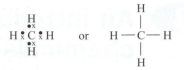

Figure 3.1 Lewis structure of methane.

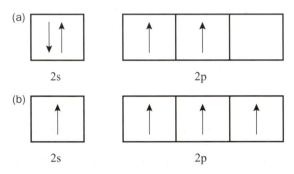

Figure 3.2 (a) The ground state outer electron shell of carbon. (b) Energetically excited state of carbon.

How are the carbon–hydrogen bonds formed? The electronic configuration of the carbon atom in its ground state is $1s^2 2s^2 2p^2$. The outer electronic shell of the carbon atom in its ground state can be represented as shown in Figure 3.2(a). If one of the paired electrons in the 2s sub-shell is promoted to the 2p, then the carbon atom is now in an energetically excited electronic state (Figure 3.2b).

If the 2s and 2p sub-shells are treated as atomic orbitals, then the four orbitals can be 'mixed' to form four energetically equivalent sp^3 hybridised atomic orbitals (Figure 3.3a). In order to minimise electron repulsion, these atomic orbitals are arranged in the shape of a tetrahedron with angles of 109° 28′.

Because of the tetrahedral nature of the carbon atom, the structure of methane is not a two-dimensional one, although it is often represented as such for convenience. Another

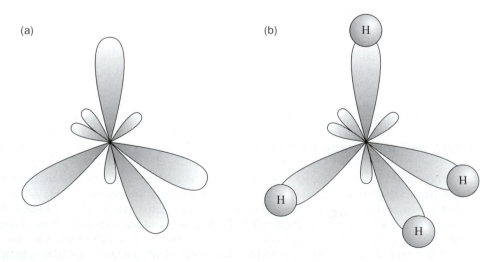

Figure 3.3 (a) sp^3-hybridised carbon. (b) Methane, CH_4.

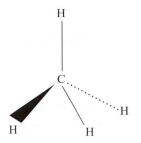

Figure 3.4 The tetrahedral structure of methane.

representation of methane is shown in Figure 3.3(b) where hydrogen 1s-atomic orbitals have overlapped with the four sp^3-atomic orbitals of the central carbon to produce four carbon–hydrogen single bonds (Figure 3.4).

The enthalpy change that takes place when a covalent bond is broken is called the **bond enthalpy**, and is a measure of the strength of that bond. In the case of a polyatomic molecule such as methane, the presence of all the atoms has an effect on the bond strength of the C—H bond. Thus the **average bond enthalpy** is used.

Ethane gas is the next simplest hydrocarbon and arises from the combination of two carbon atoms via a single carbon –carbon bond. It has the **molecular formula** C_2H_6 or,

$$\text{ethane, } C_2H_6 \qquad \underset{\displaystyle\underset{H}{|}}{\overset{\displaystyle\overset{H}{|}}{H-C}}-\underset{\displaystyle\underset{H}{|}}{\overset{\displaystyle\overset{H}{|}}{C}}-H \qquad \text{or } CH_3CH_3$$

When the formula of an organic molecule is represented in this way, it is referred to as a **structural formula**. The above structural formula can also be represented by the **abbreviated** molecular formula CH_3—CH_3 or CH_3CH_3. Other hydrocarbons can be formed by adding carbon atoms in succession to form chains. Thus the next two hydrocarbons are propane and butane,

$$\text{propane, } C_3H_8 \qquad H-\overset{\displaystyle\overset{H}{|}}{\underset{\displaystyle\underset{H}{|}}{C}}-\overset{\displaystyle\overset{H}{|}}{\underset{\displaystyle\underset{H}{|}}{C}}-\overset{\displaystyle\overset{H}{|}}{\underset{\displaystyle\underset{H}{|}}{C}}-H \qquad \text{or} \quad CH_3CH_2CH_3$$

$$\text{butane, } C_4H_{10} \qquad H-\overset{\displaystyle\overset{H}{|}}{\underset{\displaystyle\underset{H}{|}}{C}}-\overset{\displaystyle\overset{H}{|}}{\underset{\displaystyle\underset{H}{|}}{C}}-\overset{\displaystyle\overset{H}{|}}{\underset{\displaystyle\underset{H}{|}}{C}}-\overset{\displaystyle\overset{H}{|}}{\underset{\displaystyle\underset{H}{|}}{C}}-H \qquad \text{or} \quad CH_3CH_2CH_2CH_3$$

All of the molecules considered so far are hydrocarbons which contain single C—C bonds. No further hydrogen atoms can be added to these structures, so they are called **saturated** hydrocarbons. Although the tetrahedral nature of this single-bonded carbon atom gives a three-dimensional shape to the molecule, two-dimensional **straight chain** structures are often used. Methane, ethane, propane and butane can be represented by the general formula C_nH_{2n+2} where n is the number of carbon atoms. The family of hydrocarbons having this general formula is called the **alkanes**. Thus, methane is a saturated hydrocarbon belonging to the alkane family. Since each successive member of the family differs by a (CH_2) unit, it is also referred to as a **homologous series**.

After butane, the alkanes are named according to the number of carbon atoms – pentane is C_5H_{12}, hexane C_6H_{14}, heptane C_7H_{16}, etc. The formula for hexane is, hexane, C_6H_{14}

$$H-\overset{\overset{\displaystyle H}{|}}{\underset{\underset{\displaystyle H}{|}}{C}}-\overset{\overset{\displaystyle H}{|}}{\underset{\underset{\displaystyle H}{|}}{C}}-\overset{\overset{\displaystyle H}{|}}{\underset{\underset{\displaystyle H}{|}}{C}}-\overset{\overset{\displaystyle H}{|}}{\underset{\underset{\displaystyle H}{|}}{C}}-\overset{\overset{\displaystyle H}{|}}{\underset{\underset{\displaystyle H}{|}}{C}}-\overset{\overset{\displaystyle H}{|}}{\underset{\underset{\displaystyle H}{|}}{C}}-H$$

or $CH_3CH_2CH_2CH_2CH_2CH_3$

The molecular formula for hexane can also be written $CH_3(CH_2)_4CH_3$. On both ends of this molecule, like so many other compounds, there is a CH_3 group. This is called the **methyl** group after its parent molecule methane.

As the number of carbon atoms increases in a hydrocarbon, then so does its size and mass. This is paralleled by increases in boiling points and melting points resulting in a gradation from the gaseous phase to the solid phase. For example, methane is a gas whilst hexane is a low boiling point liquid (Table 3.2). The wide variations in melting and boiling points of organic compounds are also a function of the types of inter-molecular forces of attraction that exist between molecules, i.e. van der Waals-type forces, dipole–dipole interactions and hydrogen bonding. In general, the stronger these forces, the higher a compound's melting point and boiling point.

There is more than one way of arranging the six carbon atoms in hexane, which will still retain the same molecular formula and its single bond nature. In naming compounds like these, the longest chain of carbon atoms is selected as the 'base' name. Each of the carbon atoms is numbered so that the position of the side group/chain is indicated by the lowest number. In the structure shown below, the longest chain is composed of five atoms, therefore the base name is pentane,

i.e.

$$C^5 - C^4 - C^3 - \overset{\displaystyle C^2}{\underset{\underset{\displaystyle C}{|}}{}} - C^1$$

The carbon atoms could be numbered 1, 2, 3 . . . from left to right or from right to left. The numbering system is conventionally adopted that gives the position of the methyl group the lowest number, so in this case it is done from right to left, i.e. the methyl group is attached to carbon atom number 2. The compound is thus called 2-methylpentane. Note that the structure,

$$C - C - \overset{\displaystyle C}{\underset{\underset{\displaystyle C}{|}}{}} - C - C$$

is the same as the previous numbered structure. It is just a case of how the structure is viewed.

All of the structures of molecular formula C_6H_{14} are shown in Figures 3.5 and 3.6. Note that the longest carbon skeleton in two of the examples is composed of four carbon atoms, and so the base name here is butane. To indicate the presence of two methyl groups, 'dimethyl' is used and their respective positions along the carbon chain indicated.

All of the structures shown in Figures 3.5 and 3.6 have the same molecular formula but the atoms are arranged in a different order. Such structures are called **structural**

Table 3.2 *The melting points and boiling points of the first nine straight chain alkanes*

Name	Chemical formula	Structural formula	Melting point/$^\circ$C	Boiling point/$^\circ$C
Methane	CH_4	CH_4	−182	−164
Ethane	C_2H_6	CH_3CH_3	−183.3	−88.6
Propane	C_3H_8	$CH_3CH_2CH_3$	−189.7	−42.1
Butane	C_4H_{10}	$CH_3CH_2CH_2CH_3$	−138.4	−0.5
Pentane	C_5H_{12}	$CH_3CH_2CH_2CH_2CH_3$	−130.0	36.1
Hexane	C_6H_{14}	$CH_3CH_2CH_2CH_2CH_2CH_3$	− 95.0	69.0
Heptane	C_7H_{16}	$CH_3CH_2CH_2CH_2CH_2CH_2CH_3$	− 90.6	98.4
Octane	C_8H_{18}	$CH_3CH_2CH_2CH_2CH_2CH_2CH_2CH_3$	− 56.8	125.7
Nonane	C_9H_{20}	$CH_3CH_2CH_2CH_2CH_2CH_2CH_2CH_2CH_3$	− 51.0	150.8

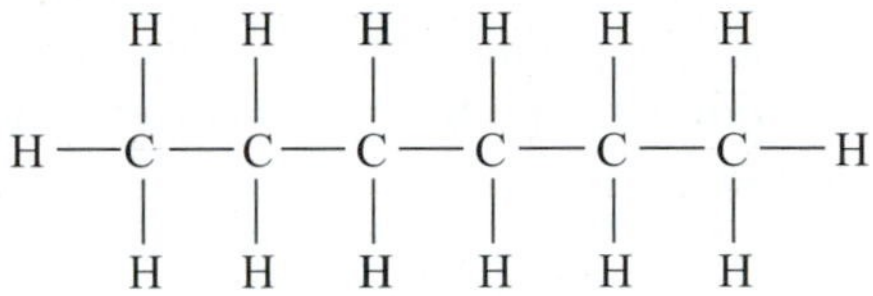

(a) hexane

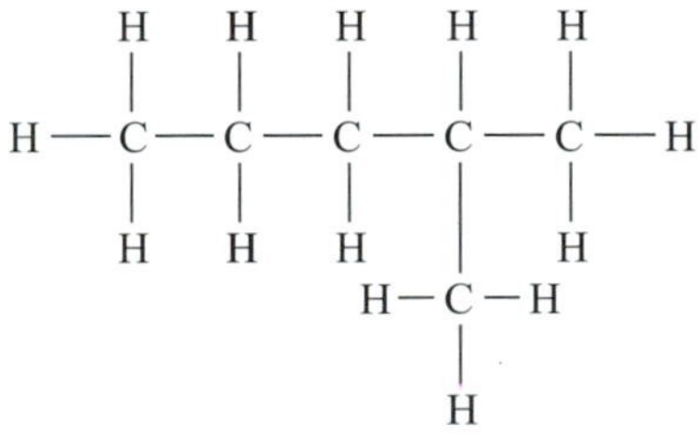

(b) 2-methylpentane

Figure 3.5 *(above)* Two structures of molecular formula C_6H_{14}.

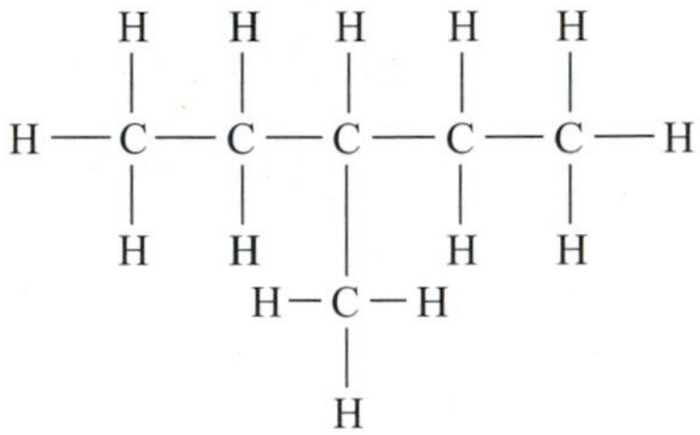

(a) 3-methylpentane

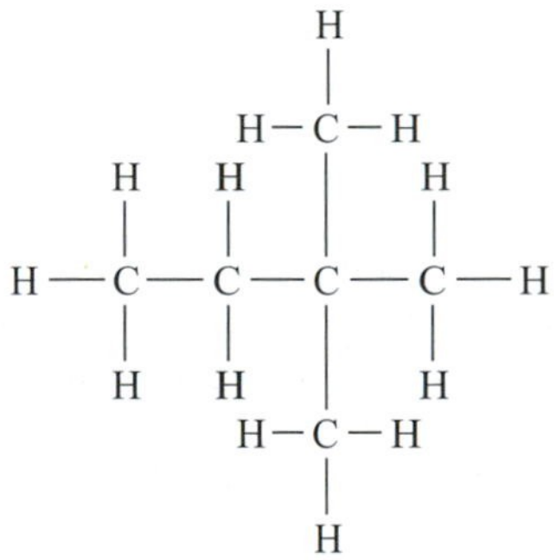

(b) 2,2-dimethybutane

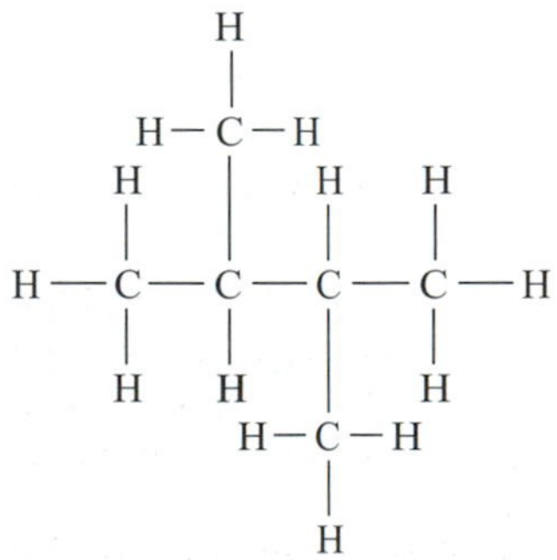

(c) 2,3-dimethylbutane

Figure 3.6 *(right)* More skeletal isomers of molecular formula C_6H_{14}.

isomers. Because they differ only in the way the carbon skeletons have been arranged, they are also called **skeletal isomers**. All five compounds exist and have different boiling points and melting points, though their chemical reactions are similar. Carbon can also form bonds with other atoms such as fluorine, chlorine, bromine, iodine, oxygen and nitrogen. For example, if the hydrogen atoms in methane are successively replaced by monovalent chlorine atoms, then the compounds corresponding to the molecular formulas CH_3Cl (chloromethane), CH_2Cl_2 (dichloromethane), $CHCl_3$ (trichloromethane or chloroform) and CCl_4 (tetrachloromethane, or carbon tetrachloride) are formed. The addition of chlorine atoms imparts to each of these compounds their own peculiar physical and chemical properties.

The methyl group, CH_3— can also combine with monovalent groups such as the hydroxyl group —OH or amine group —NH_2 producing the molecular formulae CH_4O and CH_5N, respectively. The structures of these compounds are shown in Figure 3.7. The presence of the OH group or the NH_2 also produces two new chemicals which have their own peculiar properties.

CH_3OH is called methanol and CH_3NH_2 methylamine. The Cl, OH and NH_2 groups are called **functional groups** because each organic compound's chemical reactions depend upon the type and number of these groups present. The hydrocarbon part of a molecule tends to remain unchanged during a reaction.

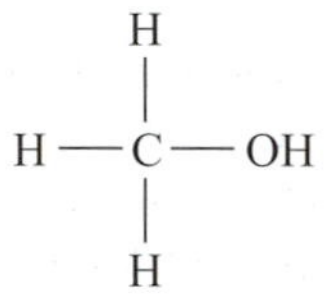

In the cases of CH_3Cl, CH_3OH and CH_3NH_2 each represents the first member of their respective homologous series. The OH group gives rise to a series of compounds called the alcohols, of which the first six linear members are shown in Table 3.3. Note that each name ends in **-ol**, which indicates that they are members of the homologous series known as the alcohols. The numbering system indicates the carbon atom on which the OH group is found.

In the case of chlorine and the amino group, the first six compounds formed are shown in Table 3.4.

The systematic naming of organic compounds like this is preferred but old names are still in common use, e.g. methyl chloride, n-butyl alcohol. Because of their complexity, many molecules are often referred to by trivial names, e.g. testosterone, Vitamin C.

H
|
H —— C —— OH   methanol
| boiling point 65.2 °C
H

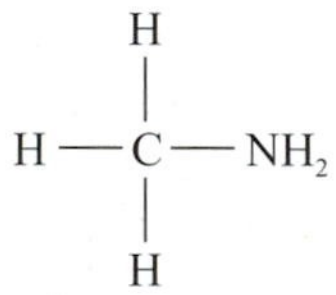

H
|
H —— C —— NH₂   methylamine
| boiling point –6.3 °C
H

Figure 3.7 The structure of methanol and methylamine.

Table 3.3 *The first six straight chain alcohols*

Name	Chemical formula	Structural formula	Melting point/°C	Boiling point/°C
Methanol	CH_4O	CH_3OH	− 93.9	65.2
Ethanol	C_2H_6O	CH_3CH_2OH	−117.3	78.5
Propanol	C_3H_8O	$CH_3CH_2CH_2OH$	−126.5	97.4
Butan-1-ol	$C_4H_{10}O$	$CH_3CH_2CH_2CH_2OH$	− 89.5	117.2
Pentan-1-ol	$C_5H_{12}O$	$CH_3CH_2CH_2CH_2CH_2OH$	− 79	137.3
Hexan-1-ol	$C_6H_{14}O$	$CH_3CH_2CH_2CH_2CH_2CH_2OH$	− 46.7	158

Table 3.4 *The first six straight chain chloroalkanes and alkanamines*

Name	Chemical formula	Structural formula	Melting point/°C	Boiling point/°C
Chloromethane	CH_3Cl	CH_3Cl	− 97.1	−24.2
Chloroethane	C_2H_5Cl	CH_3CH_2Cl	−136.4	12.3
1-Chloropropane	C_3H_7Cl	$CH_3CH_2CH_2Cl$	−122.8	46.6
1-Chlorobutane	C_4H_9Cl	$CH_3CH_2CH_2CH_2Cl$	−123.1	78.4
1-Chloropentane	$C_5H_{11}Cl$	$CH_3CH_2CH_2CH_2CH_2Cl$	− 99.0	107.8
1-Chlorohexane	$C_6H_{13}Cl$	$CH_3CH_2CH_2CH_2CH_2CH_2Cl$	− 94.0	134.5
Methanamine	CH_3NH_2	$CH_3\ NH_2$	− 93.5	−6.3
Ethanamine	$C_2H_5NH_2$	$CH_3CH_2NH_2$	− 81.0	16.6
Propan-1-amine	$C_3H_7NH_2$	$CH_3CH_2CH_2NH_2$	− 83.0	47.8
Butan-1-amine	$C_4H_9NH_2$	$CH_3CH_2CH_2CH_2NH_2$	− 49.1	77.8
Pentan-1-amine	$C_5H_{11}NH_2$	$CH_3CH_2CH_2CH_2CH_2NH_2$	− 55.0	104.4
Hexan-1-amine	$C_6H_{13}NH_2$	$CH_3CH_2CH_2CH_2CH_2CH_2NH_2$	− 19.0	130

Structural isomerism – a revisit

Consider the molecular formula $C_4H_{10}O$. The oxygen atom could either be connected to a hydrogen atom and a carbon atom, as in C—O—H, thus indicating the presence of an alcohol functional group, or it could be attached to two carbon atoms, i.e. C—O—C. This latter group is known as the **ether functional group** and again gives rise to a new family of compounds whose chemical properties are very different from the alcohols. Thus with structural isomerism, it is necessary to identify the functional groups present, then, for each functional group, to identify the possible carbon skeleton arrangements (skeletal isomers), and finally to identify the positions along the carbon skeleton where the functional groups can be placed.

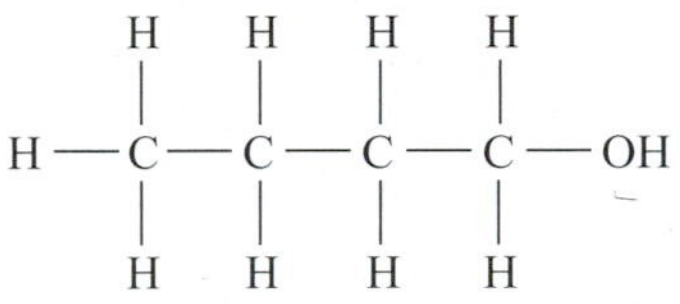

(a) butan-l-ol

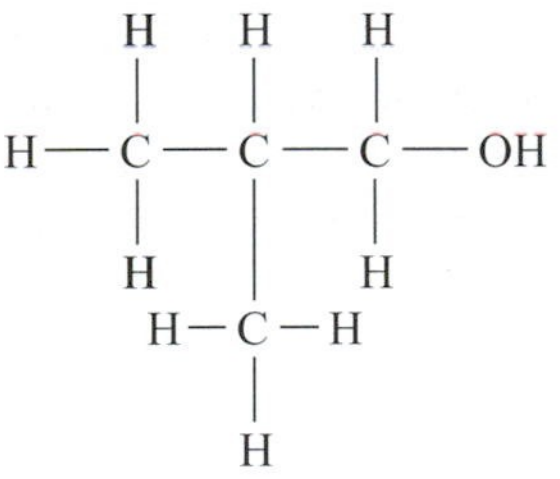

(b) 2-methylpropan-l-ol

Figure 3.8 Molecular formula $C_4H_{10}O$, the OH functional group and skeletal isomerism.

If $C_4H_{10}O$ is written as C_4H_9OH, then the skeletal isomers are as shown in Figure 3.8. The two alcohols maintain the same position of the hydroxyl group but differ in the carbon skeleton structure. The positional isomers are shown in Figure 3.9.

For the ether functional group, C—O—C, the possible skeletal isomers are shown in Figure 3.10. The way in which the ethers are named is again to select the longest carbon chain as the base name, and the remaining CH_3O or CH_3CH_2O are called methoxy and ethoxy respectively. Thus in Figure 3.10 (a) is called ethoxyethane, (b) 1-methoxypropane and (c) 2-methoxypropane.

Carbon can also form rings of carbon atoms and still be saturated with hydrogen, e.g. cyclohexane, C_6H_{12}. In this molecule, the carbon atoms are again sp^3 hybridised leading to a hexagonal ring of carbon atoms that are not planar. It is convenient to represent cyclohexane as shown in Figure 3.11.

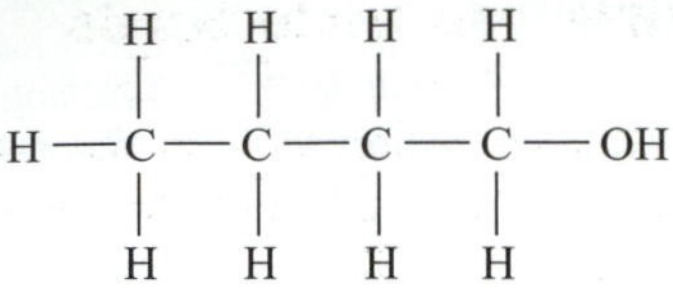

(a) butan-1-ol

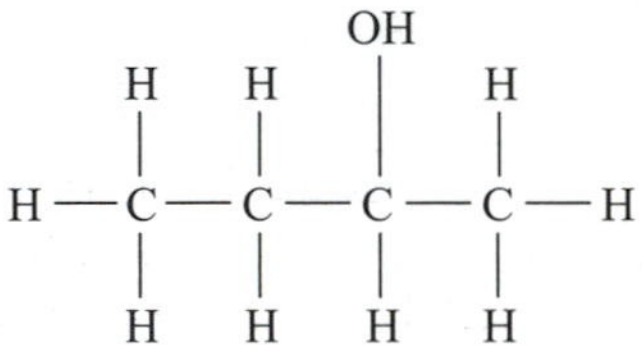

(b) butan-2-ol

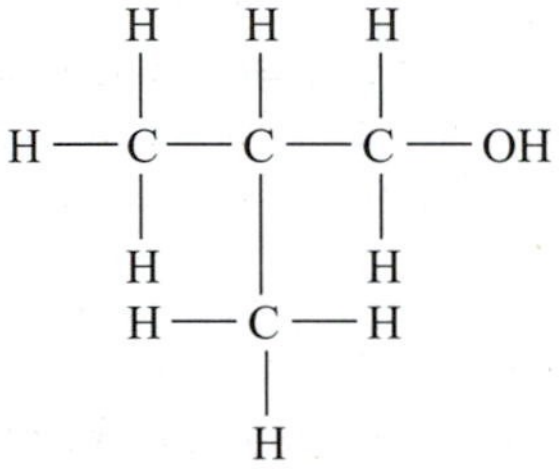

(c) 2-methylpropan-1-ol

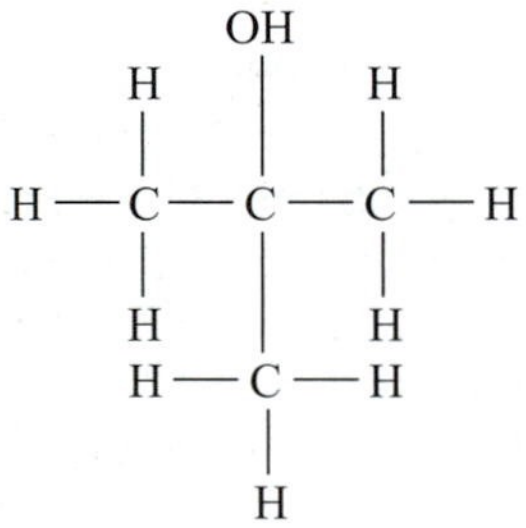

(d) 2-methylpropan-2-ol

Figure 3.9 *(left)* Positional isomers of the alcohols of molecular formula $C_4H_{10}O$.

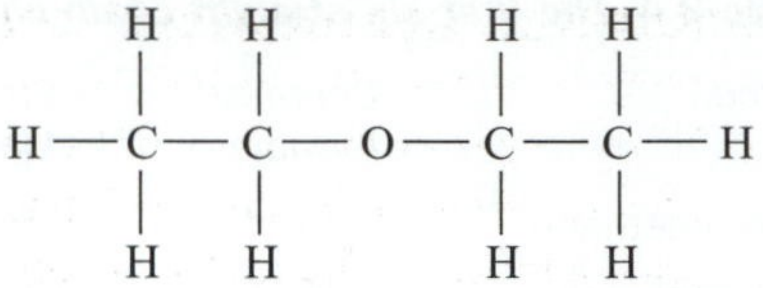

(a) ethoxyethane

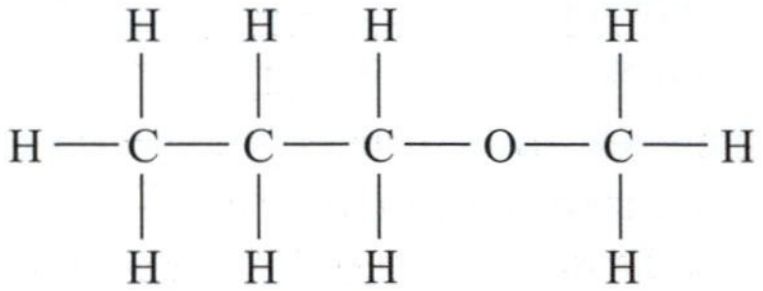

(b) 1-methoxypropane

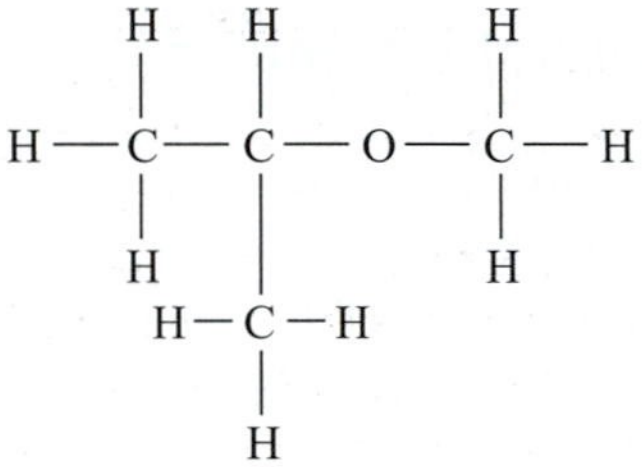

(c) 2-methoxypropane

Figure 3.10 The skeletal isomers of the ethers of molecular formula $C_4H_{10}O$.

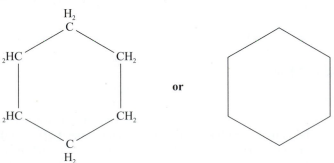

or

Figure 3.11 Cyclohexane.

Unsaturated hydrocarbon compounds – double and triple bonds

When a carbon atom joins twice or three times to another carbon atom at the expense of hydrogen, it is said to be **unsaturated**. The simplest example has the molecular formula C_2H_4. The Lewis structure for this molecule is,

$$\begin{array}{c} \text{H} \quad\quad \text{H} \\ \text{C::C} \\ \text{H} \quad\quad \text{H} \end{array}$$

When the carbon atom is in the excited electronic state as depicted in Figure 3.2(b), it is possible that one 2s- and two 2p-atomic orbitals hybridise on each carbon atom leading to three hybrid sp^2-atomic orbitals. Two of these orbitals, one from each carbon atom, overlap to form a carbon–carbon single bond (called a σ-bond or sigma bond). The other two on each carbon atom overlap with the 1s-atomic orbital of a hydrogen atom to form carbon–hydrogen single bonds. The two remaining 2p-atomic orbitals overlap to produce the second carbon–carbon bond (called a pi or π-bond).

The compound has a carbon–carbon **double** bond to which hydrogen could be added. The double bond gives the compound its own characteristic chemical reactions, and is an example of another functional group. Compounds that possess a carbon–carbon double bond are referred to as the **alkenes**.

The name of the above compound is thus eth**ene**, 'eth' after the parent saturated compound 'ethane' and 'ene' to indicate the presence of the double bond.

The simplest compound containing the carbon–carbon **triple** bond has the molecular formula, C_2H_2. The Lewis structure is,

$$\text{H:C:::C:H} \quad \text{or} \quad \text{H—C}\equiv\text{C—H}$$

If the carbon atoms are sp-hybridised, then two of the resulting 2sp-hybrid atomic orbitals will be able to overlap to form a carbon–carbon sigma-bond. The other two 2sp-hybrid atomic orbitals will overlap with the 1s-atomic orbital of hydrogen atoms to form two carbon–hydrogen σ-bonds. The remaining two pairs of 2p-atomic orbitals will overlap to produce two π-bonds between the carbon atoms.

The triple bond is yet another functional group and gives a new family called the **alkynes**. The name of the compound is eth**yne**.

Again, it is necessary to introduce a number system to indicate the position of a double or triple bond in the naming of a compound. For example, $CH_3CH_2CH{=}CH_2$ is an alkene

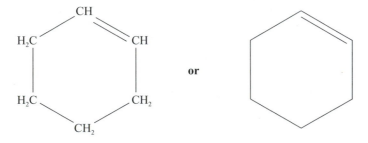

Figure 3.12 Cyclohexene.

called but-1-ene, and $CH_3CH\!=\!CHCH_3$ one called but-2-ene. Similarly $CH_3CH_2C\!\equiv\!CH$ is but-1-yne and $CH_3C\!\equiv\!CCH_3$ but-2-yne. In the case of cyclic alkenes, the simplest one is the liquid cyclohexene, C_6H_{10} (Figure 3.12).

Compounds of the type just described in which the backbone is composed either of a chain, branched chain or rings of saturated and unsaturated carbon atoms are called **aliphatic** compounds. This generic term allows them to be distinguished from those compounds that contain the benzene ring and are referred to as **aromatic compounds**.

Aromatic hydrocarbons

Benzene, C_6H_6, is a colourless liquid at room temperature. It has a melting point of 7 °C and a boiling point of 80 °C. It is a dangerous, carcinogenic substance. The smell or aroma of benzene caused it and its family members to be given the generic name aromatic compounds. The compounds are now called **arenes**, derived from **ar**omatic and alk**ene**. This latter name gives an *indication* of the bonding in benzene!

The Lewis structure of benzene is shown in Figure 3.13. These structures suggest a ring of alternate carbon–carbon single and double bonds. However, the bond lengths of all of the carbon–carbon bonds are the same, 139 pm, and intermediate between that of the carbon–carbon single bond and the carbon–carbon double bond. The average bond enthalpy of the carbon–carbon bonds in benzene is 518 kJ mol^{-1}, which again is intermediate between that of a carbon–carbon single bond and a carbon–carbon double bond (see Table 3.1). One view of benzene is that its structure is a blend of the above structures, each equivalent structure being called a **resonance hybrid**, thus the structure is represented by a set of equivalent structures (Figure 3.14).

If each carbon atom in the benzene ring is viewed as being sp^2 hybridised, then a hexagonal structure results. Two of the sp^2-carbon atomic orbitals overlap to form σ-bonds with other carbon atoms whilst the third sp^2-carbon atomic orbital overlaps with the 1s-atomic

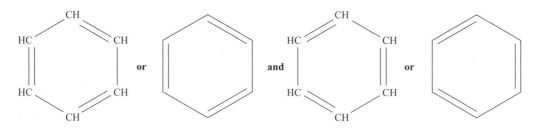

Figure 3.13 The Lewis structures of benzene.

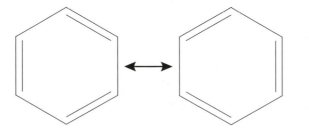

Figure 3.14 Some resonance hybrids of benzene.

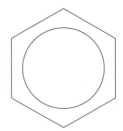

Figure 3.15 Benzene.

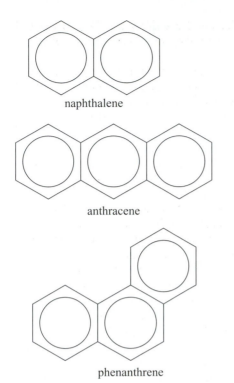

naphthalene

anthracene

phenanthrene

Figure 3.16 Some fused ring compounds of benzene.

orbital of a hydrogen atom. The remaining non-hybridised carbon 2p-atomic orbital of each carbon atom can overlap to form a π-bond. However, these π-electrons become delocalised and are spread over the entire ring of carbon atoms, which gives rise to the same bond lengths between carbon atoms and to the stability of the benzene ring. The structure in this case can be represented as in Figure 3.15.

The result of this delocalisation of electrons over the entire ring gives it a particular stability towards chemical reaction such that its integrity is maintained. Thus benzene chemistry is dominated by substitution reactions where the hydrogen atoms on the ring are replaced by other atoms or groups of atoms. The benzene ring does not normally undergo addition reactions, which are a characteristic of carbon–carbon double and triple bonds.

Benzene can form fused ring systems to give polycyclic or polyaromatic compounds. The two simplest ones are naphthalene and anthracene (Figure 3.16).

Naphthalene is the only benzene fused ring system that is produced on a commercial scale. It is used as an insecticide for moths, and is responsible for the smell of 'moth-balls'. Anthracene and phenanthrene occur naturally in petroleum deposits, and are formed in the coking of coal and in the combustion of coal- and oil-based fuels. Some polycyclic aromatic/benzoid compounds are again very dangerous.

The hydrogen atoms on the benzene ring can be substituted by other monovalent atoms or groups, for example methylbenzene and phenol (Figure 3.17).

If more than one group/atom is substituted, then isomerism occurs and a number system is used to indicate the position of the substituents (Figure 3.18). When a common or historical name is used, e.g. xylene, then the position of substitution is indicated by the prefixes *o-* (ortho), *p-* (para), *m-* (meta).

The compound shown in Figure 3.19 has explosive characteristics! It is called 1-methyl-2,4,6-trinitrobenzene, or trinitrotoluene (TNT).

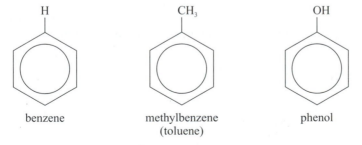

benzene   methylbenzene   phenol
           (toluene)

Figure 3.17 Some monosubstituted benzene compounds.

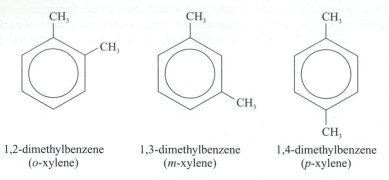

1,2-dimethylbenzene
(*o*-xylene)

1,3-dimethylbenzene
(*m*-xylene)

1,4-dimethylbenzene
(*p*-xylene)

Figure 3.18 Some disubstituted benzene compounds.

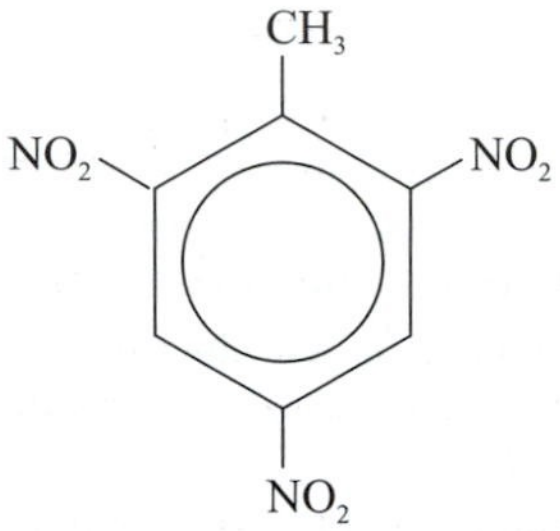

Figure 3.19 1-methyl-2,4,6-trinitrobenzene.

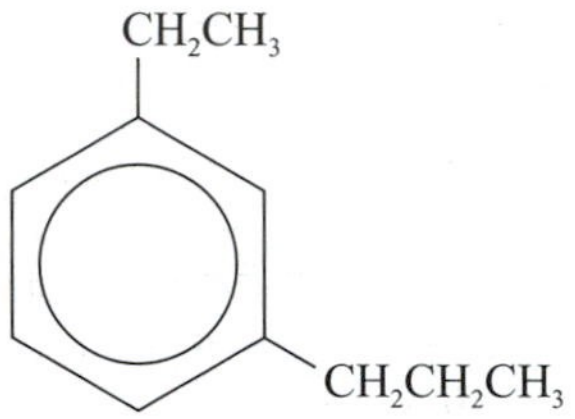

Figure 3.20 1-ethyl-3-propylbenzene.

The compound 1-ethyl-3-propylbenzene has the abbreviated molecular structure shown in Figure 3.20.

The ethyl group CH_3CH_2 is derived from ethane, CH_3CH_3, and the propyl group $CH_3CH_2CH_2$ from propane, $CH_3CH_2CH_3$.

If the benzene ring forms part of a molecule then the C_6H_5 group is known as the **phenyl** group.

Heterocyclic organic compounds

Heterocyclic chemistry is the study of a vast number of cyclic organic chemicals in which one or more of the carbon atoms making up the ring(s) have been replaced with a nitrogen, oxygen, sulphur or some other hetero-atom. For example, when one carbon atom in cyclopentane is replaced by a nitrogen atom, the resulting compound is pyrrolidine,

cyclopentane

pyrrolidine

Pyrrolidine has chemical properties closely resembling those of an amine.

If a carbon is replaced by an oxygen atom, then tetrahydrofuran is the result, which has chemical properties of an ether, whilst the introduction of a sulphur atom gives tetrahydrothiophene,

tetrahydrofuran   tetrahydrothiophene

Heterocyclic compounds can also contain double bonds, e.g.

furan   pyrolle   thiophene

Furan, pyrrole, and thiophene are all important heterocyclic compounds. These compounds react in ways that indicate aromaticity, with furan being the least and thiophene the most aromatic in behaviour. Derivatives of pyrolle are widespread in nature, e.g. in haemoglobin, the chlorophylls and Vitamin B_{12}.

When a heterocyclic compound fuses with a benzene ring, then a bicyclic compound is formed, e.g. indole,

indole

Again derivatives of indole are widespread in nature, e.g. tryptophan is an essential amino acid, which upon degradation produces skatole found in the faeces,

tryptophan   skatole

Indole-3-ethanoic acid is a plant growth hormone and many plant alkaloids contain the indole ring.

Carbohydrates include sugars and starches, and are composed of one or more sugar units (monosaccharides). These sugars are polyhydroxyaldehydes or polyhydroxy-ketones, which frequently exist as furanoside, a cyclic five-membered furanose ring, e.g. fructofuranose (Figure 3.21).

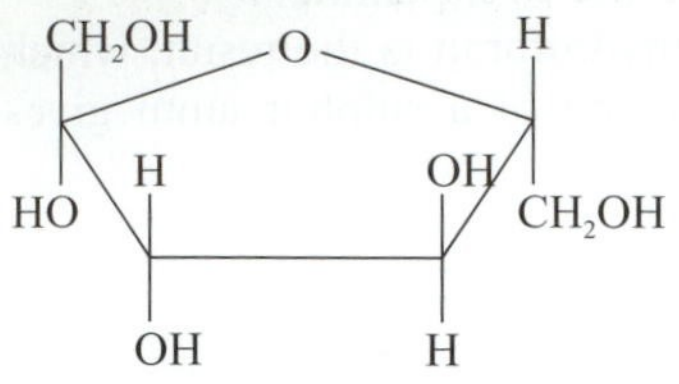

Figure 3.21 Fructofuranose.

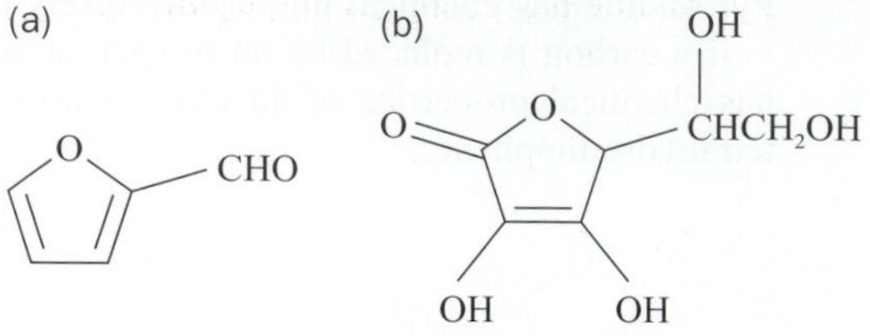

Figure 3.22 (a) Furfural. (b) Ascorbic acid (Vitamin C).

If some carbohydrates are dehydrated, then furan derivatives may result. One derivative known as furfural is used as a solvent in the manufacture of plastics and in the preparation of other furan derivatives (Figure 3.22a). Vitamin C is also related to furan (Figure 3.22b).

Pyridine is the most important nitrogen-containing, six-membered heterocyclic compound. It is used on a large scale as a solvent and as the starting compound for other heterocyclic compounds. Pyridine has the structure,

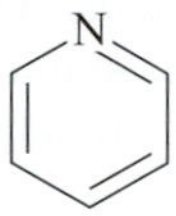

Pyridine derivatives are of great biological significance, e.g. the vitamins B_6 and nicotinamide,

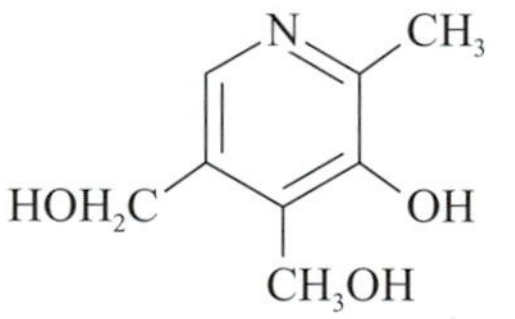

vitamin B_6 (pyrodoxine) 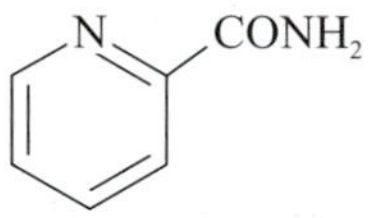   nicotinamide

Many plant alkaloids are derived from pyridine and piperidine, for example nicotine found in tobacco and piperine found in pepper (Figure 3.23).

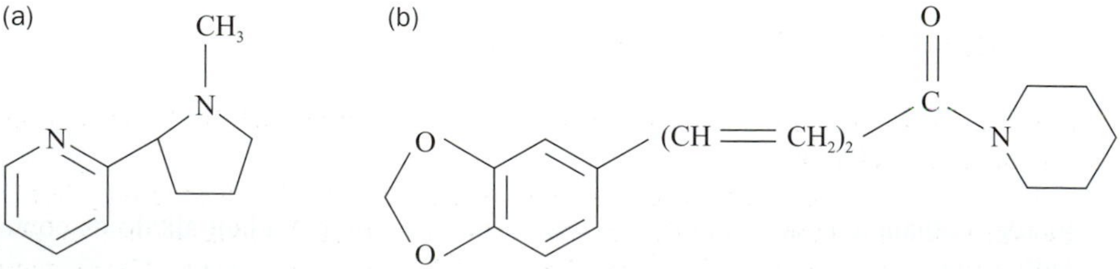

Figure 3.23 (a) Nicotine. (b) Piperine.

The heterocyclic ring containing the saturated nitrogen atom is known as piperidine,

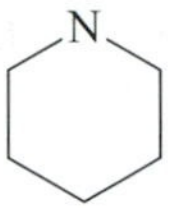

The combination of pyridine with a benzene ring yields the parent compounds for a class of alkaloids, quinoline and isoquinoline,

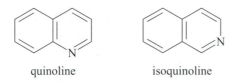

quinoline   isoquinoline

Unsaturated compounds containing oxygen – the carbonyl group

The carbonyl group is a very important group in organic chemistry,

i.e. $\diagdown_{\diagup}C{=}O$

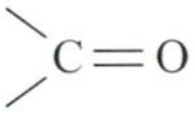

It is found linked with different atoms/groups, which give rise to a variety of different functional groups (Table 3.5).

Isomerism and the carbon–carbon double bond

Structural isomerism also occurs in molecules containing double bonds. For example, for the molecular formula C_4H_8 there are three compounds possible, all of which contain the same $C{=}C$ functional group, i.e.

$CH_3CH_2CH{=}CH_2$ but-1-ene

$CH_3CH{=}CHCH_3$ but-2-ene

$CH_2{=}C(CH_3)_2$ 2-methylpropene

But-1-ene and but-2-ene are examples of positional isomers, whilst 2-methylpropene is a skeletal isomer of but-1-ene.

The carbon–carbon double bond is not only characterised by its reactivity but also by its 'stiffness', i.e. there is no free rotation allowed about the axis of the double bond. This gives rise to **geometric isomerism**, which is a form of **stereoisomerism**. For example, but-2-ene can be represented by the structures shown on p. 80.

The atoms are connected in the same order for both structures, i.e. they have the same structural formula. Stereoisomers are pairs of compounds with the same molecular formula and the atoms connected in the same order. The atoms though are arranged

Table 3.5 *The carbonyl group*

Functional group	Homologous series	Example
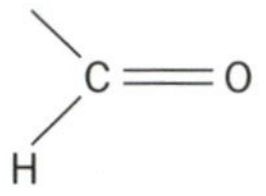 C=O (with \ above to C and / below to H)	Alkanals or aldehydes	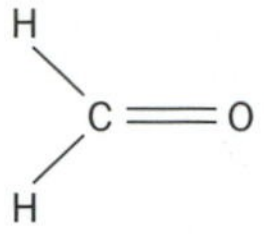 H–C=O (with H below) — Methanal or formaldehyde
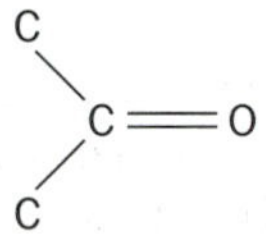 C=O (with C above and C below)	Alkanones or ketones	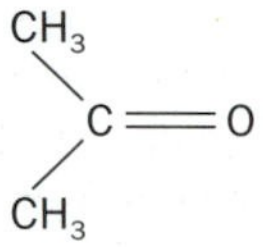 CH₃–C=O (with CH₃ below) — Propanone or acetone
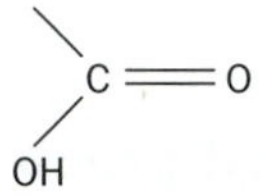 C=O (with \ above and OH below)	Carboxylic acids	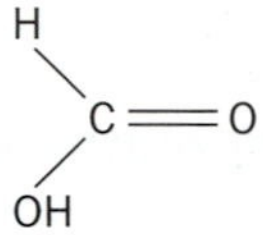 H–C=O (with OH below) — Methanoic or formic acid
		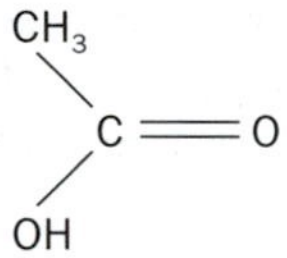 CH₃–C=O (with OH below) — Ethanoic or acetic acid

differently in three-dimensional space. In the case of but-2-ene, the methyl group or hydrogen atoms are arranged differently at the ends of the double bond, i.e.

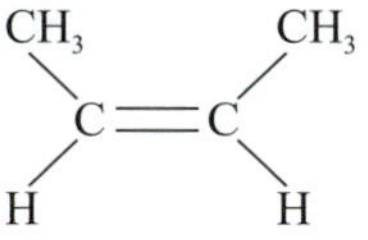

$$CH_3 \diagdown C = C \diagup CH_3 \quad \text{or} \quad CH_3 \diagdown C = C \diagup H$$

(left structure: CH_3 and H on one carbon, CH_3 and H on the other; right structure: CH_3 and H on one carbon, H and CH_3 on the other)

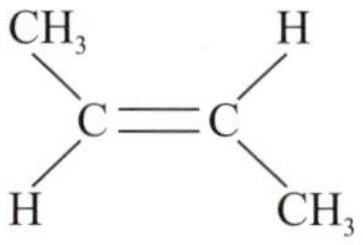

Where the methyl groups (or hydrogen atoms) are on opposite sides of the double bond, the structure is called the ***trans*-isomer**, and when both are on the same side the ***cis*-isomer**,

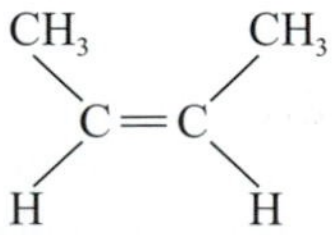

$$CH_3 \diagdown C = C \diagup CH_3 \quad \text{or} \quad CH_3 \diagdown C = C \diagup H$$

cis-but-2-ene
boiling point 3.7 °C
melting point −138.9 °C

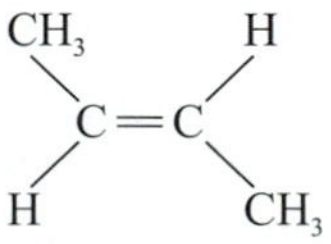

trans-but-2-ene
boiling point 0.9 °C
melting point −105.5 °C

cis/trans

The test for geometric isomerism is to determine if, and where, a double bond occurs, and what atoms/groups are present on the ends of that double bond. The general case can be represented by,

If geometric isomerism is present, then the atoms/groups represented by G_1 and G_2 must be different, *and* the atoms/groups R_1 and R_2 must be different. This can easily be seen in the case of but-2-ene if $G_1 = CH_3$ and $G_2 = H$, and $R_1 = CH_3$ and $R_2 = H$. In contrast, if propylene is examined,

then R_1 and R_2 are both the same and thus geometric isomerism is not possible.

The molecule penta-1,3-diene has two double bonds (**-diene**) in the five carbon atom backbone in the positions indicated (**1,3**), i.e.

$$CH_3CH=CHCH=CH_2$$

Here *cis*- and *trans*-isomerism about the double bond on the third carbon atom is possible,

For 3-methylpenta-1,3-diene, the isomers are,

In this last example, whether an isomer is called *cis* or *trans* is based upon which is the larger of the two groups at either end of the double bond.

Geometrical isomerism and insects

The insecticides have caused problems as well as being very useful. One way of control-ling insects that are pests is to 'design' compounds that the insects themselves produce to help in their destruction. Insects produce chemicals called **pheromones** which they use to attract a mate (sex pheromones), signal danger (alarm pheromones) or lay a trail to a food source (trail pheromones). Aphids are quite difficult to get rid of because the insects tend to collect under the leaves of plants. When an insecticide is applied, most aphids are 'missed' so the insecticide has to be applied again and in large quantities. If it is possible to alarm the aphids so that they are caused to scatter and seek other parts of the plant, then it may be that a single, smaller application of insecticide would be more effective. Investigations on aphids have shown that they secrete the alarm pheromone β-farnesene. This molecule has the formula,

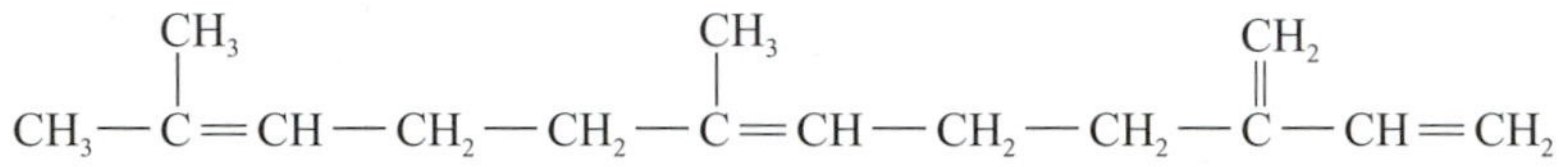

$$CH_3\!-\!\underset{\underset{CH_3}{|}}{C}\!=\!CH\!-\!CH_2\!-\!CH_2\!-\!\underset{\underset{CH_3}{|}}{C}\!=\!CH\!-\!CH_2\!-\!CH_2\!-\!\underset{\overset{CH_2}{\|}}{C}\!-\!CH\!=\!CH_2$$

The double bond in the centre of this molecule gives rise to geometrical isomerism since the other two double bonds have two identical atoms or groups on one of their ends. It is the larger of the groups at each end of a double bond that determines whether or not it is the *cis-* or *trans-*isomer. Thus the isomers of β-farnesene are:

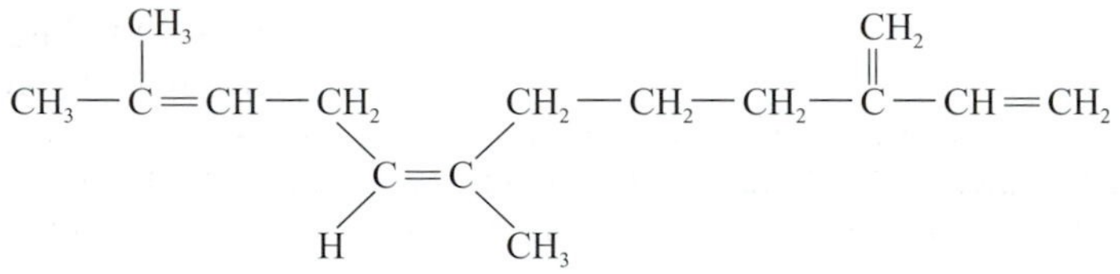

cis-β-farnesene

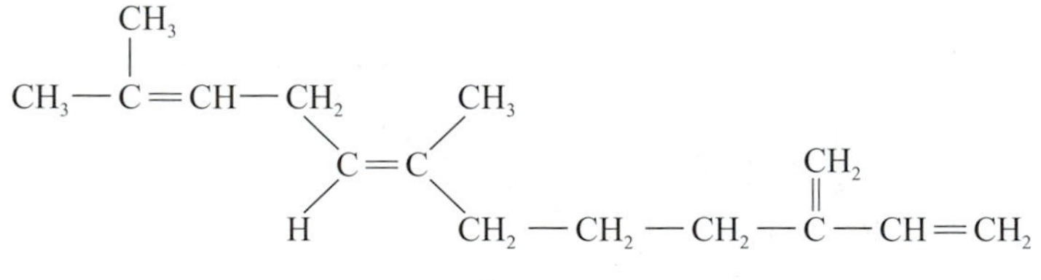

trans-β-farnesene

When β-farnesene was synthesised and applied to aphids, only one of the above struc-tures was found to be the effective alarm pheromone, i.e. *trans*-β-farnesene. The *cis*-form was inactive. Thus application of *trans*-β-farnesene causes aphids to move to more exposed places where they can be dealt with by an appropriate insecticide.

Optical isomerism and chirality

If an unsaturated, tetrahedral carbon atom (sp³-hybridised) is attached to four different atoms or groups then two three-dimensional structures are possible and one is the mirror image of the other (like a left hand appearing as a right hand when viewed in a plane mirror).

For example, the compound butan-2-ol, $CH_3CH_2CH(OH)CH_3$, would produce the structures shown in Figure 3.24 (in three dimensions). Such pairs of molecules are called **optical isomers**. The reference to right-handedness and left-handedness is termed **chirality** and the two structures are called **chiral**. The test for chirality is whether or not one structure is super-imposable on the other. The atom, which is attached to the four different atoms/groups, is called the **chiral centre**.

Optical isomers are iden-tical in their rates of chemical reactions, in the products they produce, and in their melt-ing points and boiling points. In order to distinguish them, **plane-polarised light** is used. Visible light is composed of mutually perpendicular electric (E) and magnetic (H) fields that vibrate in a sinusoidal fashion at right angles to the direction of propagation, i.e. light is a transverse wave. In normal light these electric and magnetic fields vibrate in all directions (Figure 3.25).

Some substances have the ability to 'block out' many of these vibrations allowing only electric fields through that vibrate in just one direction. Such a substance is called a **polarising material**. If this occurs, then the amount of light passing through the material is greatly reduced. The light emerging from the polarising material is said to

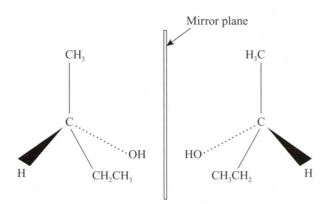

Figure 3.24 Optical isomers of butan-2-ol.

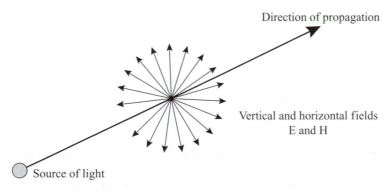

Figure 3.25 Normal visible light before polarisation.

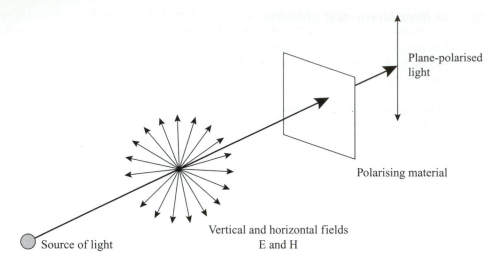

Figure 3.26 The production of plane-polarised light.

be plane-polarised (Figure 3.26). If a second sheet of polarising material is placed in line with the first, and matches its orientation, then the plane-polarised light will pass through it unimpeded. If this second sheet is rotated through 90°, it will prevent the plane-polarised light from passing through and darkness will be observed when viewing in the direction of the light source.

If the second polarising material is orientated somewhere between 0 and 90°, then some light will be seen to pass through. In the laboratory a **polarimeter** is used to investigate plane-polarised light. A schematic diagram of such an instrument is shown in Figure 3.27. Here, the normal source of light is a sodium lamp. The first polarising material, the **polariser**, is kept in a fixed position and allows plane-polarised light to travel to a second polarising material called the **analyser**. The analyser is rotatable to either the left or to the right, and is initially set so that the maximum amount of light is seen.

If a substance with a chiral centre is placed in a sample tube/cell in-between the polariser and analyser, then a much reduced light intensity may be seen. This means that the substance has rotated the plane of polarised light either to the left or to the right as viewed. In order to maximise the light intensity again, the analyser has to be rotated in the appropriate direction. A chemical that possesses a chiral centre thus has the ability to rotate the plane of polarised light passing through it.

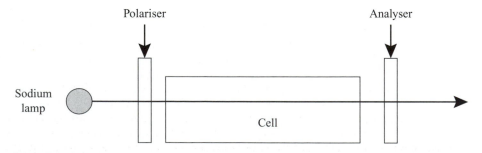

Figure 3.27 A schematic diagram of the polarimeter.

If the polariser has to be rotated clockwise (right) then the agreed sign (+) is allocated to the angle measured; if the rotation is anti-clockwise (left), then the angle is designated (−).

Substances that can rotate the plane of polarised light are described as being **optically active**. The degree of rotation depends upon the substance itself, its concentration when dissolved in some optically inactive solvent, and the path length of the cell.

If pure (+)butan-2-ol is tested, then it will rotate the plane of polarised light by +13.9°, whilst (−)butan-2-ol will rotate it by −13.9°. Hence, the magnitude of rotation of plane polarised light is the same for both the (+) and (−) optically active molecules, but occurs in opposite directions.

The amount of rotation is often measured under standard conditions and is called the **specific rotation**, $[\alpha]_D$, where D stands for the source of light – the bright yellow D-lines of the atomic spectrum of sodium. Now $[\alpha]_D$ is given by,

$$[\alpha]_D = \alpha / lc$$

where α = angle through which the analyser is turned (degrees)
   l = cell length (cm)
   c = concentration (g cm^{-3})

It is relatively easy to draw the shapes of the optical isomers *but* very difficult to determine which optical isomer is the (+) or (−) form.

Living organisms and chirality

Proteins are made from about 20 naturally occurring amino acids. Amino acids have a hydrogen atom, an amino group —NH$_2$, and a carboxylic acid group, —COOH, attached to an sp^3-hybridised carbon atom. Two examples of amino acids are alanine and serine:

H
|
H$_2$N — C* — COOH
|
CH$_3$

alanine

H
|
H$_2$N — C* — COOH
|
CH$_2$
|
OH

serine

Both examples have a chiral centre (*). The *only* naturally occurring optical isomers are (+)alanine and (+)serine – they occur in only one of the two possible chiral forms. They also have the same **handedness**. That is, all of the natural amino acids have the same **configuration**. This is denoted by the use of a small upper case (L). Optical isomers that are un-natural are denoted by a small upper case (D).

In terms of handedness, (L) means laevus or left, and (D) means dexter or right. This should not be confused with (+) and (−), which indicate what happens to plane polarised light when it interacts with the optical isomers. Some naturally occurring amino acids are (+) and others (−) but they all have the (L) configuration.

The configuration of a substance is very important in, for example, what it taste likes and whether it will be metabolised by a living organism. (+)glucose is the natural optical isomer of glucose which has the (L) configuration. This tastes sweet to humans and can be metabolised. (−)glucose tastes just as sweet but cannot be metabolised and is therefore used as a non-fattening sweetener. The two isomers of carvone produce different odours. (+)carvone occurs in caraway and (−)carvone in mint. It is clear that the configuration of an optically active isomer is important in how it interacts with a living organism such as a human. The molecule has to be able to 'dock' with the naturally occurring **receptor site**!

If a chiral compound is to be synthesised, unless some agent is present that 'directs' the configuration, then a 50–50 mixture of both optical isomers will be produced, and there will be no effect on plane polarised light. One way of obtaining a required configuration is to use natural enzymes. Enzymes are made of amino acids and are therefore chiral. Because an enzyme is chiral, it will ensure that, if a reaction is carried out where a reactant is a racemic mixture, it will react with the right isomer, leaving the other one intact.

Some chemical reactions of organic compounds

There are many chemical reactions that organic molecules can undergo that depend upon the functional groups present. Table 3.6 (reactions 1–14, see pp. 88–9) is a summary of some of the more major types.

Amongst the commonest reactions in the environment are the oxidation and reduction of organic chemicals. Oxidation can be viewed as the addition of oxygen, or the removal of hydrogen, whilst reduction is the addition of hydrogen or the removal of oxygen. The alkanes are rather chemically unreactive because of the strengths of the carbon–carbon and carbon–hydrogen single bonds, together with the inability of carbon to act as a Lewis acid. Alkanes, the unsaturated alkenes and alkynes, and aromatic hydrocarbons, readily undergo oxidation/combustion reactions (Table 3.6, reaction 1). However, the right amount of oxygen is required to avoid very dirty, smoky flames due to incomplete combustion. In addition, side reactions may occur, which produce unwanted chemicals.

Oxidation can be carried out using reagents such as acidified potassium dichromate(VI), e.g. the oxidation of ethanol to ethanoic acid. (This oxidation of ethanol also happens when the *Acetobacter* bacteria from the air is allowed into wine.) If conditions are carefully controlled, then partial oxidation to ethanal (a ketone) can be achieved. Alcohols like ethanol can be represented by RCH_2OH so that oxidation can be described by,

$$RCH_2OH + [O] \longrightarrow RCHO + H_2O$$

$$RCHO + [O] \longrightarrow RCOOH$$

Alcohols are called primary, secondary or tertiary depending upon how many carbon atoms are connected to the one that the OH group is attached to,

$$\begin{array}{ccc}
\text{H} & \text{H} & \text{C} \\
| & | & | \\
\text{C} - \text{C} - \text{OH} & \text{C} - \text{C} - \text{OH} & \text{C} - \text{C} - \text{OH} \\
| & | & | \\
\text{H} & \text{C} & \text{C}
\end{array}$$

primary alcohol   secondary alcohol   tertiary alcohol

The oxidation of a primary alcohol produces first an aldehyde and then a carboxylic acid. The oxidation of a secondary alcohol produces only a ketone (Table 3.6, reactions 2, 3).

Reduction of aldehydes to primary alcohols and ketones to secondary alcohols can be done using sodium borohydride (Table 3.6, reactions 4, 5). However, a tertiary alcohol cannot be reduced by this method. Alkenes and alkynes may be reduced by H_2 and a suitable catalyst (Table 3.6, reactions 6, 7).

Electrophilic addition reactions are common in the case of double bonds.

These reactions involves an initial attack by an electrophilic reagent, or 'electron-seeking' species on the π-electrons of the double bond. This gives rise to the formation of a positively charged carbon atom called a carbonium ion. This is then followed by the addition of a negatively charged species. In these reactions two atoms from a reactant are added across the double bond (Table 3.6, reaction 8). Alkynes react in a similar way to alkenes in undergoing addition reactions.

Compounds that have double bonds can also undergo polymerisation addition reactions. That is, single molecules (the monomer) add together to give long chain compounds called polymers. Ethene, for example, undergoes addition reaction to form polyethene.

The carbonyl group found in aldehydes and ketones also readily undergo nucleophilic addition reactions. Here the reagent has a lone pair of electrons and is an 'electron donating' species.

A condensation reaction occurs when two molecules join with the consequent elimination of a simple molecule such as water, ammonia or hydrogen chloride (Table 3.6, reaction 9).

Hydrolysis is the term describing the reaction of water with an amide or an ester (Table 3.6, reactions 12, 13). Heat and an acid catalyst are required.

If an atom or group is replaced by another atom or group, then a substitution reaction has occurred. The benzene ring readily undergoes electrophilic substitution reaction; the electrophilic reagent attacks the π-electrons spread out over the benzene ring. Note that addition across a double bond does not occur and the reaction requires a catalyst. In this reaction the iron(III) bromide acts as the carrier of polarised dibromine, allowing the positive end of the dibromine molecule to attack the benzene ring (Table 3.6, reaction 13). The presence of one or more groups on a benzene ring can cause direction of substitution (Figure 3.28) and affect the rate of reaction.

Note that in Figure 3.28 the hydrogen atoms in the 3 and 5 positions are not substituted by chlorine.

Alkanes will undergo substitution with halogens in the presence of sunlight. However, they are chemically unreactive towards polar or ionic reagents, i.e. electrophiles or nucleophiles.

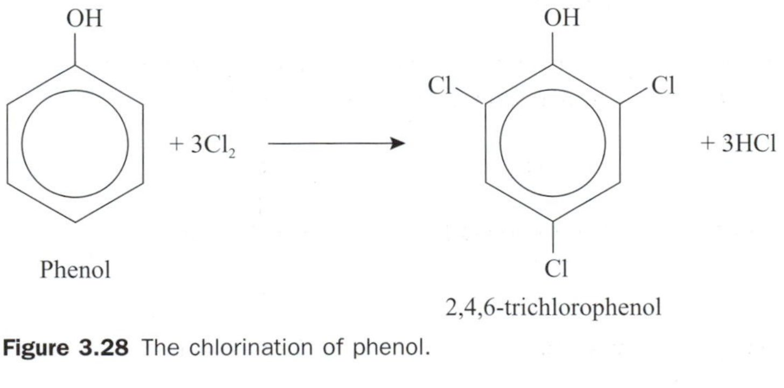

OH   OH

$+ 3Cl_2$ $\longrightarrow$ Cl   Cl $+ 3HCl$

Phenol   Cl

2,4,6-trichlorophenol

Figure 3.28 The chlorination of phenol.

Table 3.6 Some chemical reactions of some common functional groups

Functional group	Homologous series	Reactions	Example reaction	Comments
R—H	Alkanes	1 Oxidation/combustion with $O_2(g)$	$CH_4 + O_2 \rightarrow CO_2 + H_2O$	Alkenes and alkynes similarly react
CH_2—OH	Primary alcohols	2 Oxidation using acidified potassium dichromate(VI)	$CH_3CH_2OH + [O] \rightarrow CH_3CHO + H_2O$ ethanal $CH_3CH_2OH + [O] \rightarrow CH_3COOH + H_2O$ ethanoic acid	Tertiary alcohols are not oxidised
R_1R_2CH—OH	Secondary alcohols	3 Oxidation using acidified potassium dichromate(VI)	$(CH_3CH_2)(CH_3)CH$—$OH + [O] \rightarrow$ $(CH_3CH_2)(CH_3)C$=O methylpropanone	
RCHO	Aldehyde	4 Reduction using sodium borohydride, $NaBH_4$	$CH_3CH_2CHO \rightarrow CH_3CH_2CH_2OH$ primary alcohol	
R_1R_2CO	Ketone	5 Reduction using sodium borohydride	$(C_2H_5)(CH_3)CO \rightarrow (C_2H_5)(CH_3)CHOH$ secondary alcohol	
R_1R_2C=R_3R_4	Alkenes	6 Reduction with $H_2(g)$ and catalyst	$(C_2H_5)(CH_3)C$=$C(CH_3)_2 \rightarrow$ $(C_2H_5)(CH_3)CHCH(CH_3)_2$	
R_1C≡CR_2	Alkynes	7 Reduction with $H_2(g)$ and catalyst	CH_3C≡$CC_2H_5 \rightarrow CH_3CH_2CH_2C_2H_5$	

Structure	Functional group	Reaction	Equation	Notes
$R_1R_2C{=}R_3R_4$	Alkenes	8 Addition of dibromine	$CH_2{=}CH_2 + Br_2 \rightarrow CH_2BrCH_2Br$	Used as a test for an alkene, red brown colour of dibromine is lost
R_1COOH R_2OH	Carboxylic acid Alcohol	9 Condensation reaction between the acid and an alcohol using thionyl chloride, SO_2Cl_2	$CH_3COOH + CH_3OH \rightarrow$ $CH_3COOCH_3 + H_2O$ an ester	Condensation reactions eliminate simple molecules such as H_2O, NH_3 or HCl
R_1COCl R_2NH_2	Acid chloride Amine	10 Condensation reaction between acid chloride and an amine	$CH_3COCl + C_2H_5NH_2 \rightarrow$ $CH_3CONH_2 + HCl$ an amide	
R_1COOR_2	Ester	11 Hydrolysis (or reaction with H_2O)	$CH_3COOCH_3 + H_2O \rightarrow$ $CH_3COOH + CH_3OH$	
$RCOCl$	Amide	12 Hydrolysis (or reaction with H_2O)	$CH_3COCl + H_2O \rightarrow$ $CH_3COOH + HCl$	
$C_6H_5{-}H$	Benzene ring	13 Electrophilic substitution, bromination using $FeBr_3$ as a catalyst	$C_6H_6 + Br_2 \rightarrow C_6H_5Br + HBr$	Other groups present on the ring can direct where substitution takes place on the ring
RCH_2X	Halogenoalkanes where X can be Cl, Br, I	14 Nucleophilic substitution	$CH_3CH_2CH_2Br + H_2O \rightarrow$ $CH_3CH_2CH_2OH + HBr$	

Halogenoalkanes undergo characteristic nucleophilic substitution, where the reagent is a nucleophile such as water, ammonia and the hydroxyl group, which carry lone pairs of electrons (Table 3.6, reaction 14). In halogenoalkanes the halogen atom is more electronegative than the carbon atom it is attached to, and thus causes polarisation of the bond. The partial positive charge that is left on the carbon atom is therefore somewhat electron deficient and thus will tend to attract groups with electrons.

Summary

- Catenation is the formation of long chains, branched chains and rings of carbon atoms.
- Saturated hydrocarbons contain only carbon–carbon single bonds together with carbon–hydrogen bonds.
- A molecular formula shows the ratio of atoms present in a molecule.
- A structural formula is a simplified two-dimensional arrangement of a molecular structure in which the bonds are shown.
- Structural isomers have the same molecular formulae but the atoms are arranged in a different order.
- Skeletal isomers differ in the way the carbon skeleton is ordered.
- A functional group is an atom or group of atoms that gives an organic compound its peculiar physical and chemical properties.
- Unsaturated hydrocarbons contain double bonds and/or triple bonds.
- Compounds that contain a backbone of saturated or unsaturated carbon atoms organised in chains, branched chains or rings are termed aliphatic compounds.
- Aromatic compounds contain the benzene ring, C_6H_6.
- The hydrogen atoms attached to the ring can undergo substitution.
- Benzene can form fused ring systems.
- Heterocyclic ring compounds have one or more carbon atoms replaced by a heteroatom.
- The important carbonyl group occurs in the alkanals, alkanones, carboxylic acids, esters, etc.
- The rigid carbon–carbon double bond can lead to a form of stereoisomerism called geometric isomerism. The geometrical isomers are termed *cis* and *trans*.
- A chiral centre is one containing an asymmetric carbon atom and gives rise to stereoisomerism called optical isomerism. Optical isomers are optically active.
- Optically active chemicals can rotate the plane of polarised light. Optical isomers are non-superimposable on each other. By convention $(+)$ isomers rotate the plane of polarised light clockwise, and $(-)$ isomers rotate it to the left.
- Organic molecules can undergo a range of chemical reactions according to the functional groups present, e.g. oxidation, reduction, addition, condensation, hydrolysis and substitution reactions.

Questions

1   From the *pairs* of molecules listed below, which

(a)   belong to the same homologous series

(b)   are position isomers

(c)   are functional isomers

(d)   are skeletal isomers

(e)   are identical molecules

(f)   are none of the above.

(i)	$CH_3CH_2CH_2OH$	(ii)	$CH_3CHOHCH_3$
(iii)	$CH_3CHOHCH_3$	(iv)	$(CH_3)_2CHOH$
(v)	CH_3OCH_3	(vi)	CH_3CH_2OH
(vii)	$(CH_3)_2CHCH_3$	(viii)	$CH_3CH_2CH_2CH_3$
(ix)	$CH_3CH_2CH_2OH$	(x)	CH_3OCH_3
(xi)	$(CH_3)_2CHCH_2Br$	(xii)	$CH_3CH_2CHCH_3CH_2Br$

2 An organic material, containing one carbon–carbon double bond has the molecular formula C_3H_5Cl. Draw all the possible structural formulae.

3 Using the compound from (2) above, identify the one compound that can exhibit geometric isomerism.

4 An ester, X, has the molecular formula $C_4H_6O_2$. It is suspected to have one carbon–carbon double bond in its structure.

(a) Suggest a test that could be used to show that this compound contains a carbon–carbon double bond.

(b) There are five possible structures for ester X. One of these structures exists as a pair of geometric isomers. Draw the five structures, showing which one exists as a pair of geometric isomers.

(c) If the five possible esters are hydrolysed, what five alcohols will be formed? Which pair of alcohols shows positional isomerism?

5 (a) Draw a Lewis-type structure to show the bonding in ethanol, C_2H_5OH.

(b) Draw the structural formula for the functional group isomer of ethanol.

(c) Write down the chemical reaction that would occur between ethanol and ethanoic acid if these two reagents were to be heated with an acid catalyst.

(d) The molecule shown below is an insecticide. It can be made in the laboratory by reaction between an alcohol and a carboxylic acid. Name and draw structures of the alcohol and carboxylic acid that can be used to make the insecticide shown below.

$$CH_3CH_2C\equiv CCH_2CH_2CH_2COOCH_2CH_3$$

6 (a) The compound **A**,

$$CH_3CH_2C\equiv CCH_2CH_2CH_2COOCH_2CH_3$$

can be reduced using suitable reagents to form compound **B**,

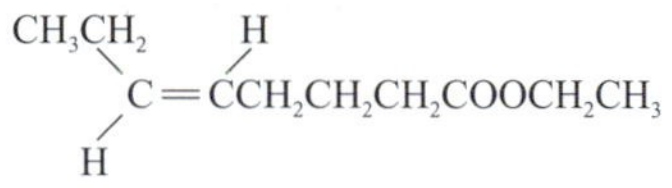

It can also be reduced using hydrogen to compound **C**,

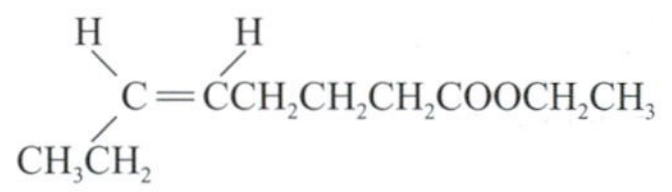

Comment on the structures of the above new products **B** and **C**.

(b) Compound **B** undergoes an addition reaction with hydrogen bromide. When it does so, it forms two positional isomers. Draw the structures of these two isomers.

(c) Both of the two positional isomers in (b) are chiral. Identify the chiral centres on these molecules.

(d) Compound **C** also undergoes an addition reaction with hydrogen bromide. Do the products resemble those obtained for **B** in part (b) above?

Further reading

Both the texts listed below were recommended reading in Chapter 1. As basic introductions to organic chemistry the reader will find them easy to follow.

Atkins, P.W. and Beran, J.A. (1992) *General Chemistry*, 2nd edn. Scientific American Books, W.H. Freeman, New York.

Glugstone, M. and Flemming, R. (2000) *Advanced Chemistry*. Oxford University Press, Oxford.

There are many good textbooks on the market that cover organic chemistry. The following are two which presume a prior knowledge of basic organic chemistry.

Clayden, J., Greeves, N., Warren, S., Wothers, P. (2000) *Organic Chemistry*. Oxford, University Press, Oxford.
A very user-friendly textbook that offers excellent explanations for organic chemical reactions. Reaction mechanisms are explained very well. This is aimed at first- and second-year undergraduates.

Solomons, G. and Fryhle, C. (2000) *Organic Chemistry*, 7th edn. John Wiley, Chichester.
This is another well-organised, very clearly formatted text. It deals with both basic and advanced concepts. The book comes with quite a useful CD.

4 Planet Earth and the origin of our environment

- The origin of the Universe and the big bang theory
- The formation of the elements via the birth and death of stars
- The origin of the solar system and the formation of the Earth via accretion
- The evidence for the internal structure of the Earth
- The formation of the Earth's atmosphere and hydrosphere
- Origin of life and the formation of dioxygen in the atmosphere
- The determination of the age of the Earth by radiometric dating

The hot big bang

Currently, many scientists support the big bang theory (Weinberg, 1993; Gribbin, 1998, 2000) as probably giving the best description of how the Universe was formed and subsequently developed.

Between 15×10^9 and -20×10^9 years ago, it is thought that the origin of all matter in the Universe was compressed into a point about the size of a hydrogen nucleus. The Universe began to be formed when this infinitely dense point of matter at a temperature of about 10^{10} K violently exploded and distributed matter throughout space. Not only was matter created in the form of fundamental sub-atomic particles and energy produced, but time and space also came into existence.

Initially, the Universe contained neutrons, protons, electrons and other fundamental particles. The neutrons began, on a very large scale, to be transformed into protons and electrons (plus particles called anti-neutrinos). The protons and electrons then began to interact with each other and other particles. Neutrons were also able to combine with protons to produce deuterium nuclei, which then combined to form helium nuclei ($^4_2$He). An excess of protons ensured that the Universe was composed of about 25 per cent helium and 74 per cent hydrogen nuclei. As the Universe continued to expand and cool electrons were able to combine with protons, and protons with protons to form hydrogen ($^1_1$H, $^2_1$H), helium ($^3_2$He, $^4_2$He) and lithium atoms ($^7_3$Li).

The birth and death of stars – the formation of the elements

[Kaler 1997; Taylor 1972; Prantzos *et al.* 1993] At very high temperatures the kinetic energy of hydrogen nuclei is greater than the repulsive energy between their positive charges, and hydrogen nuclei are able to fuse together. This process began the birth of stars. The birthplace of stars is now found in the spiral arms of galaxies (Figure 4.1), where there are dense clouds of gas and dust, dominated by the element hydrogen. Gravitational forces cause the concentration of matter resulting in stars which are born

Figure 4.1 The spiral Galaxy M81 in the northern constellation Ursa Major.

Source: Reproduced with the permission of Robert Gendler.

in clusters of a few hundred. After these clusters are formed, the remaining matter becomes dispersed to leave isolated stars or small groups of stars.

At first, the denser regions in the clouds collapse to form 'protostars'. It is the gravitational energy of the collapsing star that is the source of its energy. Once the star has contracted enough so that its central core can undergo nuclear reactions that convert hydrogen to helium (called hydrogen burning), it becomes a main sequence star. Hydrogen burning lasts nearly the entire lifetime of the star.

Hydrogen burning occurs in several steps,

$$\text{Step 1} \qquad 2{}^{1}_{1}\text{H} + 2{}^{1}_{1}\text{H} \quad \overset{10^{9} \text{ years}}{=} \quad 2{}^{2}_{1}\text{H} + 2{}^{0}_{+1}\beta + 2\nu$$

$$\text{Step 2} \qquad 2{}^{2}_{1}\text{H} + 2{}^{2}_{1}\text{H} \quad \overset{1 \text{ second}}{=} \quad 2{}^{3}_{2}\text{He} + \gamma$$

$$\text{Step 3} \qquad 2{}^{3}_{2}\text{He} \quad \overset{10^{6} \text{ years}}{=} \quad {}^{4}_{2}\text{He} + 2{}^{1}_{1}\text{H}$$

Adding these equations gives,

$$4{}^{1}_{1}\text{H} = {}^{4}_{2}\text{He} + 2{}^{0}_{+1}\beta + 2\nu + \gamma$$

Since it is the slowest, step 1 is the rate-determining step. However, there are so many hydrogen atoms in the core undergoing this proton–proton chain reaction that the time involved loses its importance.

The Sun is a main sequence star. It also contains other elements besides hydrogen and helium, and therefore must be a secondary star, which has incorporated elements from other sources.

A star's lifetime depends upon its mass. A star of one solar mass (the mass of the Sun) has a lifetime of about 10^9 years, a star of half a solar mass one of about 2×10^{11} years, whilst the lifetime of a star 15 times the solar mass is about 1.5×10^7 years. For a main sequence star, the greater the mass, the larger it is and the faster it uses up its hydrogen. Its interior and exterior temperatures are also high. Stars from one to eight solar masses will sooner or later become red giants with diameters 10–100 times greater than they had as a main sequence star. As the hydrogen is used up, the temperature of the core of the star falls and its exterior expands. The core, which at this stage is mainly helium, continues to contract under gravitational forces causing the density and temperature to markedly increase. The outer shell becomes hotter and the remaining hydrogen in it begins to burn. Helium burning begins in the nucleus and the lighter elements start to be formed. Typical nuclear fusion reactions occurring are,

$$ {}^4_2\text{He} + {}^4_2\text{He} = {}^8_4\text{Be} + \gamma $$

$$ {}^8_4\text{Be} + {}^4_2\text{He} = {}^{12}_6\text{C} + \gamma $$

What happens next depends upon the size of the star. If the star is small, i.e. less than 1.4 solar masses, then, as the helium fuel becomes exhausted, the star becomes a red super-giant. The star loses mass because of the generation of strong stellar winds. Eventually all of the matter in its outer regions will be lost, and a hot core of carbon will be left centred in a nebula of expelled gas. The star remnants will shrink to about the size of the Earth and cool to become a white dwarf.

$$ {}^{12}_6\text{C} + {}^1_1\text{H} = {}^{13}_7\text{N} + \gamma $$
$$ {}^{13}_7\text{N} = {}^{13}_6\text{C} + \nu + \beta^+ $$
$$ {}^{13}_6\text{C} + {}^1_1\text{H} = {}^{14}_7\text{N} + \gamma $$
$$ {}^{14}_7\text{N} + {}^1_1\text{H} = {}^{15}_8\text{O} + \gamma $$
$$ {}^{15}_8\text{O} = {}^{15}_7\text{N} + \nu + \beta^+ $$
$$ {}^{15}_7\text{N} + {}^1_1\text{H} = {}^4_2\text{He} + {}^{12}_6\text{C} $$

$$ \overline{4\,{}^1_1\text{H} = {}^4_2\text{He} + 2\beta^+ + 3\gamma + 3\nu} $$

Figure 4.2 The carbon–nitrogen cycle.

$$ 2\,{}^{12}_6\text{C} = {}^{20}_{10}\text{Ne} + {}^4_2\text{He} $$
$$ 2\,{}^{16}_8\text{O} = {}^{28}_{14}\text{Si} + {}^4_2\text{He} $$
$$ 2\,{}^{16}_8\text{O} = {}^{31}_{16}\text{S} + {}^1_0\text{n} $$

Figure 4.3 Carbon and oxygen burning.

If the star is large, i.e. bigger than about 1.4 solar masses, then the higher temperatures (about 6×10^8 K) ensure alternative hydrogen burning processes that occur in a carbon–nitrogen cycle (Figure 4.2). Note that carbon-12 is initially involved but is not consumed overall, and thus acts as the catalyst for the cycle.

Carbon and oxygen burning also takes place (Figure 4.3).

At higher temperatures (about 3×10^9 K) γ-rays can also cause disintegration of a nucleus, e.g. ${}^{28}_{14}\text{Si} + \gamma = {}^{24}_{12}\text{Mg} + {}^4_2\text{He}$, which initiates silicon burning and the formation of heavier elements. This continues up to the formation of ${}^{56}_{26}\text{Fe}$, which has a particularly stable nucleus.

All of the nuclear reactions listed so far are exothermic reactions. However, if the temperature of the star is high enough, endothermic nuclear reactions can occur which break down nuclei and cause the emission of neutrons, protons, helium nuclei and neutrinos. Neutrinos are very fast and therefore take a considerable amount of energy away from the star which cools down. Contraction of the core occurs until the density becomes so high that further collapse is resisted and any more incoming matter 'bounces' off the core. This sudden 'bounce' causes a **supernova** explosion. The supernova explosion throws out carbon, oxygen and other elements into interstellar space.

The heavier elements, i.e. with masses greater than $^{56}_{26}$Fe, are formed by the capture of neutrons formed in stellar processes such as those occurring in red giants.

There are two neutron capture processes. The first is slow neutron capture, which takes place from 1 to 10^4 years. Here, one neutron is captured by a nucleus so that the neutron to proton ratio is increased. This is then usually followed by β-decay, for example,

$$^{58}_{26}\text{Fe} + ^1_0\text{n} = ^{59}_{27}\text{Co} + \beta + \gamma$$

The heaviest element produced by this process is $^{209}_{83}$Bi.

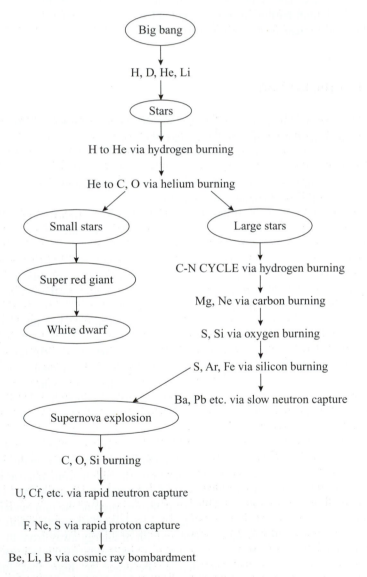

Figure 4.4 The formation of the elements.

The second process is fast neutron capture which occurs between one-hundredth of a second and about 10 seconds,

e.g. $^{58}_{26}Fe + {}^{1}_{0}n = {}^{59}_{27}Co + \beta + \gamma$

Some isotopes such as $^{112}_{50}Sn$ and $^{114}_{50}Sn$ are believed to have been formed by rapid proton capture. However, the low abundance of such isotopes shows that neutron capture is more favourable than proton capture.

Some isotopes of low abundance that would not be expected to survive within a star are found in the cosmos. Amongst such isotopes are $^{9}_{4}Be$ and $^{11}_{5}B$. These are thought to be formed by a process called spallation, where heavy particles found in cosmic rays have collided with elements such as carbon and nitrogen and broken them up into lighter nuclei.

Matter that is expelled by stars at the end of their lives is returned to the interstellar medium where it can take part again in the birth of a new star. A cosmic cycle thus exists (Figure 4.4).

The origin of the solar system

[Beatty *et al.* 1999] The planets in our solar system can be grouped into three classes according to their size, density and chemistry. The first group contains the large, low density 'gas giants' Jupiter and Saturn. The second group contains the medium-sized, high density 'rocky' planets, i.e. Mercury, Venus, Earth, the Moon and Mars, the asteroids and satellites. The last group contains the 'icy' giants Uranus and Neptune. Pluto is classified as a planet but is only the size of Neptune's satellite Triton.

One explanation of the origin of the solar system is the Nebula Hypothesis, which focuses on the gravitational collapse of a dust and gas cloud. It assumes that all members of the solar system condensed from the same solar nebula about 4.7×10^9 years ago. The initial nebula was formed in the spiral arm of the Milky Way from dust and gas which had become disturbed in some way, e.g. a shock wave from a nearby supernova. The disturbance probably caused variations in the density of matter within the cloud that resulted in sufficient amounts of it being able to aggregate under the influence of gravitational forces. Since the cloud was itself revolving around the centre of the Milky Way, then the outer edge of the nebulous cloud was moving more slowly than its inner parts. As the nebula rotated, it continued to collapse in on itself. In order to conserve angular momentum, as the cloud became smaller the speed of rotation increased. As its size diminished, the nebula started to flatten. This was because matter, attracted by gravity, was able to move more easily towards the centre of the nebula along its axis of rotation rather than perpendicular to it. Thus a flattened disc resulted, which contained a central, denser, condensed region. As the gas and dust concentrated more and more in this central region, potential energy was converted to kinetic energy and so the temperature increased. Eventually, the temperature became great enough for nuclear fusion reactions to start, and a star, the protosun, began to form. If the disc uniformly rotated, then particles on the rim would have rotated faster. As a consequence of this, matter continued to flow inwards and, again because of the conservation of angular momentum, angular momentum was transferred outwards. This is why the planets move in an almost common plane and why they have a higher spin than the Sun.

As the protosun was being formed, very small 'solid' bodies also began to take shape and grow in the cooler parts of the nebula. Near to the very hot protosun, as the nebula cloud cooled, temperatures became low enough to allow metals to condense first followed by the rock forming silicates. However, temperatures remained too high to allow water to freeze or to allow the retention of volatile compounds such as ammonia and carbon dioxide. These small bodies collided with each other to form larger bodies, ranging in size from a few metres to about the size of Mars. Such large objects are called **planetisimals**. How big each individual planetisimal became depended upon their distance from the protosun and the density of the nebula material. Asteroids are believed to be surviving examples of these planetisimals.

The larger planetisimals acquired further matter because their increased gravity caused inter-planetisimal collisions. This process is termed **accretion**, and eventually led to the formation of the inner planets Mercury, Mars, Venus and Earth. Further away, at distances equal to that of Jupiter and greater, ice could form. Ice is made of the elements hydrogen and oxygen which are the most reactive and abundant elements in the Universe. Rocks are made mainly of silicon and oxygen. Because of the much larger abundance of oxygen compared to silicon, there was more than enough oxygen to make both rocky material and massive amounts of water. Hence, objects forming at distances from the protosun, and thus cool enough, could amass much more solid matter in the form of ice than those close to it.

After about a million years of cooling and accreting, very strong solar winds blew the remaining gas and dust out of the system. Once an object acquires sufficient mass, it will exert enough gravity to attract and keep the hydrogen and helium present in the solar nebula. This may explain how the outer planets became so large, why they have the chemical composition they do, and why there is no planet between Mars and Jupiter. Larger unaccreted pieces of matter have remained in the solar system as asteroids and comets.

The timescale involved from the collapse of the gas and dust until the formation of the planets was very short (about ten million years). Collisions between planetisimals transferred energy to the forming planets, which caused vaporisation and melting. The composition of the resulting objects was thus different from the original materials, which first condensed from the nebula.

At some time during the accretion period, collision between some large body and the infant Earth took place which led to the formation of the Moon. The mantle of that body was ejected to form the Moon whilst its core was assimilated by the Earth. Other collisions caused Mercury to lose its mantle and Venus to rotate clockwise. This bombardment of the forming planets lasted until 3.8×10^9 years ago and resulted in the cratered appearance of the Moon, Mercury and Mars, and the destruction of our Earth's earlier crust.

The formation of the Earth

As the Earth accreted material, its temperature increased. This was due to three reasons. First, collision of planetisimals with the primitive Earth caused the transference of kinetic energy from the former to the latter. Second, as the mass of the Earth increased, its gravity also increased causing its material to compress and increase in density. This resulted in a rise in temperature to about 1,300 K. The increase in mass also increased the gravitational attraction of the Earth for planetisimals. This increased their kinetic energy and the impacts became more energetic. The third reason was the collision with other matter of particles emitted by radioactive substances within the Earth. The kinetic energy transferred caused a rise in temperature to about 2,300 K. At this temperature, iron will melt.

During accretion, an incoming planetisimal could interact with a primitive Earth in several ways. Some of its impact energy could have caused the heating up of any existing atmosphere, some of the energy would be dissipated as heat throughout the Earth's body, and some would cause the vaporisation of some its surface material. The structure of the Earth at this stage would then be a hot, outer shell surrounding a cooler interior. The surface itself would cool down a little due to the loss of energy by radiation. Most of the acquired impact energy would have stayed in the Earth raising its temperature to such a level that it melted its rocks and metals. Thus, the surface of the Earth was covered by an ocean of molten material known as magma.

The variety of planetisimals/meteorites striking the Earth consisted of metals (e.g. iron and nickel), sulphides, oxides, silicates and volatile materials such as water, ammonia and other trapped gases.

Any volatile materials would have been hot enough to become part of the primitive atmosphere. Further impaction caused this atmosphere to increase in temperature and its component gases to be lost form Earth. Our current atmosphere is therefore believed to be a secondary one.

Whilst the less volatile components remained as part of the Earth, they would have partially melted to form part of the magma ocean. The metals, because of their higher melting points and densities, sank in the molten ocean to eventually become part of the Earth's metallic core. Sulphides behaved in a similar way but, because of their lower densities, formed a layer between the mantle and the core. This process of the cooling and sinking of various materials has produced four major layers or shells. These correspond in elemental composition to the three major types of meteorites observed to land on Earth at the present time. Observations made on these meteorites together with the study of earthquake waves has enabled a structure to be proposed for the present Earth (Box 4.1).

Thus, Earth started its life as a fairly homogeneous primitive planet which became hotter. The temperature rise was enough to melt iron and nickel which, because of their higher density, sank towards the centre of the planet to form the core. Lighter material floated upwards and gradually cooled forming a primitive crust. The Earth became a zoned or layered planet with a lighter, less dense, crust separated from the dense core by a mantle of medium density material. As the outer layers cooled and solidified, large cracks developed because of thermal stress which left the outermost layer, the lithosphere, broken up into large blocks or plates.

The formation of the Earth's atmosphere and hydrosphere

It is possible that, during its early period of formation, the Earth may have grown big and cool enough to have acquired an atmosphere. The Earth's gravity may have been sufficient to have caused the **direct capture** of some of the original gas cloud from which it was made. However, the composition of that atmosphere would have depended upon the gravity that existed at that time. A gaseous atom or molecule can escape the gravitational pull of a planet if its velocity is high enough. The higher the gravity, the higher the velocity of a particle must be in order to escape from the surface of a planet. Once the solid Earth began to form, any gaseous atom or molecule with a velocity lower than the escape velocity could have contributed to the formation of the first atmosphere.

The velocity of a gaseous particle depends upon its temperature and its relative molecular/atomic mass. The hotter a gas is the faster its particles will move. In addition, particles having a low mass will move faster. The temperature of the Earth's

Box 4.1

The internal structure of the Earth

The Earth has an average radius of 6,370 km. Figure 4.5 shows a summary of the zoned structure of the Earth.

Inner core

Between a depth of 5,150 and 6,370 km is the inner core, which is believed to be composed of a solid mixture of iron and nickel. Its density is about 13.5×10^3 kg m^{-3} and it contributes about 1.7 per cent of the total mass of the Earth.

Outer core

The outer core is found between a depth of 2,900 and 5,155 km. It contributes some 30.8 per cent towards the total mass of the Earth. It is a hot, electrically conducting liquid composed of iron and sulphur. Its density ranges from 12.3×10^3 kg m^{-3} at its greatest depth to about 9.9×10^3 kg m^{-3} at its shallowest. Convection currents also exist in this liquid outer core. Its ability to conduct electricity combined with the Earth's rotation produces a dynamo effect that leads to electrical currents that, in turn, produce the Earth's magnetic field. Its movement also leads to the subtle jerking evident in the Earth's rotation.

Outer core–mantle interface

At the outer core–mantle interface, there is a change in chemical composition from an iron-rich core to a silicate-based mantle.

Lower mantle

The lower mantle starts at a depth of 650 km and extends to a depth of 2,900 km. It constitutes 52.2 per cent of the Earth's mass. The density changes from about 4.3×10^3 to 5.4×10^3 kg m^{-3}. It is probably composed mainly of silicon, magnesium and oxygen, together with some iron, calcium and aluminium. Evidence for this assumption comes from measurements made on the abundance of elements in the Sun and in primitive meteorites.

Transition zone

Between a depth of 400 and 650 km is a transition zone or mesosphere. It constitutes 7.5 per cent of the Earth's mass. Here, there is a phase change from high to low density minerals. It is the source of magmas, and contains calcium, aluminium, silicon and oxygen.

Upper mantle

From about 10 to 400 km deep lies the upper mantle. It forms 10.3 per cent of the Earth's mass. It is composed of three layers. The deepest layer between 250 and 400 km has a density of about 3.4×10^3 kg m^{-3} and is believed to be composed of a solid rock, peridotite. Between depths of 50 and 250 km, there is a low seismic speed layer, which is partly molten peridotite. This layer exhibits phase changes from solid to liquid and liquid to solid. It is referred to as the asthenosphere. The upper layer from about 10 to 50 km deep has a variable peridotite composition and a density of 3.3×10^3 kg m^{-3}.

The crust

Between the upper mantle and the crust is the Moho discontinuity which separates the mantle from the crust. Here, there is a chemical composition change from peridotite to basalt and granite. The crust varies in thickness according to whether it is continental or oceanic crust in nature. Oceanic crust has a depth of between 0 and 9 km and forms about 0.1 per cent of the Earth's mass. It is essentially basaltic in nature, with a density of some 3.0×10^3 kg m^{-3}. Continental crust has a thickness of between 0 and 90 km and forms about 0.4 per cent of the Earth's mass. Continental crust is granitic in nature.

atmosphere was much higher during its infancy than now. The escape velocity for Earth at its *current* temperature is 1.13×10^4 m s^{-1}. Atoms and molecules with a relative molecular/atomic mass greater than 10 at higher temperatures have an escape velocity below this figure for the gravity that *now* exists. Hence, the earliest atmosphere must have had gases present with relative molecular/atomic masses greater than 10 since its temperature was so high. Dihydrogen and helium would therefore not be expected to be main components of this atmosphere.

Since hydrogen is the major component of the Universe and the forming solar system, compounds of hydrogen would be expected to dominate an atmosphere. In an early atmosphere created by direct capture, gases such as methane, ammonia and water vapour and other volatile compounds would be expected to be its main components, together with neon, krypton and xenon. The current compositions of the atmospheres of the four outer giant planets Jupiter, Saturn, Uranus and Neptune support this suggestion. Since these planets are much more massive and colder than Earth, their atmospheres also contain dihydrogen and helium. If the Earth ever had this kind of primary atmosphere, it has since been lost.

A second way that the Earth may have acquired an early atmosphere is via outgassing (volcanic-type activity) from its solid interior. During early accretion, solid objects involved in creating the planetisimal would have contained absorbed and adsorbed volatile materials. As the planetisimal began to grow, these volatile materials would have become trapped and incorporated into its internal structure. Hence, even if the early Earth had no atmosphere, there may well have been enough volatile material 'entombed' within it to form one at a later date. During the time that the interior of Earth began to undergo internal heating and differentiation, it is highly likely that the release of volatile materials through outgassing would have taken place. If the volatile materials had been retained until the Earth started to form zones, then the amount of outgassing would have been immense.

The atmosphere caused by outgassing would have contained only those gases which had been readily absorbed/adsorbed by the internal solid material. In addition, methane and ammonia would not be present in anything but trace amounts because they would have been destroyed by the high temperatures associated with outgassing. The Noble Gases are known not to strongly adhere to surfaces and therefore would be expected to be present in the atmosphere in low abundance. Carbon monoxide and/or carbon dioxide together with dinitrogen would also be expected to exist in an atmosphere caused by outgassing (see Figure 4.5).

A third way in which the Earth's atmosphere may have developed was by the addition of volatile materials carried by impactors toward the end of the accretionary period. For example, some of the matter that contributed to the formation of the initial planetisimal had compositions very similar to the carbonaceous chondrite meteorites, which

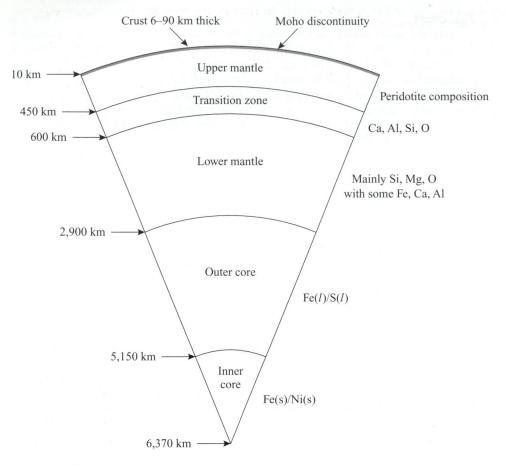

Figure 4.5 A schematic diagram of the zoned structure of the Earth (not to scale).

contain up to 20 per cent by volume of water. Generally, the solid impacting bodies would have been much richer in volatiles because they would have been amongst the last bodies to be formed. Some of these bodies may have been similar in composition to the comets that can be currently observed in the solar system.

It is generally agreed that the early atmosphere that existed before life appeared on Earth, the **prebiotic atmosphere**, was formed primarily by outgassing. This atmosphere would have been rich in carbon dioxide, water vapour and dinitrogen, with traces of dihydrogen, ammonia and methane. Evidence for such an atmospheric composition has been found in the analysis of the gases emitted from volcanoes and of the gases trapped in old solidified igneous rocks. Table 4.1 shows typical compositions, together with that of the current Earth's atmosphere. It can be seen that the gases emitted from Hawaiian volcanoes, compared with the current atmosphere, contain no dioxygen. This is because it is a very reactive gas at high temperatures. If it had been initially present in hot magma, then it would have taken part in chemical reactions to produced sulphur dioxide, carbon dioxide and water vapour. Any iron(0) or iron(II) present would also have reacted with oxygen.

At the temperatures of outgassing, water, hydrogen chloride, carbon dioxide and other volatile materials would have all been initially too hot to exist in other than the gaseous

Table 4.1 *Some typical gas composition data for volcanic eruptions, igneous rocks and the current Earth's atmosphere*

Chemical	Composition of emissions from Hawaiian volcanoes % by volume	Composition of gases trapped in basalt % by volume	Composition of gases trapped in andesite % by volume	Composition of current atmosphere of Earth % by volume
Water	73.5	83.1	92.9	0–3
Carbon dioxide	11.8	8.1	2.0	0.03
Dinitrogen	4.7	2.0	1.2	78.01
Sulphur dioxide	6.6	1.1	0.2	–
Sulphur trioxide	2.3	–	–	–
Difluorine	–	3.8	2.3	–
Carbon monoxide	0.5	0.2	0.5	–
Dihydrogen	0.4	1.2	0.4	–
Dichlorine	0.05	0.5	0.5	–
Argon	0.2	<0.12	<0.1	0.93
Dioxygen	–	–	–	20.94

state. As this atmosphere cooled to the **critical temperature** of water (374 °C), then water would start to condense and an early hot ocean formed. The hydrogen chloride would have readily dissolved in this water. Most of the carbon dioxide would have remained in the atmosphere with some dissolving in the ocean water. Both the dissolution of hydrogen chloride and carbon dioxide would have made the water acidic, causing it to react vigorously with the crustal rocks, dissolving silica (SiO_2) and cations such as Na^+. Support for this is seen in that the large amounts of sulphur dioxide, sulphur trioxide and dichlorine in volcanic emissions are rapidly removed from the *current* atmosphere by dissolution in the seas and oceans. Dihydrogen, because of its low mass and high escape velocity, is lost to space. Carbon monoxide is oxidised to carbon dioxide, and ultimately enters the seas either dissolved in its water or precipitated in sediments. The vast amount of water emitted to the atmosphere as steam is condensed and joins the normal water cycle. Since they are so chemically unreactive, dinitrogen and argon have gradually accumulated in the atmosphere via volcanic activity to give the present-day totals. Thus, the development of our atmosphere is linked to the cooling of the Earth's surface, emissions of gases from volcanic activity, and the dissolution of some of those gases in condensed waters.

Table 4.2 shows the present-day composition of the oceans compared with that of the crust. Present-day oceans consist on average of 96.7 per cent water and 3.3 per cent dissolved salts. Over 99 per cent of all dissolved substances are formed from the cations Na^+, Ca^{2+}, Mg^{2+}, K^+, and the anions Cl^{-1} and SO_4^{2-}. This table shows that four of the most abundant elements in seawater, namely chlorine (Cl^{-1} ions), sulphur (as SO_4^{2-}), bromine (as Br^{-1}) and carbon (as $CO_2(aq)$, HCO_3^{-1} and CO_3^{2-}) do not appear as such in the Earth's crust. The presence of sodium, magnesium, calcium, potassium and strontium ions can all be accounted for by the interaction of the components of the initial atmosphere with continental rocks (erosion). However, chlorine, sulphur, bromine and carbon originate from volcanic sources. The concentration of chemically combined carbon in seawater is much lower than that of chlorine and sulphur, though carbon dioxide is a much more abundant gas in volcanic emissions than dichlorine or sulphur

Table 4.2 *The concentrations of elements dissolved in seawater and in the Earth's crust*

Concentration in seawater % by mass		Concentration in Earth's crust % by mass	
Chlorine (as Cl^-)	1.900	Oxygen	46.60
Sodium (as Na^+)	1.060	Silicon	27.70
Magnesium (as Mg^{2+})	0.127	Aluminium	8.13
Sulphur (as SO_4^{2-})	0.088	Iron	5.00
Calcium (as Ca^{2+})	0.040	Calcium	3.63
Potassium (as K^+)	0.038	Sodium	2.83
Bromine (as Br^-)	0.007	Potassium	2.59
Carbon (as CO_2, HCO_3^-, CO_3^{2-} and organic matter)	0.003	Magnesium	2.09
Strontium (as Sr^{2+})	0.001	Titanium	0.44

dioxide. An explanation of this lies in the equilibria that are formed when carbon dioxide, sulphur dioxide and dichlorine react with water upon their dissolution.

The equilibria that exist when water and carbon dioxide react are given by,

$$CO_2(g) + H_2O(l) \rightleftharpoons H^+(aq) + HCO_3^-(aq)$$

$$HCO_3^{-1}(aq) + H_2O(l) \rightleftharpoons H_3O^+(aq) + CO_3^{2-}(s)$$

The equilibria that exist for the reactions of chlorine and sulphur dioxide with water can be written,

$$2SO_2(g) + 2H_2O(l) + O_2(g) \rightleftharpoons 4H^+(aq) + 2SO_4^{2-}(aq)$$

$$Cl_2(g) + H_2O(l) \rightleftharpoons 2H^+(aq) + OCl^-(aq) + Cl^-(aq)$$

If the concentration of calcium ions in seawater is high enough, then calcium carbonate, calcium sulphate and calcium chloride could all be precipitated from solution and, according to le Chatelier's Principle, the above equilibria shifted to the right-hand side.

However, the solubility of these salts differs considerably. Calcium carbonate is by far the least soluble and will be preferentially precipitated (to form limestone) and hence carbon will be removed from the sea. Calcium sulphate would then be precipitated if the seawater were to be confined long enough for evaporation to occur, whilst calcium chloride would only be precipitated if large volumes of seawater were to be evaporated.

The oldest rocks known are about 3.8×10^9 years old and show evidence that they have been changed by an environment which contained water. As yet, there is no direct evidence concerning the role of water in Earth's development in the period between 4.6×10^9 and 3.8×10^9 years.

Outgassing depends upon internal heating within the Earth caused primarily by the radioactive decay of isotopes. This process would have been much more vigorous in the infancy of the Earth than it is now due to the presence of more radioactive isotopes that were yet to decay. The atmosphere, free of dioxygen, formed at an exponentially decreasing rate.

Where did the oxygen in the Earth's atmosphere come from? The answer to this question lies in the origin of life and its link with the seas and oceans.

The atmosphere, life and the appearance of oxygen

The chemical theory of life is based on the assumption that simple organic compounds were first made from simple inorganic molecules in a reducing atmosphere devoid of dioxygen. The atmosphere formed from outgassing contained mainly carbon dioxide, water vapour and nitrogen. This atmosphere would have been subjected to electrical discharge (lightning) and ultraviolet (UV) radiation from the Sun. The amount of UV radiation passing through the atmosphere would have been much greater than now because of the lack of atmospheric dioxygen, which would have prevented the formation of a protective ozone layer. It is believed that there would have been enough energy available to enable chemical reactions leading to the formation of elementary organic molecules, including simple amino acids and sugars.

Various laboratory experiments have been devised to provide supporting evidence for the chemical theory of life. Mixtures of CO_2, H_2O and N_2 subjected to electrical discharges have been found to lead primarily to the formation of nitric(III) and nitric(V) acids, with much smaller amounts of methanal (HCHO) and hydrogen cyanide (HCN) (Miller 1974; Miller and Orgel 1974). The latter two compounds have the potential to form more complex organic compounds. When aqueous solutions of iron(II), Fe^{2+}, and CO_2 have been irradiated with UV radiation, methanal has also been formed (Miller and Orgel 1974). This suggests that reduced iron may have been important in pre-biotic Earth. Experiments on the effects of UV radiation on water containing suspended clay particles and dissolved CO_2 have resulted in the formation of methanol (CH_3OH) and other simple organic compounds. The atmospheric conditions that prevailed at the time were, however, probably too severe for organic compounds to have survived for very long; thus, it is probable that the development of more complex molecules occurred in the oceans.

Over vast periods of time, any organic compounds would have gradually accumulated in the oceans. In those places where evaporation allowed them to become concentrated, molecular interactions may have occurred which caused larger, more complex molecules to form. It has been suggested (Dillon 1978) that clay particles may have provided the surfaces where HCN, HCHO and other organic molecules could have been adsorbed and concentrated. Some large organic molecules could then have been produced by polymerisation via surface catalysed reactions on these clays.

If HCN in mildly reducing conditions undergoes electrical discharge, low yields of amino acids are produced. If these amino acids are heated, then thermal polymerisation is known to occur (Dillon, 1978). If the resulting molten material is allowed to cool, polymeric molecules similar to proteins form. It was also discovered that, when these polymeric molecules or protenoids were formed, an unexpectedly small number were obtained from a large number of amino acids in an unexpectedly ordered way. The synthesis was thus not a random one. When the protenoids were placed in warm water and cooled, it was found that they formed spherical structures of a uniform size. These structures or protenoid microspheres have been suggested as a possible model for prototype cells. It appears from these experiments that protein-type structures are readily obtainable.

In order to be recognised as 'living', organic molecules need to feed and be self-replicating. It has been suggested (Cairns-Smith 1985) that the template for replication originally existed on the electrically charged surfaces of clay particles. Some organic molecules would be adsorbed and therefore concentrated and reordered on these surfaces. Genetic information would thus be stored in the form of the distribution of electrical charges on the clay surfaces.

The only molecules in the modern world that can replicate themselves are the nucleic acids DNA and RNA. How nucleic acids such as these were synthesised on clay surfaces, how they took over the role of providing a template for replication, and how cells evolved is not, as yet, known.

If evolving life in an aqueous environment did form protocells, then it is possible that they possessed hydrophobic (water-hating) external surfaces. These could have been the precursors of cell membranes. Once discrete protocells were formed, simple division could begin to take place. This may have been a result of some physical process. For example, as a sphere increases in size, its surface area to volume ratio changes, and it is possible that a critical ratio may have been reached at which the structure became unstable. The outer membrane may have split to form two or more units.

Very little is known about the metabolism of the protocells, but it is thought that they were **heterotrophic** in that they obtained the necessary energy and matter for building more complex organic molecules from simpler ones. These protocells would have had very limited synthetic ability and probably consisted of only a few complex macro-molecules that replicated themselves from organic molecules not much simpler than themselves. How the energy of the macromolecules was liberated and harnessed in the protocells is not known. If the first organisms were mainly **heteromorphs**, then over a long period of time the accumulated organic matter in the hydrosphere would have been reduced, thus ultimately limiting their growth. Heteromorphs would have needed to develop more elaborate synthetic abilities in order to replicate themselves from simpler molecules. Organisms with more complicated biochemical pathways would be at an advantage over those with less elaborate ones, thus competition and natural selection began to operate.

It is possible that the first organisms were not only **autotrophic** (self-feeding) but also **chemo**autotrophic. If so, then the energy necessary for this type of metabolism would have been derived from light or chemical reactions, and the molecules used in synthesis ones like carbon dioxide, hydrogen sulphide and dinitrogen.

The early true autotrophs probably carried out reactions of the form,

$$6H_2S + 6CO_2 \longrightarrow 6S + C_6H_{12}O_6 + 6H_2O$$

i.e. hydrogen sulphide reduces carbon dioxide to carbohydrate. This still occurs in some bacteria today.

As further mutation and selection took place, water became the reducing agent,

$$6H_2O + 6CO_2 \longrightarrow C_6H_{12}O_6 + 6O_2$$

The first organisms to perform this were probably the predecessors of the blue-green bacteria. Oxygen produced by this process would dramatically change the biosphere. For example, it would react with any hydrogen sulphide in the atmosphere thus,

$$2H_2S + O_2 \longrightarrow 2H_2O + 2S$$

Eventually, the autotrophs that used water in photosynthesis would have replaced the more primitive organisms using hydrogen sulphide, and the last traces of hydrogen sulphide would disappear from the atmosphere. This suggestion is supported by the existence of bacteria in red and white tube worms, which are found around hydrothermal vents where water heated by the Earth's crust emerges on the ocean floors. Here, large mounds are observed which are streaked with green, yellow and purple colours of metal sulphides, and water samples taken from these regions smell of hydrogen sulphide.

The tubeworms have sulphur oxidising bacteria inside them. During the oxidation process, the bacteria release chemical energy from hydrogen sulphide and sulphides. This energy is then used by the bacteria to reduce carbon dioxide to organic molecules such as carbohydrates. Chemosynthesis is currently taking place according to the equation,

$$6CO_2 + 6H_2S + 6O_2 + 6H_2O \xrightarrow{\text{chemical energy}} C_6H_{12}O_6 + 12H^+ + 6SO_4^{2-}$$

It is possible that some very early life forms used a similar pathway in their metabolism.

There is still a range of metabolic types in existence today. For example, one group of organisms, the desulphovibrios, appear to be intermediate between heteromorphs and automorphs and may be similar to some of the early, very primitive organisms. The desulphovibrios are sulphate-reducing bacteria that can use an inorganic reaction to generate energy that is used in the formation of complex organic materials from simple organic ones,

$$CaSO_4 + 4H_2 \longrightarrow CaS + 4H_2O + \text{energy}$$

Photoautotrophs would have had a huge advantage over heterotrophs and chemo-autotrophs because light energy from the Sun was readily available. The first photo-autotrophs were probably very different from modern-day plants and blue-green bacteria. Modern photosynthetic organisms generally use chlorophyll as a catalyst to convert carbon dioxide and water to carbohydrate and dioxygen,

$$6CO_2(g) + 6H_2O(l) \xrightarrow[\text{chlorophyll}]{\text{light}} C_6H_{12}O_6(s) + 6O_2(g)$$

The early photosynthesisers may have used purple pigments for photosynthesis. Some evidence for this is found in the present-day organism *Halobacterium halobium*, which grows only in *very* salty conditions, e.g. salt lakes. Virtually nothing else can live in such saline environments, so this bacterium has developed in an environment free of competition. Thus, it is thought that selection pressure has been low and the organism still possesses a primitive structure. Its photosynthesis is also unusual. The photosynthetic pigment is the protein bacteriorhodospin. This pigment absorbs visible light in the middle region of the visible spectrum and therefore appears purple. The primitive seas were possibly purple-coloured. *Bacteriorhodosporin* is an organic compound belonging to a family called the terpenoids. Terpenoids have been found in rocks 3.0×10^9 years old (Cairns-Smith 1985) and may be the remains of the earliest photosynthesisers.

Although purple bacteria like *Halobacterium* would have been successful, they would have been eventually outcompeted by organisms that could use carbon dioxide. The quantity of carbon dioxide in the atmosphere was almost certainly higher than today because of (i) outgassing, and (ii) carbon dioxide produced by fermentation reactions of the primitive heterotrophs.

Later organisms developed the green pigment chlorophyll, which absorbs strongly in the blue and red parts of the visible spectrum. The great advantage of the chlorophyll-containing organisms was that they could convert carbon dioxide to organic matter using light, i.e. they were true autotrophs. These organisms were the precursors of modern photosynthesisers and had certainly evolved 2.3×10^9 years ago.

Evidence for the existence of early autotrophic life is to be found in the fossilised layered structures called stromatolites that exist in sedimentary limestones throughout the world. The thickness of these deposits can be as much as 1 kilometre and they can be hundreds of kilometres across. Some rocks bearing stromatolites have been dated as 3.5×10^9 years old. Stromatolites appear to have been formed from organic matter composed of thread-like structures that grew as sticky surface mats on the surface of water. These mats probably trapped sediments and the layered structure resulted from repeated cycles of growth and sediment capture. Such structures resemble that of the colonies of modern blue-green **cyanobacteria**. These also form felt-like microbial mats, which cover the surface of shallow pools of water, and are composed of alternating layers rich and poor in organic matter.

As the oxygen-producing autotrophs began to dominate the Earth, the atmosphere gradually became less reducing and more oxidising. Initially, there would have been very little free oxygen in the atmosphere because the dioxygen molecules released would have been 'mopped up' by the reduced rock strata that contained iron compounds. In the original rock strata, iron occurred as iron(II), Fe^{2+}. Compounds that contain iron(II) are dark blue or grey in colour, whereas those that contain iron(III), Fe^{3+}, are red, brown or yellow. Many other minerals became oxidised and weathering reactions accelerated, producing new minerals.

Sedimentary strata containing abundant oxidised iron compounds are termed **redbeds** and began to appear about 2.0×10^9 years ago. After this time, they became increasingly more common. Redbeds older than 2.2×10^9 years have not been found. Hence 3.0×10^9 years ago reducing conditions prevailed, and 2.2×10^9 years ago oxidising conditions prevailed.

The dioxygen liberated by the photoautotrophs was initially toxic to many of the organisms present. Their survival depended upon their ability to develop new ways of metabolising or disposing of the dioxygen in some other way, or to avoid it.

Sedimentary **banded ironstone deposits** made up of alternating layers of iron(III) oxide, Fe_2O_3, and chert, SiO_2, plus organic matter, are useful indicators of the transition period between reducing and oxidising conditions. These rocks are thought to have been formed because the oceans contained organisms that produced dioxygen by photosynthesis. The dioxygen released reacted with the soluble reduced iron(II) to give insoluble hydrated iron(III) oxide, which precipitated out of solution. The banding suggests some kind of balance between the photosynthetic oxygen-producing organisms, and the supply of reduced iron. Banded ironstone deposits first appeared around 3.8×10^9 years ago and stopped being deposited about 2.0×10^9 years ago. This geological evidence suggests it is probable that dioxygen-producing organisms began to appear about 3.5×10^9 years ago, but there was little free atmospheric dioxygen before about 2.2×10^9 years ago.

The age of the Earth – radiometric dating

The currently accepted age of the Earth is taken to be 4.6×10^9 years. How was this figure determined?

Because of the dynamic nature of the Earth's crust and the existence of a rock cycle, it is probable that the oldest of Earth's rocks have been recycled and destroyed. Thus it has not been possible for scientists to determine the *exact* age of the Earth by dating rocks. Primordial rocks have yet to be found. An approximate age has been determined and, by assuming that the Earth and the rest of our solar system were formed at about the same time, so has the age of the solar system.

The ages of rocks on Earth, lunar rocks and meteorites have all been determined by radiometric methods. However, such methods only date rocks from the last time they

Table 4.3 *Common parent/daughter isotopes used in radiometric dating*

Parent isotope	Daughter isotope	Half-life/ 10^6 years
$^{235}_{92}U$	$^{207}_{82}Pb$	704
$^{238}_{92}U$	$^{206}_{82}Pb$	4,467
$^{232}_{90}Th$	$^{208}_{82}Pb$	1,400
$^{40}_{19}K$	$^{40}_{18}Ar$	1,193
$^{147}_{62}Sm$	$^{143}_{60}Nd$	106,000
$^{87}_{37}Rb$	$^{87}_{38}Sr$	48,800

were melted or disturbed in some other way. Radiometric dating relies upon the radioactive decay of long-lived isotopes that occur naturally in rocks and their component minerals. As seen in Chapter 1, a radioactive isotope will exponentially decay at a constant known rate. It is from this latter fact that calculations can be done to determine the age of a rock. Clearly, because of the vast time scales involved, very long-lived isotopes are needed, together with the identity of their daughter product. Those most commonly used are listed in Table 4.3.

The easiest rocks to date are igneous rocks. When an igneous rock solidifies, then all of its component crystalline minerals can be assumed to form at the same time. Hence, any radioactive element present at the time of solidification will be encapsulated in the rock. At this moment, it has become a 'radiometric clock', that has just started to tick! If a mineral in the rock is assumed to contain uranium in its crystalline structure and no lead at its formation, then any lead that is found to be present will have resulted only from the decay of the uranium. The relative amounts of uranium and lead, i.e. $^{235}_{92}U:^{207}_{82}Pb$ and $^{238}_{92}U:^{206}_{82}Pb$, can be determined using a mass spectrometer. Zircon is a mineral often used in dating because its crystalline structure excludes lead but includes uranium when it is formed. Individual zircon crystals found in Australia have been determined as being 4.2×10^9 years old. Complete rock dating is possible, which uses $^{40}_{19}K:^{40}_{18}Ar$ and $^{87}_{37}Rb:^{87}_{38}Sr$. The former parent:daughter relationship is useful because potassium is an element found in three of the most common rock forming minerals, mica, feldspars and hornblendes. Unfortunately, the leakage of argon from the rock can be a problem, particularly if the rock has undergone heating. Typically, complete rock analysis has resulted in the granites of Greenland being dated as 3.8×10^9 years old.

So far we have seen that it is possible to determine the age of events that have led up to the formation of the rocks and minerals of Earth. How then has the age of the Earth itself been determined? This had been done by analysis of deep ocean sediments. It is assumed that the sediments have been formed from all kinds of rocks from all over the planet, and therefore gives an accurate reflection of the mean crustal composition. By analysing the uranium and lead present in whole rock samples taken from deep sea sediments, the age at which all of the sedimentary sources were part of a uniform magma early in the Earth's history can be determined. Ages between 6.7×10^9 and 5.5×10^9 years old have been determined using this approach. However, these figures give a maximum age because the daughter lead isotopes were also present at the formation of the Earth and so too much lead would be contributing to the calculation.

In order to surmount this last problem, the isotope of lead $^{204}_{82}Pb$ which is not derived from a decay series is used. This isotope is a common one that is found both on Earth and in meteorites. Its abundance has remained constant throughout time. Some rock samples formed at the beginning of the Earth's existence contain no uranium. It is possible to measure the ratios of the amounts of $^{204}_{82}Pb$ and $^{206}_{82}Pb$ to $^{207}_{82}Pb$, and to determine the amounts of $^{206}_{82}Pb$ and $^{207}_{82}Pb$ present at the time of the Earth's formation. The ratios have been found to be $^{204}_{82}Pb:^{206}_{82}Pb = 1:8$ and $^{204}_{82}Pb:^{207}_{82}Pb = 1:9$, thus if the amount of $^{204}_{82}Pb$ in deep sea sediments is known then it is possible to calculate how much $^{206}_{82}Pb$ and $^{207}_{82}Pb$ would have been present when the isotopic clock was started. Since both isotopes of uranium occur naturally together, then there are two decay series that can

Box 4.2

Methods of dating rocks

From Chapter 1,

$$N/N_0 = e^{-\lambda t}$$

where N_0 = number of original parent atoms, N = number of parent atoms left after time t, t = time from start of decay, λ = decay constant = $0.693/t_{1/2}$, where $t_{1/2}$ = half-life of the parent isotope.

Rearranging this equation gives,

$$N_0 = N/e^{-\lambda t} = Ne^{\lambda t}$$

If the number of daughter atoms is D, then $D = N_0 - N = Ne^{\lambda t} - N = N(e^{\lambda t} - 1)$

Hence, $D/N = (e^{\lambda t} - 1)$

therefore, $e^{\lambda t} = D/N + 1$

and $\lambda t = \ln(D/N + 1)$

Hence, $t = (1/\lambda)\ln(D/N + 1)$

If we know, for example, that the parent:daughter ratio is 1:25, i.e. for every 25 daughter atoms (D) there is 1 parent atom, and $t_{1/2}$ for the parent isotope is 7.04×10^8 years, then the age of the material is given by,

$$t = (t_{1/2}/0.693)\ln(25/1 + 1)$$
$$= (7.04 \times 10^8/0.693)\ln 26$$
$$= 3.31 \times 10^9 \text{ years}$$

The method described above relies upon four important conditions. The rock must be a closed system in that daughter and parent isotopes have never been added or subtracted. The half-life/decay constant must be accurately known. The number of daughter and parent isotope atoms must also be accurately known. Finally, there should be some method for correcting for the presence of atoms identical to the daughter atoms already present when the rock or mineral was formed.

In those cases where daughter isotopes were already present, since some were formed at the start of the creation of the elements, the **isochron** method is used. Hence, if $D = N(e^{\lambda t} - 1)$ and if there were D_0 daughter atoms at the start already present, then the correct number of daughter atoms would be $D = D_0 + N(e^{\lambda t} - 1)$.

If we divide both sides of this last equation by the amount of stable isotope S of the daughter element to relate N and D abundances to the initial background, we get $D/S = (D/S)_0 + N/S(e^{\lambda t} - 1)$. Hence, a plot of D/S vs N/S would yield a straight line of slope $(e^{\lambda t} - 1)$ and intercept of $(D/S)_0$. Since $t_{1/2}$ is known, then t can be calculated, providing many points are used to obtain the line.

There are various ways in which dating can be achieved, e.g. fission track dating, disequilibrium dating. Different pairs of parent and daughter isotopes can be used depending upon the situation being investigated. For further details, the reader should consult the appropriate references (Dickin 1997; Lewis 2000).

complement each other in determining the age of the Earth. Both sets of calculations give a value of 4.6×10^9 years.

Similar calculations on lunar rocks and meteorites have resulted in values of their age of 4.6×10^9 and between 4.4×10^9 and 4.6×10^9 years, respectively (Box 4.2).

Summary

- The Universe was probably formed from a very dense point of matter that exploded somewhere between 15 and 20 billion years ago.
- Galaxies were probably formed by gravitational forces between matter that was not uniformly distributed through space.
- The occurrence of the elements is a consequence of the births and deaths of stars in a cosmic cycle of events.
- The solar system was formed from a nebula of gas and dust that collapsed under gravity to form a rotating disc of matter. Planetisimals grew by accretion to become planets.
- As the Earth cooled, matter according to its melting point and density settled out to form a zoned internal structure.
- The composition of the Earth's first atmosphere is unknown, but it was too hot to hold lighter molecules and atoms.
- The secondary atmosphere was probably formed by outgassing as the Earth's zoned structure developed. This atmosphere may also have been formed by direct capture or as a result of impactation during the late accretionary period.
- Electrical discharge through the secondary atmosphere is believed to be one way that simple inorganic molecules were changed to precursor organic molecules necessary for the start of life.
- Living organisms are capable of replicating themselves. RNA and DNA are the only molecules known to be able to replicate themselves.
- The first cells or protocells developed a necessary cell membrane. Their metabolism is thought to have been heterotrophic. The first living organisms are believed to be autotrophic and chemoautotrophic.
- The development of chlorophyll in cells led to the early photosynthesisers and therefore dioxygen which appeared in the atmosphere. The initial atmosphere of the Earth was a reducing atmosphere which developed into an oxidising atmosphere.
- The age of the Earth by radiometric dating is approximately 4,600 billion years.

Questions

1 Compare the compositions of the atmospheres of the planets in our solar system, and comment on the likelihood of them sustaining life. How is the search for life in the Universe being conducted?

2 (a) Elements heavier than $^{56}_{26}Fe$ can be formed by the capture of slow neutrons. For example, the isotope iron-58 can capture a slow neutron to become iron-59. This isotope decays to become cobalt-59 accompanied by the emission of β rays. Write out this series of events in the form of balanced nuclear reactions.

  (b) Man has made a number of elements not found in nature. For example, the bombardment of americium-241 with helium nuclei produces berkelium-243 plus two neutrons. Write out the balanced equation for this nuclear reaction. Why do you think neither americium or berkelium are found in nature?

3   In what kind of extreme environments (high pressures, high temperatures, low temperatures, etc.) does life on Earth now exist? How are these life forms adapted for these extreme conditions?

4   Devise a flow chart showing how unicellular life may have started and developed.

5   A mineral sample taken from a rock was found to contain a $^{40}_{19}$K:$^{40}_{18}$Ar of 2:17. The half life of $^{40}_{19}$K to $^{40}_{18}$Ar is 1193 Ma. Calculate the age of the rock.

References

Beatty, J.K., Petersen, C.C. and Chaikin, A. (eds) (1999) *The New Solar System*, 4th edn. Sky Publications, New York.

Cairns-Smith, A.G. (1985) Seven clues to the origin of life. *Nature*, **338**, 217–23.

Dickin, A.P. (1997) *Radiogenic Isotope Geology*. Cambridge University Press, Cambridge.

Dillon, L.S. (1978) *Evolution Concepts and Consequences*. 2nd edn. Mosby & Co, St Louis, MO.

Gribbin, J. (1998) *In Search of the Big Bang*. Penguin Books, London.

Gribbin, J. (2000) *The Birth of Time – How We Measure the Age of the Universe*. Phoenix Press, London.

Kaler, J.B. (1997) *Cosmic Clouds: Birth, Death and Recycling in the Galaxy*. Scientific American Library, W.H. Freeman, New York.

Lewis, C. (2000) *The Dating Game*. Cambridge University Press, New York.

Miller, S.L. (1974) The atmosphere of the primitive Earth and the prebiotic synthesis of amino acids. *Origins of Life*, **5**, 139–51.

Miller, S.L. and Orgel, L.E. (1974) *The Origins of Life on Earth*. Prentice Hall Inc., Englewood Cliffs, NJ.

Prantzos, N., Vangioni-Flam, E. and Casse, M. (eds) (1993) *Origin and Evolution of the Elements*. Cambridge University Press, Cambridge.

Taylor, R.J. (1972) *Origin of the Elements*. Wykeham Publishers, London.

Weinberg, S. (1993) *The First Three Minutes: A Modern View of the Origins of the Universe*. Flamingo, London.

Further reading

Francis, P. and Dise, N. (1997) *Atmosphere, Earth and Life, S269 Earth and Life*. Open University, Milton Keynes.
This is a module study book for the Open University second level course S269 Earth and Life. It is very readable and well illustrated. Useful summaries complete each chapter. Chapters deal with the evolution of the atmosphere, the appearance of oxygen and its importance in the start of life. Fossil and rock records for the evolution of the atmosphere are examined in detail. The roles of ozone and carbon dioxide are also included.

Levy, D. (ed.) (2000) *The Scientific American Book of the Cosmos*. Macmillan, London.
This book is written for the lay person by experts in their respective fields. Topics range from the big bang theory, the formation of galaxies, the birth and death of stars, the creation of the solar system, etc. No previous scientific knowledge is required.

Liddle, A.R. (1999) *An Introduction to Modern Cosmology*. John Wiley, Chichester, New York.
A very readable introduction to cosmology. No previous knowledge of astronomy is assumed, but a level of basic physics is!

Scientific American (March 1998) Special Edition: 'Magnificent Cosmos'.
A range of topics written by distinguished cosmologists. Topics range from the search for new planets, life in our and other solar systems, the evolution of the Universe, dark matter, etc. Very readable, very understandable.

Silk, J. (1997) *A Short History of the Universe*. Scientific American Library, W.H. Freeman, New York.

This is a very good introduction to cosmology. It is well illustrated and delivers a wide range of up-to-date information and ideas ranging from the occurrence of dark matter to the formation of the elements and galaxy formation.

Skelton, P., Spicer, R. and Rees, A. (1997) *Evolving Life and Earth, S269 Earth and Life.* Open University, Milton Keynes.

Again a well-illustrated and very readable module study book for S269 Earth and Life. It is about the history of the interactions of evolving life with the Earth, with emphasis on the impact of the eukaryotes and their diversification. (Please note that all of the books for S269 are well worth a look!)

Strickberger, M.W. (2000) *Evolution*, 3rd edn. Jones & Bartlett, London.

Everything you would like to know about evolution! Although not an easy book to read, it contains much material and a vast range of references to original papers. You do need an understanding of basic organic chemistry to be able to cope with some chapters. Well worth the effort in the end!

5 The Earth as a finite resource

- The meaning of renewable and non-renewable natural resources
- The three main classes of reserves – proven or measured, probable or indicated and possible or inferred
- Types of mineral resources
- Solid waste, its production and treatment
- Life cycle analysis
- The Precautionary Principle and innovation
- Economic and environmental sustainability
- The extraction of metals and their environmental problems

Introduction

Worldwide increases in population, industrialisation and availability of transport have been paralleled by very rapid increases in the demand for, and production of, material goods and services. Some of the consequences of this are that the use of natural resources, production of waste materials, and pollution of the environment have all increased. Pollution of the environment and the observation that some materials are being used up faster than they are being found and mined, are giving rise to fears concerning the future of both the Earth's resources and the quality of the environment. This has led to demands for reductions in the amount of waste produced in processes, and for ways to be found that will increase the amount of materials and/or energy that can be recovered from waste. Such recovery would slow down the exhaustion of non-renewable resources and might even lower the use of renewable resources to the rate of their replenishment.

In the past, the air, land and the oceans have been used as sinks, and as a means of diluting and dispersing all kinds of waste products. When the waste has been at low concentrations, then the environment has been able to absorb its effects with little or no change occurring in environmental conditions. Unfortunately, as more and more waste has been released into the environment, this has led to some major disturbances in environmental quality.

Renewable and non-renewable resources

Resources are materials that are useful to mankind, and are described as being renewable or non-renewable according to how long they took to form. Renewable resources can, within certain limitations, be replaced in a matter of days or years. For example, the availability of solar energy is only restricted by the amount of sunlight available (weather conditions) and by the day/night cycle, whilst wood can be replaced in the time it takes a tree to grow to a usable size. Given the appropriate regeneration time some renewable resources would be limitless. If they are not managed properly, then the rate at which such resources are used can exceed that at which they are renewed.

Non-renewable resources cannot be replaced during times comparable with the life-span of humans. It took millions of years for coal and oil to form, whereas all the copper and iron that exists on Earth were formed when the Universe came into being. Minerals were formed some 4.5 billion years ago and what now exists on Earth is all that is available. How long these resources last will depend upon the rate of use, degree of recycling, use of substitution materials, how much of each is left and innovation.

In this chapter the focus will be upon mineral resources. Minerals can be solids, liquids or gases. Hence, coal, petroleum and natural gas are all minerals. Unfortunately, minerals are rarely found in the pure state. Many solid minerals usually exist with unwanted materials or 'gangue', but together they form an ore. An ore is thus an aggregation containing one or more useful minerals in concentrations that are worth economically exploiting, together with waste matter.

One of the difficulties in assessing the amounts of resources left in the environment lies in the terms 'resource' and 'reserve' and how they are used. To avoid this confusion, the US Bureau of Mines (1980) has established a common classification and nomenclature which will be used here.

A mineral resource is defined as 'a concentration of naturally occurring solid, liquid, or gaseous material in or on the Earth's crust, in such form and amount that economic extraction of a commodity from the concentration is currently or potentially feasible' (US Bureau of Mines 1980). Mineral resources can be divided into those that have been identified and those that are yet to be discovered (Figure 5.1). In the case of identified

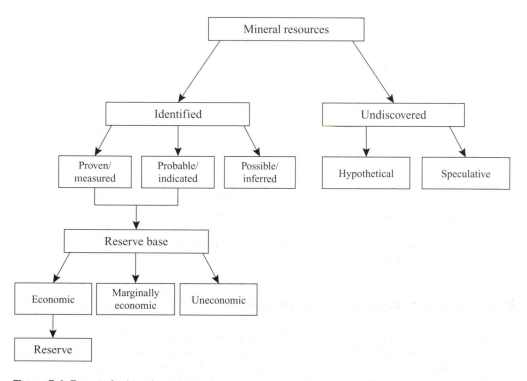

Figure 5.1 Types of mineral resources.

resources, their location, grade and quantity are known or have been estimated from geological evidence. Those that are known can be placed into three categories, **proven or measured**, **probable or indicated** and **possible or inferred**. Proven or measured mineral resources are those whose volumes and tonnage have been well established by exploration and detailed sampling. Both the geology and boundaries of the resource are also well known. In the case of probable or indicated resources, the volume and tonnage have also been calculated but from less precise data. Possible or inferred mineral resources are those based on the assumption that there is a degree of continuity beyond the measured or indicated resource boundaries. This is largely based on geological evidence. There may not be the evidence available based on sampling and measurements.

Undiscovered resources have a postulated existence. They are referred to as being hypothetical or speculative. Hypothetical resources are expected to be found existing in the same regions where there are known resources under the same geological conditions. Speculative resources are even less probable in that their existence is predicted to occur in those regions where mineral discoveries have not been made or in sites of unknown economic potential.

Mineral resources are also classified according to whether they are currently economic to extract, are marginally economic, or uneconomic. Reserves of ores then become 'that part of a resource that can be economically and legally extracted at any given time' (US Bureau of Mines 1980). The status of a resource can shift almost overnight by changes in the law, in economic factors, or in improvements in technology.

The expression **reserve base** of a resource refers to the sum of the amounts of measured and indicated resource. These can include resources that are currently economic, marginally economic and occasionally those that are uneconomic. The term **reserve** is then the part of the reserve base that can be economically extracted or produced at the time of determination. It refers only to recoverable materials.

Types of mineral resources

There are three main types of mineral resources: those that supply metals, those that supply non-metals and those that supply energy.

Metals have been separated from their ores for many years because of their very useful physical and chemical properties. The uses of pure metals are much more limited than 'designed for a job' alloys. These materials can be further subdivided on the basis of their abundances into *abundant* and *scarce* metals. To be described as abundant (see Tables 5.1 and 5.2), a metal must constitute 0.01 per cent or greater by mass of the Earth's crust. (*Note*: Mineral data depicted in all tables and mentioned elsewhere in this chapter are provided by the US Geological Survey, Geological Division, Mineral Resources Programme.)

Although abundant metals are found in large quantities in a wide variety of forms, they are not uniformly distributed over the Earth's crust. As can be seen from Table 5.1, those countries that have the largest land areas, China, Canada, Australia, Russia and the US, all have the largest metal reserves. For the rest, there is usually no correlation between the size of a country and the size of its metal reserves. Metal reserves depend upon the geology of a country, the extent of soil cover and whether or not it has been explored extensively or mined out. Ethiopia is a large country with low reserves because it has not been mineralogically extensively explored and a large proportion of the land is covered by young sedimentary deposits. Great Britain was once a source of lead and tin but economic deposits have been mined out. Jamaica is a very small country

Table 5.1 *The geochemically abundant metals – crustal concentrations, principal ores and main ore producing countries*

Metal	Mean crustal abundance %	Principal minerals	Main ore producing countries
Iron	5	Haematite, Fe_2O_3 Magnetite, Fe_3O_4 Limonite, $FeO(OH)$	Australia, Brazil, Canada, China, India, Russia, S. Africa, Sweden, Ukraine, US
Aluminium	8	Diaspore, $AlO(OH)$ Gibbsite, $Al(OH)_3$	Australia, Brazil, China, Guinea, Jamaica, Venezuela, Suriname, Russia, India, US
Magnesium	2.8	Dolomite, $CaCO_3.MgCO_3$ Magnesite, $Mg CO_3$ Epsomite, $MgSO_4.7H_2O$ Carnallite, $MgCl_2.2KCl.6H_2O$	Australia, Austria, Brazil, China, Greece, India, N. Korea, Russia, Slovakia, Spain, Turkey, US
Titanium	0.06	Ilmenite, $FeTiO_3$ Rutile, TiO_2	Australia, Canada, India, Norway, S. Africa, Ukraine
Chromium	0.01	Chromite, $FeCrO_4$	Albania, Finland, Turkey, Cuba, Japan, India, Pakistan, Philippines
Manganese	0.09	Pyrolusite, MnO_2	Australia, Brazil, China, Gabon, India, Mexico, Ukraine

Table 5.2 *The geochemically abundant metals – reserves and production*

Metal	World reserves 1999 metric tonnes	World mine production 1999 metric tonnes	World reserve base 1999 metric tonnes
Iron	As crude ore 14×10^9	992×10^6	30×10^9
Aluminium	As bauxite and alumina 2.5×10^9	123×10^6	34×10^9
Magnesium	As magnesium compounds (Mg content) 2.5×10^9	2.91×10^6	3.4×10^9
Titanium	Contained as TiO_2 370×10^6	4.01×10^6	640×10^6
Chromium	Gross weight 3.6×10^9	12.8×10^6	7.5×10^9
Manganese	Gross weight (metal content) 680×10^6	6.74×10^6	5.0×10^9

with a high reserve of aluminium in the form of bauxite because of the rock type and its climate. Concerns have been expressed about the scarce metals (Tables 5.3 and 5.4) in that they are not as environmentally or economically sustainable as the abundant metals.

Non-metallic minerals are used mainly in the form of chemical compounds or mixtures. For example, halite, NaCl, or common salt (Figure 5.2), can be used directly for both its own direct chemical value or to provide a feedstock for the manufacture of derived chemicals.

The world mining production of salt in 1999 was 200×10^6 tonnes. There are very large reserves and reserve base. Amongst the many producers of salt are Australia, Brazil, Canada, China, Germany, Iceland, Mexico, Spain, US and the UK. Because of its presence in seawater (some 2.7 per cent by mass) and the existence of vast marine

Table 5.3 *Some of the geochemically scarce elements – crustal concentrations, principal ores, and main ore producing countries*

Metal	Mean crustal abundance %	Principal minerals	Main ore producing countries
Copper	0.005	Chalcopyrite, $CuFeS_2$ Chalcocite, Cu_2S Covellite, CuS Malachite, $Cu_2CO_3.(OH)_2$	Chile, US, Indonesia, Canada, Australia, Peru, Russia
Zinc	0.007	Sphalerite/zinc blende, ZnS Calamine/Smithsonite, $ZnCO_3$	Australia, Canada, China, Peru, US
Lead	0.001	Galena, PbS	Australia, China, US, Peru
Tin	0.0002	Cassiterite, SnO_2	China, Peru, Indonesia, Brazil, Australia, US
Silver		Argentite, Ag_2S	US, Australia, Canada, Mexico, Peru
Platinum metals	Pt 0.000001 Pd 0.0000015	Platinum as Sperrylite, $PtAs_2$ Braggite PtS_2 Cooperite, PtS Palladium as Froodite and Michenerite, $PdBi_2$ Arseopalladininite, Pd_3As	US, Canada, Russia, S. Africa
Gold	0.0000004	Native Calaverite, $AuTe_2$ Krennerite, $(Au,Ag)Te_2$ Sylvanite, $(Au,Ag)Te_4$ Petzite, $AuAg_3Te_2$	US, Australia, Canada, S. Africa, Russia, Brazil
Nickel	0.007	Pentlandite, $(Ni,Fe)_9S_8$ Garnierite, $(Ni,Mg)_6Si_4O_{10}(OH)_8$ Limonite, $(Fe,Ni)O(OH).nH_2O$	US, Canada, New Caledonia, Russia, Australia, Brazil, Cuba, S. Africa
Molybdenum	0.0001	Molybdenite, MoS_2	US, Chile, China, Canada, Russia, Peru, Mexico, Iran, Armenia
Cobalt	0.00029	Linnaeite, Co_3S_4 Smaltite, $CoAs_2$ Cobaltite, CoAsS	US, Australia, Canada, Congo, Cuba, Russia, Zambia
Tungsten	0.0001	Wolframite, $(Fe,W)O_4$ Scheelite, $CaWO_4$	Austria, Russia, China, Bolivia, Burma, Brazil, N. Korea, Portugal, Uzbekistan
Mercury	0.000008	Cinnabar, HgS	Algeria, Kyrgyzstan, Spain, US

evaporite deposits, there is practically an inexhaustible amount of salt available. In the UK large quantities of halite are mined in Cheshire and Shropshire. Salt is also mined in Northern Ireland. Brine is produced in Teesside, whilst salt is a by-product in potash production in Cleveland. In 1999, the UK production of salt was some 6.6×10^6 tonnes.

When brine is produced, it is done by dissolving underground deposits of salt in water and pumping out the solution/suspension. To avoid one of the major problems of underground mining, i.e. subsidence, the cavities produced are filled with water with enough salt left to support the overlying rock. The brine is either used directly by the chemical

Table 5.4 *Some of the geochemically scarce elements – reserves and production*

Metal	World reserves 1999 metric tonnes	World mine production 1999 metric tonnes	World reserve base 1999 metric tonnes
Copper	As copper content 340×10^6	1.26×10^6	650×10^6
Zinc	As zinc content 190×10^6	7.64×10^6	430×10^6
Lead	As lead content 64×10^6	3.04×10^6	143×10^6
Tin	As tin content 7.7×10^6	2.1×10^5	12×10^6
Silver	As silver content 2.80×10^5	1.59×10^4	4.20×10^5
Platinum metals	As platinum group of metals 7.1×10^4	Platinum 150 Palladium 125	As platinum group of metals 7.8×10^4
Gold	As gold content 49	2.33	77
Nickel	As nickel content 4.6×10^4	1.14×10^4	1.40×10^5
Molybdenum	As molybdenum content 5.5×10^6	129	12×10^6
Cobalt	As cobalt content 4.5×10^6	2.83×10^4	9.6×10^6
Tungsten	2.0×10^6	3.13×10^4	3.20×10^6
Mercury	1.2×10^5	2,300	2.4×10^5

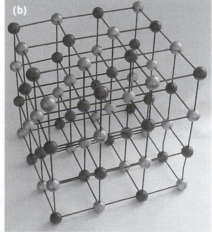

Figure 5.2 (a) Rock salt and (b) its internal structure (note the interlocking face-centred cubic structures of chloride and sodium ions).

industry or evaporated to produce white salt. In the form of rock salt, salt is mined by conventional underground mining techniques. In hot countries, e.g. Tunisia, seawater is collected in lagoons and allowed to evaporate using solar energy. Salt is one of the products of that evaporation. Although an essential component of our diets, very little salt is actually consumed worldwide by humans. The main use of salt is in the chloro-alkali industries. Salt is used to make sodium, chlorine, sodium carbonate (soda ash), sodium hydroxide (caustic soda) and hydrochloric acid. Many other compounds also involve salt in their manufacture.

There are two main problems associated directly with salt. The first is the environmental consequences of saline water being released during industrial processes. The second is caused by run-off on to agricultural land from the de-icing of roads during the winter. Water containing salt can damage roadside vegetation and can be the cause of high levels of sodium ion in groundwater. In addition, salt is known to cause the corrosion of motor vehicle bodies, metal structures and structures containing metals. There are no economic substitutes for salt, for example potassium chloride is a substitute for salt in 'low salt' products but at a much higher cost.

Energy resources are those from which useful energy can be obtained. These include non-renewable fossil fuels and uranium, and the renewable resources such as wind, wave and tidal energy.

Waste and environmental pollution – the explosion at Loscoe, Derbyshire

Waste is unwanted material that is produced, in the main, by human activities (Table 5.5). It is described as being useless, often hazardous and of no economic or other value. How it is discarded often makes waste a source of pollution. Industrial, commercial and domestic wastes usually all contain the same materials that were used to make the useful products from which the waste originated. One way of dealing with waste might be to restore some kind of value to it, thus removing it from the waste stream and thereby reducing its environmental impact. The major problem in doing this often lies in the complex mixture of waste. Unfortunately, much waste was often dumped in the environment with no records being kept of either where it was dumped or what it contained. A large expenditure of money would thus be necessary to finance the identification of the components of waste and to effectively separate them. Some waste materials such as the gaseous pollutants found in the atmosphere are technically impossible to collect and concentrate. The relationship between the composition of the waste, how concentrated the components are and their value is a very important property of waste and can dictate how it is handled and treated.

Environmental pollution is the discharge of any material (e.g. oil, asbestos, NO_x) or energy into the biosphere which may cause acute (short-term) or chronic (long-term) damage to the Earth's ecological balance, or which lowers the quality of life and even causes death. Pollutants may cause damage which is immediate and easily identifiable, or damage of an insidious nature, which is only detectable over a long period of time.

The indiscriminate discharge of untreated industrial and domestic wastes into the biosphere, the 'throwaway' attitude towards solid wastes in particular, and the use of chemicals without considering potential consequences have resulted in some major environmental disasters.

Such a situation occurred in Loscoe, south-east Derbyshire, when in the early hours of the morning of 24 March 1986 a bungalow at 51 Clarke Avenue was completely demolished by a large explosion. The occupants, a husband and wife and their son, were

Table 5.5 The classification of waste

Waste classification	Examples
By physical state, i.e. gas, solid or liquid	Wine bottle – a solid CFCs – gases Sewage effluent – gases, liquids and solids
How the original material was used	Wine bottle – packaging waste Apple peelings – food waste
By type of material	Wine bottle – glass Newspaper – paper Aluminium can – metal
Chemical properties	Paper – burns easily Apple peelings – compostable
According to health hazard	CFCs – not toxic to humans Cadmium – toxic to humans
Where it was used	Apple peelings – domestic waste Wine bottle – domestic waste CFCs – domestic and industrial waste

trapped under the rubble. Fortunately, when rescued it was found that the wife had only suffered a broken arm and a fractured pelvis, whilst the husband and son escaped with minor injuries.

It was first thought that the cause of the explosion was due to a methane gas leak from either a faulty gas appliance or the mains supply. Tests carried out in surrounding buildings showed the presence of high concentrations of methane gas in two other bungalows. When it was chemically analysed, the gas was found to have a composition that was different from that of the domestic methane gas supply (Table 5.6). The latter was not to blame, so where then had the gas come from?

Close to where the explosion occurred there was a landfill site that had been used for a number of years for the dumping of waste materials. It was known that such sites could be a source of gas emissions containing methane and so further investigation of this site was suggested.

By 26 March 1986 the mains gas supply had been conclusively ruled out as the source of the explosion. Scientific evidence supported the suggestion that it was caused by the movement of methane gas from the landfill site (Box 5.1). The explosion was

Table 5.6 Typical gas analysis of landfill gas, natural gas and the atmosphere

Gas	Landfill gas	Natural gas	Atmospheric air
Oxygen	1.8% v/v	0.0% v/v	21% v/v
Carbon dioxide	37% v/v	0.3% v/v	0.2% v/v
Nitrogen	18% v/v	1.8% v/v	79% v/v
Helium		Trace	
Methane	54% v/v	94% v/v	
Methane/carbon dioxide ratio	1.5:1	313:1	
Ethane	0.002% v/v	3.25% v/v	
Methane/ethane ratio	27,000:1	29:1	

Box 5.1

The nature and origin of methane in a landfill site

Pure methane is a colourless, odourless, tasteless gas. It has a lower explosive limit (LEL) of 5.3 per cent by volume with air. This means, for example, that 100 m^3 of a mixture of air and methane containing 94.7 m^3 of air and 5.3 m^3 of methane would be an explosive combination. Its upper explosion limit (UEL) is 15.0 per cent by volume with air. It is therefore a very serious fire and explosion hazard when exposed to a source of heat.

Methane is probably best known as the major component of North Sea gas used in UK homes. For domestic use this is mixed with a special chemical to give it a characteristic smell for safety reasons. It is also trapped in coal seams and is often released during the mining of coal – here it is known as firedamp and is often the cause of deep mine explosions.

In landfill sites, methane is produced by the decomposition of the biodegradable materials some metres below the surface of the landfill site. The actual composition and rate of emission of this 'landfill gas' are variable and depend on the composition of the materials contained within the site, compaction, and surface and underground treatments of the dumped waste. Materials can remain unchanged for decades.

Domestic waste consists mainly of paper, wood, food, plastics, metal, glass and textiles in varying proportions. The organic constituents in this waste are broken down by natural bacterial action in two stages. In the first stage carbohydrates (e.g. paper), proteins (e.g. meat, fish) and fats are broken down into simple sugars, amino acids, glycerol and long-chain fatty acids, respectively. This is then followed by the production of carbon dioxide, hydrogen and short-chain fatty acids that lead to the second stage production of methane and more carbon dioxide. A typical composition by volume of landfill gas, together with those of normal air and natural gas, is given in Table 5.6.

Apart from the material present there are five factors that affect the production of landfill gas:

1   An increase in water content of the waste material, which is paralleled by an increase in the rate of decomposition and therefore gas generation.
2   The pH of the landfill material will dictate the survival of the methane-producing bacteria, the optimum pH range lying between 6.4 and 7.4.
3   The temperature, which should lie between 2 °C and 37 °C, necessary for the existence of successful anaerobic digestion conditions. The core of a landfill will have temperatures well within this range.
4   The presence or absence of oxygen. As gas leaves the landfill site, it is replaced by air, particularly in the surface area. Oxygen in this air is toxic to anaerobic bacteria and hence it inhibits the production of methane. Thus the depth of the site, the rate of gas extraction and the permeability of the covering material are all important.
5   Time. There is no set time pattern but typically a steady rate of production of gas is achieved in one to two years, containing 55–65 per cent volume for volume of methane.

believed to have been the result of a sudden, unusually large fall in atmospheric pressure that allowed a large surge of gas from the landfill site to travel along cracks and fissures in the sub-surface rock strata and through drains to eventually accumulate under the suspended floorboards of 51 Clarke Avenue. When the central heating boiler came on at about 6.00 a.m., it also ignited the methane and air mixture contaminating the property.

In 1988, a report concluded that the explosion had been caused by the release and migration of landfill gas as a result of abnormal atmospheric conditions. A fall in atmospheric pressure of 35 millibars had been recorded in the ten hours preceding the time of the explosion. Because the landfill site was insufficiently ventilated, the methane gas found its way out of the site by the easiest routes. A system of induced ventilation was established by means of boring holes for the collection and removal of the landfill gas. It was suggested that better co-ordination between the various authorities responsible for waste management should be implemented. A clear identification of those responsible for a site after it was no longer operational and the terms of that responsibility were both seen as being essential. The report also concluded that the potential environmental and public health problems associated with the complete removal of the tip were too great for this to be a viable solution. It also stated that refuse which was biodegradable could not properly be dumped on a landfill site close to domestic properties unless there was a thorough understanding of the site geology, and a rigorous application of appropriate methodology to ensure the satisfactory removal of landfill gas from the site.

Although the Loscoe explosion stands out because of its disastrous consequences, surveys have shown that there are many other landfill sites that are potential sources of danger. In 1988, Her Majesty's Inspectorate of Pollution identified some 600 sites in the United Kingdom that needed the retrospective fitting of gas control methods. A further 700 sites near to domestic residences have been identified as potentially hazardous should gas migrate off site. Typically, the City of York in North Yorkshire has over 26 disused landfill sites within the city boundary. Indeed, in 1998, the York evening paper reported that action had to be taken to stop potentially dangerous methane gas from seeping from a former York rubbish tip towards a new housing estate.

The problem with older sites is that the waste content is not known, and is usually a mixture of both industrial and domestic wastes in varying proportions. If such a site were to be opened, the result could be the release of dangerous chemicals and other materials. It is situations like this, and accidents of the type described above, that have led to concerns about human impacts upon the environment.

Worries about the production of waste and the use of chemicals lie in six main areas:

1   Exposure of humans to toxic materials which can cause living cells to mutate, leading to physical abnormalities in a developing embryo or foetus, the formation of cancers, and other slow latent effects on health.
2   Tipping of waste which has the potential of causing acute poisoning of living organisms.
3   Contamination of groundwater.
4   Marring or destruction of amenities because of the presence of waste or hazardous materials.
5   Irreversible damage caused by the disposal of long-lived pollutants on land or in marine environments.
6   Waste of resources which could be recovered from rubbish (Box 5.1).

Pollution can be controlled by the adoption of one or more of four general approaches:

1   Temporarily stop or reduce the output from a process that is producing a high degree of pollution, thus allowing time for the environment to recover. This would only allow short-term respite for the environment and cause serious financial problems to an industrial concern.
2   Dilute the pollutants by the more effective dispersal of pollutants in air or water.
3   Remove the pollutants produced during a process before they are emitted to air and water. Unfortunately, such pollutants may need to be treated before being buried or stored on land where they can again pose a potential hazard.
4   Change the process to reduce pollution. This can involve the production of lower amounts of pollutants because the process has been improved, the substitution of a less hazardous material, or the separation and reuse of materials from the waste stream.

Changing an industrial process is considered by many to be the most effective and efficient way to meet the demands being made to control waste and minimise natural resource use, and to meet the increase in costs of pollution control, waste disposal and raw materials. A relatively new way of doing this is based on a 'life cycle' approach to the analysis of a particular process.

The life cycle of a steel cooking pan

When reviewing the production of a steel cooking pan using a 'life-cycle' approach, broad questions need to be asked such as 'Where did the steel come from? What was involved in the pan's manufacture? How did it come to be in the shop from where it was bought? What will happen to it when it comes to the end of its useful life?' An inanimate object like this pan is treated as if it has a 'life cycle' that can be compared in some ways to a living thing.

Any manufacturing process involves both the movement of material and energy into and out of that process throughout its life cycle. If the amounts of materials and energy involved can be controlled and *reduced*, then the Earth's physical resources will last longer, more time will be available to search for possible substitute materials and ways of recycling, and there will be fewer polluting wastes entering the environment. The first step is to identify *where* materials and energy are involved at all points in the life cycle of a product.

A cooking pan starts its life (the cradle) when the raw materials are extracted from the Earth's crust by the mining of iron ore, coal and limestone (Figure 5.3). This is followed by the removal of soil, rock and other impurities to concentrate the iron ore before it can be converted into iron. Physical separations may also be necessary for the coal and limestone. Before use coal has to be converted into coke by heating it in the absence of air. This is done on site in a modern steel plant. The chemical extraction of the iron occurs in a blast furnace where the ore, mixed with coke and limestone, is subjected to a blast of very hot air. It is often necessary to transport the iron ore, coal and limestone to the blast furnace site. The iron from the blast furnace contains a small percentage of impurities, mainly carbon, which cause it to be far too brittle and hard for use in the production of items like a cooking utensil. Steel is produced by blowing oxygen gas on to the molten iron, which converts these impurities to gaseous oxides that are removed from the furnace. A cooking pan must be resistant to staining and corrosion by hot liquids and is therefore made of stainless steel, and so other elements

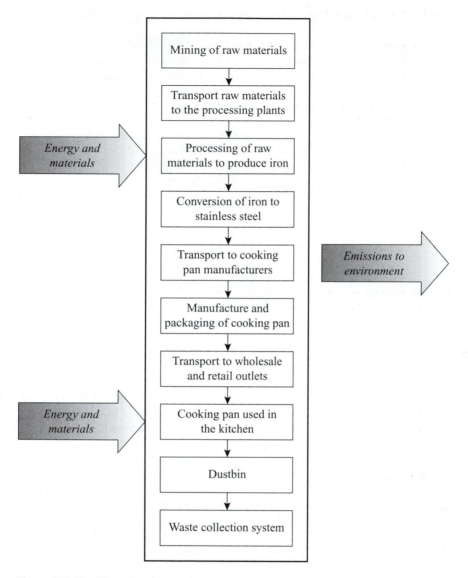

Figure 5.3 The life cycle of a steel cooking pan.

such as chromium and nickel must be added. This type of steel is made in an electric arc furnace where ordinary molten steel is mixed with the appropriate amounts of these elements in the absence of oxygen. When the steel is of the correct composition, it is run out of the furnace and allowed to cool in the form of sheets.

The steel is delivered to the manufacturers of kitchen utensils, where the shape of the pan is pressed out from the sheet steel and a handle attached. The finished pans are then distributed to the wholesale and retail industries who sell them to their consumers. Eventually, when the pan has ceased to perform its function properly, it is probably consigned to the dustbin (the grave) where it becomes part of the domestic waste stream. Throughout this 'cradle to the grave' life cycle of the pan, several mineral resources together with vast amounts of energy have been used. Various types of wastes are

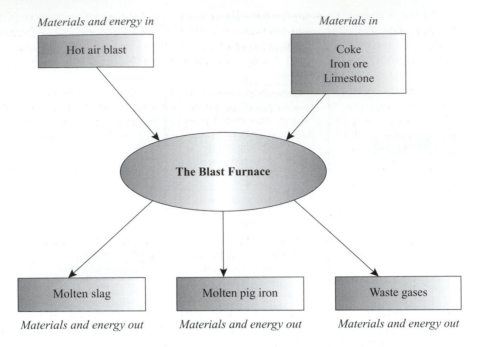

Materials and energy in   *Materials in*

Hot air blast

Coke
Iron ore
Limestone

The Blast Furnace

Molten slag   Molten pig iron   Waste gases

Materials and energy out   *Materials and energy out*   *Materials and energy out*

Figure 5.4 The blast furnace – movement of materials and energy.

produced, which have to be disposed of and treated. There are thus many points of inter-action with the environment throughout the production of the pan.

Figure 5.4 shows a very simplified diagram of the energy and material flows in just the blast furnace part of the production.

How then can manufacturers minimise environmental impacts? How can the use of resources be maximised so that the use of more expensive alternatives is delayed? How can resources be conserved? The answers to these questions lie in the concept of 'sustain-able resource management', which involves as its central idea the control of all forms of waste and the prolonging of the availability of existing resources.

In order to understand why sustainable resource management is so important, it is necessary to examine what is meant by environmental conservation. One definition of conservation is 'the careful preservation and protection, especially of natural resources, the quality of the environment, or of a plant or animal species, to prevent exploitation, destruction or neglect' (Penguin Hutchinson 1996). Conservation includes the main-tenance of such amenities as national parks, wildlife centres, sites of particular historic interest and wilderness areas. The natural resources of a country are often directly related to its economy and the standard of living of its people. Thus the wasteful use or even non-use of those resources could result both in short- and long-term social and economic damage.

A closer look at the production of iron in the blast furnace

When a non-renewable resource is used, where does the material go and is it lost forever? This question can be answered by investigating more closely one aspect of steel produc-tion, namely the way in which iron is produced in the blast furnace.

The raw materials extracted from the Earth's crust and used in the production of iron are all examples of natural, non-renewable resources. The most important ore is hematite, Fe_2O_3, which contains about 70 per cent of iron by mass (Table 5.1). To produce one metric tonne of iron requires 1.6 tonnes of this iron ore, 0.7 tonnes of coke and 0.2 tonnes of limestone.

The purpose of the blast furnace is to separate the iron from the oxygen, i.e. the reduction of the iron oxide. Blasts of very hot air are blown through small holes in the bottom of the furnace. A metric tonne of iron would require 3.6 tonnes of air at a temperature in excess of about 800 °C. The chemical reactions that go on inside the furnace are very complex but the principal ones are as follows.

Near the base of the furnace, coke burns in the very hot blast of air to produce carbon monoxide and a large quantity of heat,

$$2C(s) + O_2 = 2CO(g) + heat$$

Near the top of the furnace the carbon monoxide combines with oxygen from the iron oxide to form iron and carbon dioxide plus a smaller quantity of heat.

$$Fe_2O_3 (s) + 3CO(g) = 2Fe(s) + 3CO_2(g) + heat$$

The iron melts as its sinks through the hotter parts of the furnace and collects at the bottom of the furnace. Here it is tapped off and allowed to cool. The solid iron at this stage is called pig-iron or cast iron.

The coke used is formed when coal is heated in ovens to about 1,500 °C in a much reduced supply of air. The other products formed are coal gas, ammoniacal liquor and coal tar. Typically, one metric tonne of coal will produce 670–720 kg of coke containing about 92 per cent by mass of carbon. Coke can be used as a high energy containing fuel, and as a source of organic chemicals when it is combined with oxygen and hydrogen. It is also used, as illustrated here, on a very large scale as an industrial reducing agent in the blast furnace for the production of iron.

The function of the limestone is to help remove earthy materials such as sand, SiO_2, that can clog the furnace, react with the iron and generally cause problems. The limestone is decomposed by the heat to give calcium oxide,

$$CaCO_3 (s) = CaO(s) + CO_2(g)$$

This calcium oxide reacts with the earthy materials to give a molten slag of silicates,

$$CaO(s) + SiO_2(s) = CaSiO_3(l)$$

The slag collects on top of the molten iron and can again be tapped off. This slag is either dumped or used as concrete aggregate, railway ballast or soil conditioner.

Applying the Law of Conservation of Mass to the reactions between the coke and oxygen (the reactants) to give carbon monoxide gas (the product), and between iron oxide and carbon monoxide to produce iron and carbon dioxide, the total mass of the reactants and the total mass of the product will be the same. Thus, when the iron ore is described as a non-renewable resource, it means that during the production of iron its atoms are separated and rearranged with the consequent liberation of heat energy. Hence, the products contain less energy than the reactants. Matter has not been destroyed but has been reorganised, in this case, in a more useful way.

However, once the iron ore and the coal used to make coke are removed from the environment, they are no longer available to produce iron. Indeed, the coke ultimately becomes carbon dioxide, which can find its way into the atmosphere where it may contribute to global warming. Although iron ore and coal occur in large quantities in various parts of the world, there is not a bottomless pit of these resources. When they are used to produce iron, what took many millions of years to form are effectively 'destroyed' in hours. Fortunately, iron-based metals are recycled and reused on a very large scale.

The Precautionary Principle and innovation

In order to protect the environment from anthropogenic actions, a better understanding of the processes that go on in the environment is needed. This would enable more accurate modelling of these processes, which could result in more reliable predictions and therefore more reliable preventative action. Unfortunately, not enough data is available to allow correct action to be taken for most of the time. Accurate historical data is almost non-existent and extensive reliable monitoring over a long period of time has not been carried out. Whilst accurate observations and measurements are vital, these are of little use in a current situation or in the short/medium term. There is also a lack of good modelling because of the very complex nature of the interactions that occur in the environment between a wide range of variables. Indeed, it can be argued that what is going on in the environment is too complex to be fully understood.

Because of the inadequate nature of both data and modelling, the concept known as the Precautionary Principle has been developed. This Principle takes the view that the environment is so important that mistakes cannot be afforded. Action should be taken even if there is no agreement concerning a problem, and in advance of scientific proof of cause and effect, on the grounds of wise management and cost effectiveness – it is better to pay less now than possibly more later. The Principle suggests that all kinds of resources should not be 'worked out' since it cannot be predicted what the outcome would be if all or too much of anything were to be removed from the environment. It also places responsibility on everyone whose actions may affect the environment to prove that what they are doing will not cause harm.

The Precautionary Principle is enshrined in the basic treaty of the European Union – the Maastricht Treaty:

> Community policy on the environment shall aim at a high level of protection taking into account the diversity of situations in the various regions of the community. It shall be based on the precautionary principle and on the principles that preventative action should be taken, that environmental damage should as a priority be rectified at source and that the polluter should pay.
>
> (Article 130r, Maastricht Treaty 1992)

This would appear to be a very clear statement of intent. However, it leaves a number of important questions to be answered. How likely is it that a problem will develop and become of sufficient gravity that action should be taken? What is an acceptable degree of risk? How much are people prepared to sacrifice in order to avoid the problem?

In using the Precautionary Principle it must be accepted that it is not possible to be 'absolutely certain' that a particular problem will occur. There is a general agreement that 'beyond reasonable doubt' is sufficiently close to 'absolute' for all practical purposes. The Precautionary Principle requires evidence that serious and irreversible damage might occur and a cost-benefit analysis of the different problem-solving options available. If correctly used, the Principle will lead to the taking of responsible and

---

Box 5.2

Using plants to clean up water

Innovation 1

Reeds are wetland plants. Thus their root systems are in soil which is usually covered with water. The common reed *Phragmites australis* is used to treat domestic and industrial waste waters in the UK. A pit is dug, which is then sealed at the bottom and sides with clay and/or a synthetic liner. Soil is then added and the reeds planted in rows. Effluent is then fed through pipes onto the reed bed. This effluent percolates through the soil to the root system. There, micro-organisms degrade the organic components. Oxygen is fed to the roots via the leaves and hollow stems so that aerobic processes take place, which convert the pollutants to harmless substances such as carbon dioxide, nitrogen and water. A small amount of sludge is produced. Some anaerobic degradation also takes place.

In Rosedale, North Yorkshire, UK, in 1996, a sewage treatment plant was opened, which uses reed bed technology to treat the domestic dirty water waste of about 300 people. The reeds are about two metres high and present a pleasant view to the beholder. There is very little smell associated with the beds, which once set up have low running costs. There are no sludge disposal problems either.

Reed beds are also being used by a large chemical company in the north of England to remove organic pollutants from its effluent waste water.

Innovation 2

At Rutgers University, New Jersey, US, scientists have developed a new way of cleaning up water that has been contaminated with uranium. They have been using sunflowers. These flowers have been found to reduce the amount of uranium in such water by as much as 95 per cent. The uranium is absorbed by the roots of the plant and concentrated there. Negligible amounts of uranium are found in the stems of the sunflowers. The roots are dried and treated with acid. The acid dissolves the uranium, leaving behind a mass of plant residue which is no longer radioactive. The uranium is concentrated in the acid, which can then be easily treated. The technique has already been applied to a former nuclear processing plant in the US, where it was found to reduce the uranium in effluents to below the US Environmental Protection Agency water standard of 20 μg l^{-1}.

---

sensible precautions concerning potentially harmful situations. There are no simple answers to the questions posed above. Every case must be considered on its merits.

The Precautionary Principle, if applied incorrectly, can impede innovation. Innovation is a major force for the improvement and sustainable development of the environment. In order to improve the lives of current and future generations then significant innovation is needed, especially in the area of eco-efficiency.

Since it is impossible not to affect the environment in some way no matter what is done, new ways of reducing human impacts must be developed. In the cases of resource usage, pollution and waste management the key word is 'minimisation'. There is no such thing as zero waste, therefore innovative ways must be developed to minimise waste and treat it so that it becomes safe.

Innovation can take the form of well-thought-out processes such as the use of reed beds and other plants to make waste water and raw sewage safe for disposal (Box 5.2).

Innovation can also be about finding substitute materials for those in short supply. Copper, for example, is about the twenty-fifth most abundant metal and is regarded as a scarce element. It is used in alloys such as brass and bronze, in the pure form, because of its ability to conduct heat and resistance to corrosion, and in electrical wiring because of its excellent electrical conductivity. In 1973 (USGS 1998) copper reserves were estimated as 110×10^6 tonnes and the rate of production as 6×10^6 tonnes per year. If the trends then occurring had continued, reserves of copper would have ended by 1990 and, if the trends remained static, by 2010. Thus, in 1973 it was estimated that the availability of virgin copper would last for about 20 to 40 years. This sort of calculation initially led to fears that supplies of the metal would run out in a very short time and resulted in the rise in the price of copper. Had these figures been correct, had the search for and exploitation of lower grade ores not been attempted and had society continued to use copper in the same amounts for the same purposes, then the fears might have become reality. However, the availability of a resource like copper depends upon a range of factors such as how much in demand it is, how easy it is to extract from its ores, the economics of extraction, and the assumption that all known reserves have been found. In addition, copper is a metal that is easily recycled.

The price of copper has fallen in recent years. Indeed, the world production of copper is currently exceeding demand. One of the greatest motivators for change is an increase in costs. As soon as this happens, substitute materials are considered. So why is copper not being used since it would appear to be both available and cheap? It has been substituted by aluminium, of which there is no shortage, in certain areas of the electrical industries, and by optical fibres and microwaves for communications.

Similarly, oil, like copper, is classified as a non-renewable resource. What has happened, and continues to happen, is that the amount of easily accessible oil has decreased. Although oil prices rise and fall according to political and economic influences, in the UK the price of petrol continues to increase. Oil is also a strategic material and the demand for it continues to be high. Thus, the search for new reserves continues, and those resources once identified as being too diffuse to be worthy of extraction on economic grounds are being used. Because of the increase in costs and worries about how long it will last, people who want oil will be forced to be more economical in its use. Innovation is now occurring with the design of more efficient motor car engines, new devices for the production of useful energy, and the synthesis of substitute fuels.

Innovation can take the form of 'designing' into a packaged product ways of minimising waste that also take into account the options for the treatment of that waste. Which technique is used to dispose of the waste at the end of the life of the product depends on the nature of the packaging material. For example, packaging can be designed so that it can increasingly be recycled. Can the product itself be made to last longer (an economic problem for the manufacturer)? If durability and recycling are possible, are the steps taken to achieve these more environmentally damaging than the original packaging and contents from which the innovation stemmed? These are the kinds of questions that need to be addressed in order to make any progress in minimising environmental impacts caused by waste packaging. It is clear that, in waste management, innovation equates to adaptability. Any new waste management technology used must be able to respond to changes in both product and the composition of its waste. Equally, the options available for the treatment of waste must be able to change so that environmental impacts are reduced.

Sustainability

The most effective forms of environmental management should result in extending the longevity of natural resources, reduction of environmental impacts, and improvements in the quality of the environment. These are the corner stones of **environmental sustainability**.

The idea of sustainability began in 1972 as a consequence of the United Nations Conference on the Human Environment held in Stockholm, Sweden. Perhaps the best known definition arose as a result of the formation of the World Commission on Environment and Development in 1987. The role of this commission (the Brundtland Commission) was to identify and promote the cause of the sustainable development of the environment. The Commission defined sustainable development as 'development that meets the *needs* of the present without compromising the ability of future generations to meet their own needs'. It was appreciated by the Commission that what constituted human need varied throughout the world and depended a great deal on the attitude, culture and values of a particular society. Similarly, what could actually be achieved depended on the current state of technology, the way in which societies were organised, and on the ability of the biosphere to absorb the effects of human activities. Hence, because of these complexities sustainability was recognised as being somewhat vague in that it can mean whatever you want it to mean at any time.

In 1992, the Earth Summit held in Rio de Janeiro resulted in Agenda 21, which called for sustainable and environmentally sound development throughout the world. One of the greatest concerns recognised by this meeting was the place of waste generation and waste management in the control of impacts on our environment. Consequently, a set of objectives and actions were defined, which were aimed at waste minimisation, maximising the reuse and recycling of materials, and ensuring the safe treatment and disposal of waste. Europe had already produced independently in 1989 a waste management strategy, which was issued in the form of a framework directive in 1991. A hierarchy of waste management was listed, i.e. (i) waste prevention and minimisation; (ii) reuse, recycling and recovery; and (iii) the optimisation of final disposal. Other actions have followed, including the Maastricht Treaty of 1992 which defines EU environmental policy, the establishment of a European Environment agency and continuing EU directives aimed at managing resources and waste.

Cost is inextricably linked with environmental management. Economists have therefore played their part in trying to define what is meant by sustainability. Inherent in all definitions is that future generations should be at least as well off in terms of health and prosperity as the current generation. The economic view of our environment is that it is another form of capital. Sustainable development then becomes the maximum development that can be achieved without depleting the capital assets, i.e. the resource base, of a nation. Many economists see this resource base as being composed of four major components: man-made capital, natural capital (natural resources), human capital (a knowledge base resource) and cultural and moral capital. If the idea of sustainable development in an economic sense is accepted, then any society must behave in such a way that future generations do not have to pay a heavy price for the results of that behaviour. If this is not the case and what has happened was unavoidable, then adequate compensation for the costs incurred must be provided.

Sustainability is not just about whether a product or action can evolve in such a way so that damage limitation can be achieved and the environment sustained. Sustainability also involves cultural, social and economic dimensions. Everyone has their own view on what is valuable, worth saving or what currently needs doing. Different people have different interests according to where they live, previous experiences, economic state, etc. and will view the same problem in very different ways. What might be considered as environmentally damaging by one person is not viewed in the same way by another. The *values* of people will affect the way in which they behave towards the environment. What an individual does is crucial with respect to cumulative impacts. Environmental managers must therefore identify how they can cause people to change their ways

Economic factors clearly have a bearing on motivating people to respond to change. It is hoped that reliance can be placed on the development of the *idea* that the environment is worth paying for, that it is valuable. If this is not possible, then charges must be made, e.g. landfill tax was introduced in the UK in 1996 to discourage the overuse of landfill as a waste treatment option. It may be necessary to impose other charges to enforce the perception that the environment is valuable. Measures are being introduced, such as an increase in fuel prices and the introduction of an aggregate/quarrying tax and water pollution tax. If this should happen, then the charges must be set at a reasonable level in order to meet economic sustainability, and any law or regulation must be enforceable.

In order to determine when waste management has become sustainable, criteria or target setting could be used. However, this can only occur in the light of current scientific methods, knowledge and understanding of what is meant by sustainability. Over time, all these will change. Couple this with the complex nature of the economic, social and cultural aspects of waste management, and the achievement of sustainability will probably never be measured accurately and involve an ever changing set of targets. It is the attempt at trying to meet these targets that will be important.

A case study – the mining and extraction of copper

The countries that are major producers of copper and the main ores of copper are listed in Table 5.3. Table 5.7 shows some world mine production figures compared with world mine capacity figures. It is clear from this table that the production of mined copper is less than the mines are able to produce.

A similar table of data of refined copper production from both primary (ores) and secondary (reprocessed scrap) production of copper (Table 5.8) vs world refined production capacity also shows that, over the same period of years, less refined copper is being produced than is possible.

Table 5.7 also shows the total world consumption of copper between 1994 and 1999. The production of refined copper has gradually exceeded world demand.

Like the uneven distribution of mineral resources throughout the world, there is also an unequal distribution of consumption (Table 5.9).

The mining and extraction of copper illustrates well the techniques and problems associated with the production of a metal. The biggest copper mine in the world is the Bingham Canyon Mine. This open-cast mine is situated in the Oquirrh Mountains, some 18 miles south-west

Table 5.7 *World mine production, world mine capacity and world consumption for copper*

Year	World mine production tonne 10^6 (rounded)	World mine capacity tonne 10^6 (rounded)	World consumption tonne 10^6
1994	9.6	10.9	11.55
1995	10.1	11.4	12.20
1996	11.1	12.1	12.56
1997	11.5	12.5	13.06
1998	12.2	13.3	13.45
1999	12.6	14.0	14.03

Table 5.8 *World refined copper production and world refined copper capacity*

Year	World refined production tonne 10^6	World refined copper capacity tonne 10^6
1994	11.12	13.04
1995	11.83	13.27
1996	12.71	14.06
1997	13.54	15.36
1998	14.04	16.08
1999	14.37	16.68

Table 5.9 *World distribution of copper*

Area	Mine production tonne 10^3	Consumption tonne 10^3
Africa	490.8	111.1
America	7,728.4	4,115.8
Asia	2,197.2	5,450.6
Europe	1,307.0	4,294.2
Oceania	906.7	172.3

of Salt Lake City in Utah, US. The mine is approximately 2.5 miles wide measured across its rim, and 0.5 miles deep. Some 5.6 billion tonnes of overburden (rock, soil, etc.) and ore (chalcopyrite) have been removed since it started to operate over 90 years ago. It has produced so far 14.2 million tonnes of copper, 18 million troy ounces of gold, 150 million troy ounces of silver and substantial quantities of molybdenum, platinum and palladium.

The ore grade is 0.6 per cent of copper by mass. Currently, some 120×10^6 tonnes of overburden and ore are removed from the mine each year. It is because of the size of the deposit and the low grade of the ore that such large quantities of rocks have been removed, resulting in the very large hole.

The first step in the mining of the copper ore involves the drilling of a pattern of 50-foot-deep holes up to four times daily. Half a tonne of an explosive composed of 94.5 per cent ammonium nitrate and 5.5 per cent diesel fuel oil is inserted into each of these holes and detonated. The resulting rock is then passed on to crushers.

Any blasting that accompanies a mining process produces dangers from flying rocks, dust problems and noise pollution. There are no conurbations near to the Bingham pit so these problems do tend to be minimal. However, dust and noise pollution can be very intrusive when a mine is by built-up areas. Dust can, for example, impede photosynthesis when it coats plant leaves, and causes respiratory problems in human beings.

The second step is the primary crushing of the copper ore and is done inside the pit itself. This reduces the size of the largest boulders from about one metre in diameter to 0.2 cm. Performing this operation in the pit has removed the need to use rail transport to move the rock to a distant crushing plant, thereby reducing demand for energy. The ore is crushed to separate the copper mineral from the barren rock that surrounds it. Usually, crushing causes cracking and separation along the interfaces between the wanted mineral and the unwanted rock. The particle size produced is crucial – if the particles are not fine enough or too fine, then poor concentration will be the result in the next step. Often, secondary and tertiary crushers are required to reduce the size of the particles but these are unnecessary in the case of copper pyrite ore.

After primary crushing the ore is sent, via a five-mile-long conveyor belt system, to a semi-autogenous mill where the process of concentrating the ore is started. This type of mill relies upon larger rocks breaking smaller ones – steel balls are not used to help break the rocks. At the mill, the ore is reduced to the consistency of talcum powder. The copper ore is then concentrated by a flotation method. Flotation separation depends upon how wet the desired mineral can be made. Sulphide minerals are hydrophilic, i.e. they are wetted with water. In flotation separation the object is to make the mineral hydrophobic or water repellent, whilst the remainder of the unwanted material remains hydrophilic. In the present case, pine oil and chemicals (alcohols called xanthates and carbon disulphide) are added to the crushed copper ore mixed with water in a tank. This mixture is then agitated and air is blown through it to form a froth of oil-coated bubbles. The mineral particles are rendered hydrophobic by the added chemicals and adhere to the oil-coated air bubbles. These bubbles as they rise to the surface carry the mineral with them. This 'scum' is regularly scraped off the surface and collected. The technique is not perfect and even after several attempts the residues (now called the 'tailings') left at the bottom of the tank contain some wanted ore. This flotation separation results in a 'scum' that contains some 26 per cent by mass of copper.

The processing of the ore thus produces tailings which are essentially sands and slimes as wastes. Sands have particle sizes of less than 0.1 mm and can normally be placed on a tip. Slimes are composed of even smaller particles and water. These require time to settle in 'tailings ponds' which, in the case of Bingham, take up an area of some nine square miles. Thus tailing ponds take up a vast area of land. However, to ensure that the tailings do not enter the environment via leakage, a water collection system has been installed, which collects and recycles all groundwater.

Where waste rock has been collected in low tips, these have been revegetated, but problems of stability can occur as a result of erosion and changing internal structure. Any copper sulphide ore particles that are present can undergo oxidation in the presence of water to form soluble copper(II) ions and sulphuric acid. This increase in acidity can cause the leaching out of heavy metals and potential contamination of water tables, streams and rivers. Particulate matter in the tailings pond can be so small that colloidal suspensions can be formed. In addition, particles upon settling can trap water so that, when the sediments formed are transferred to a tip, the tip never becomes stable.

Other methods of separation and concentration that are used depend upon a variety of factors. If there are, for example, two or more minerals in an ore, it is likely that their particle sizes after crushing are different. Therefore, sieving may be used for a partial separation. If some minerals are softer than others, then during the crushing process these will be ground finer and sieving may again prove useful. If a mineral possesses magnetic properties, e.g. magnetite (Fe_3O_4) and ilmenite ($FeTiO_3$), then powerful electromagnets can be used to separate them from non-magnetic materials. Gravity separators depend upon the relative masses of minerals present in an ore. A spiral concentrator will enable lighter particles to collect at its top and the heavier ones at its base. It is important that the particle sizes are more or less the same otherwise separation will not be achieved.

After flotation separation, the concentrate in the form of slurry is sent via a 17-mile-long pipeline to the smelter. Again, the pipeline removed dependency on a railway thus saving energy. At the smelter the slurry is dried and, together with fine sand, oxygen and pre-heated air, injected into a flash smelting furnace. The furnace is maintained at a temperature of about 1,000 °C. As the sulphide ore is partially oxidised, heat is evolved in the chemical reaction. This heat sustains the temperature necessary for smelting through nearly all of the process. The chemical reaction that proceeds is,

$$2CuFeS_2(l) + 5/2O_2(g) + SiO_2(l) = Cu_2S(l) + FeS(l) + FeSiO_3(l) +$$
$$2SO_2(g) + heat$$

When necessary, additional heat is supplied using oil-fired burners. Thus the flash smelting process has the advantage that the chemical reactions occurring in the smelting process provide much of the heat necessary, thereby reducing considerably demands on non-renewable fuel sources. The molten material produced collects in a bath at the bottom of the furnace in two layers. The upper, lighter layer is 'slag' containing iron silicate and other impurities. The lower, heavier layer is 'copper matte' which contains some 65 per cent by mass of copper. Small volumes of gases are produced, which contain high concentrations of sulphur dioxide. The sulphur dioxide is conveyed to a plant where 98–99 per cent of it is converted into sulphuric acid.

The molten copper matte is then tapped from the flash melting furnace and trans-ferred to a flash converter. The molten slag, upon cooling, is dumped on to slag piles, or treated further for the removal of residual copper.

In the converter sand is added and oxygen blown through the molten material. Iron(II) sulphide is more readily converted to the oxide than is copper(I) sulphide, i.e.

$$2FeS(l) + 3O_2(g) = 2FeO(l) + 2SO_2(g)$$

This oxide then combines with the sand to form an upper layer of liquid iron silicate slag,

$$FeO(l) + SiO_2(l) = FeSiO_3(l)$$

The slag is continuously removed until the right amount of copper(II) sulphide has accumulated. More oxygen is then blasted through the melt. The copper(I) sulphide is converted to copper,

$$Cu_2S(l) + O_2(g) = 2Cu(l) + 2SO_2(g)$$

The resulting copper is 98–99 per cent pure and is referred to as 'blister' copper because of its bubbly appearance due to the gases that have been blasted through it.

The copper is cast into blocks called 'anode blocks' for further refining. The sulphur dioxide produced is again redirected to the sulphuric acid plant.

During the final purification, the copper becomes the anodes of electrochemical cells (Box 5.3). The electrolyte is a solution of copper(II) sulphate acidified with sulphuric acid. The cathodes are pure copper. As the electrolysis proceeds, pure copper is deposited on the cathodes whilst the impurities in the blister copper collect below the anodes. This 'anode slime' contains elements such as silver, gold, platinum and other valuable substances, which help to offset the cost of copper production. This electro-refining process is energy intensive, and whilst it is occurring hazardous electrolytic mists and some toxic gases are formed.

From the 120×10^6 tonnes of overburden produced per year at the Bingham pit, about 35 per cent is the required ore, which ultimately yields 2.52×10^5 tonnes of pure copper (Figure 5.6) In other words, for every tonne of copper produced, about 475 tonnes of waste is also produced.

Similarly, for every tonne of chalcopyrite that is processed, approximately 0.5 tonnes of sulphur dioxide is produced. This material is both toxic and corrosive. Until rela-tively recently, tall chimney stacks ensured the sulphur dioxide that entered the atmos-phere was placed high enough to ensure 'safe' dilution in the vicinity of the plant. However, anyone downwind of the smelters would receive their 'dose' of acid rain! More enlightened and innovative management, coupled with the need to meet environ-mental pollution regulations, has seen the installation of new enclosed flash smelters and converters. These have been coupled with efficient sulphuric acid-making plant, which retains 99.9 per cent of the SO_2 produced. Thus per 1,000 tonnes of ore processed about 0.5 tonnes of SO_2 enters the atmosphere.

Box 5.3

The electrolysis of copper(II) sulphate(VI) solution

Electrolysis is the decomposition of chemicals using a direct electrical current. Positive ions are reduced by the addition of electrons and negative ions oxidised by the removal of electrons. Electrolysis is often used to decompose chemical compounds which would otherwise be thermodynamically difficult. For example, copper(II) sulphate(VI) when dissolved in water that has been acidified with sulphuric acid forms a blue solution in which the salt is fully dissociated into its ions, $Cu^{2+}(aq)$ and $SO_4^{2-}(aq)$,

$$CuSO_4(s) \rightarrow Cu^{2+}(aq) + SO_4^{2-}(aq)$$

Water also provides a source of ions,

$$H_2O(l) + H_2O(l) \longrightarrow H_3O^+(aq) + OH^-(aq)$$

If a piece of pure copper is attached to the negative side of a direct voltage source and a piece of impure copper attached to the positive side and the two pieces immersed in the solution, then an electrical current will be seen to flow. The copper(II) sulphate(VI) gradually loses its blue colour and the piece of impure copper decreases in size. The piece of copper attached to the positive side of the DC-supply is called the **anode** and the piece attached to the negative side the **cathode**. The anode carries a positive charge and therefore the negative ions are attracted to the anode. It is because of this reason that negative ions are called **anions**. Similarly, the copper piece attached to the negative side of the DC-supply becomes negatively charged and attracts the positive ions. The positive piece of copper is thus called the **cathode** and the positive ions are called **cations**. In general, metals form cations and non-metals and radicals like sulphate(VI) anions. Pieces of material that connect a solution to a DC-supply are called **electrodes**. Figure 5.5 shows the electrolysis of copper(II) sulphate(VI) solution.

At the anode, neither hydroxyl OH^{-1} ions or sulphate(VI) ions are discharged. Copper atoms in the anode ionise via electron loss and the ions go into solution,

$$Cu(s) - 2e^- = Cu^{2+}(aq)$$

The anode thus slowly dissolves.

At the cathode, copper(II) ions are preferentially discharged,

$$Cu^{2+}(aq) + 2e^- = Cu(s)$$

As the copper anode dissolves, any insoluble matter entrapped within the impure copper settles beneath the anode at the bottom of the electrolysis cell. The direction of current flow is thus from the battery to the anode in the form of electrons, through the solution to the cathode in the form of ions, and to the battery from the cathode in the form of electrons.

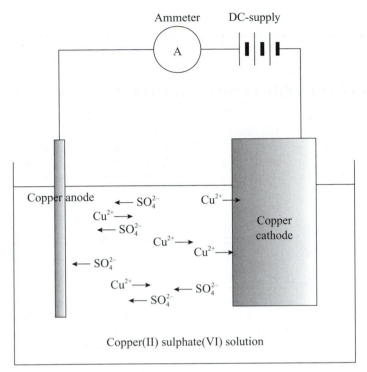

Figure 5.5 The electrolysis of copper(II) sulphate(VI) solution.

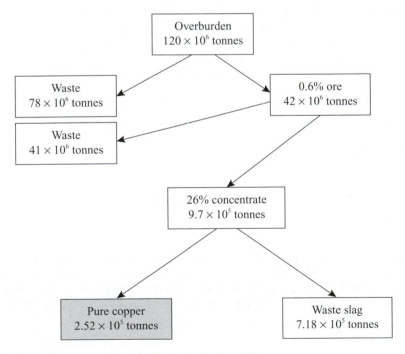

Figure 5.6 Annual outputs from the Bingham Mine.

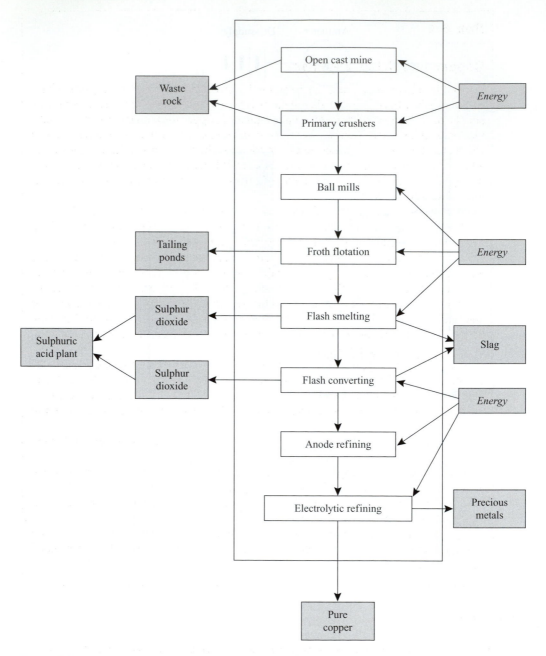

Figure 5.7 Materials and energy exchanges for the extraction of copper.

The extraction of copper at Bingham pit has involved a re-evaluation based on (i) the decision to continue and develop the use of low concentration ores, and (ii) the need to take into account both economic and environmental sustainability. An energy and materials analysis of the process is complex and involves many interactions with the environment (Figure 5.7).

Box 5.4

Copper and living organisms

Copper is an essential trace element for humans and all other living organisms. Because it is a trace element then there is a degree of 'built-in' protection against a small excess being absorbed. The main source of copper for humans is dietary intake via foodstuffs such as kidneys, liver, shellfish and nuts. In soft water regions, copper can be leached out from water pipes and therefore be taken in via drinking water. In addition, carbonated drinks may contain appreciable amounts of copper. Other sources of copper include cooking pans, coins, pigments, wine, roofing material, fungicides, algaecides and molluscides.

Prior to 1998, an EU Directive (80/778/EEC) set the mandatory upper level for the UK for copper in drinking water as 3 mg dm^{-3}. The 1998 European Drinking Water Directive set a new limit of 2 mg dm^{-3} to be achieved by 2003.

The daily intake of copper that a human adult requires ranges between 2 and 5 mg, and a child 1 and 2 mg. It is a major component of over 30 enzymes and other proteins. Of the amount of copper consumed, about 30 per cent is absorbed and the rest excreted. Copper is stored mainly in the liver but will also be found in the kidneys and intestines. How much is absorbed depends upon the amount of iron and calcium being ingested. The more of the latter two elements ingested, the lower the amount of copper that will be absorbed.

A deficiency of copper causes a variety of problems and symptoms, e.g. anaemia, baldness, reduced arterial elasticity. Some children may suffer from Minkes kinky hair disease due to a deficiency of cytochrome oxidase. This disease is fatal between the ages of three and five. The inherited Wilson's disease is caused by an excess of copper.

To cause death, copper would have to be administered in doses exceeding 250 mg a day.

If soil is deficient in copper, then there will be a serious reduction in crop yields. Sheep are very sensitive to copper toxicity and to deficiency disorders. If grass contains less than 5 μg g^{-1}, then this can lead to copper deficiency in sheep, but if it exceeds 10 μg g^{-1} it will cause toxicity.

Pollution caused by copper and its compounds arises from copper mining and smelting, brass manufacture, electroplating and excessive use of copper-based agrochemicals.

Summary

- Resources are classified as renewable and non-renewable, and can be inanimate or living.
- An ore is usually a mixture of one or more minerals and unwanted gangue.
- Resources can be described as proven or measured, probable or indicated and possible or inferred according to how well established their existence.
- Resources are classified according to their current economic value, i.e. economic, marginally economic and uneconomic.
- The reserve base of a resource is the sum of the amounts of measured and indicated resources.
- The reserve is that part of the reserve base that can be economically extracted or produced.
- Waste is unwanted material produced by anthropogenic processes. It is always produced and cannot be avoided. It must be minimised.

- Landfill sites are the ultimate destination of most wastes before or after treatment.
- Pollution is the result of the intentional or unintentional dumping of waste in the environment.
- A life cycle 'cradle to grave' approach can be adopted to help identify resource usage and wastage.
- Sustainable resource management fosters the control of waste and therefore the prolonging of available natural resources.
- The Precautionary Principle takes the view that all necessary precautions should be taken to avoid irreversible environmental damage, even though there may be no evidence of such damage being initially present.
- Innovation involves the substitution of rare materials for more common ones, the introduction of more efficient technology and the implementation of new processes.
- Environmental sustainability is ensuring that the needs of the present generation do not compromise those of future generations.
- The extraction of any material for human use must take into account both environmental and economic sustainability.

Questions

1   Find out:

(a)   How the extraction of salt has affected the environment in certain parts of Cheshire.

(b)   How salinisation has affected irrigated agricultural land.

2   (a)   How are sodium carbonate, sodium hydroxide and hydrogen chloride manufactured from salt?

(b)   What are the main uses of the compounds named in (a)?

(c)   Identify any real or potential environmental problems associated with the manufacture of these chemicals.

3   Differentiate clearly between the terms 'resource' and 'reserve'. In what ways can resources and reserves be increased?

4   Some people have grave reservations concerning the concepts of economic and environmental sustainability. Why do you think this is?

5   Outline the life cycle of a glass sauce bottle in terms of energy and material inputs and outputs (cf. Figure 5.3). What are the possible environmental consequences of glass bottle manufacture?

6   In the UK 'The polluter pays!'

(a)   What is meant by this expression and how is it enforced?

(b)   What implications in this statement are there for the environment and the use of its resources?

7   What are the uses of chromium? What properties would an innovative material have to have in order to replace chromium?

8   Why are aluminium and lead relatively easy to recycle when compared with steel?

9   Select one metallic mineral (other than copper!).

(a)   Give an account of its occurrence, distribution, consumption and reserves.

(b)   What is the normal method(s) of mining the ore and of processing it?

(c)   Outline any pollution problems that may be involved with the mining and processing.

References

Penguin Hutchinson (1996) *Hutchinson Encyclopaedia*, CD. Penguin Hutchinson Reference Library, Helicon Publishing and Penguin Books, Oxford.

US Bureau of Mines (1980) US Geological Survey Circular 831: *Principles of a Resource/Reserve Classification of Minerals*, US Geological Survey, Reston, VA.

US Geological Survey (USGS) (1998) *Copper: Statistical Compendium*. Available at http://minerals.usgs.gov/minerals/pubs/commodity/copper/stat/.

Further reading

Blunden, J. and Reddish, A. (1996) *Energy, Resources and Environment*, 2nd edn. Open University, London.
Again a very readable book! Contains chapters on mineral resources and the effects of mining and processing. Waste disposal methods are also included.

Earnshaw, A. and Greenwood, N.N. (1997) *Chemistry of the Elements*, 2nd edn. Butterworth-Heinemann, Oxford.
The abundance, distribution, extraction processes and uses for all of the metallic and non-metallic elements are fully described in an easy-to-read and understandable way. The manufacture of many important compounds is also described.

Gore, A. (2000) *Earth in the Balance*, new edn. Earthscan Publications, London.
This book is an account of a wide range of environmental problems and suggested solutions. Scientific explanations are offered, which are lucid and understandable to the lay person.

Rogers, J.J.W. and Feiss, P.G. (1998) *People and Earth: Basic Issues in the Sustainability of Resources and Environment*. Cambridge University Press, Cambridge.
Contains a very good description of sustainable development and natural resource management.

Street, A. and Alexander, W. (1998) *Metals in the Service of Man*, 11th edn. Penguin Books, London.
All you ever wanted to know about metals! The most common metals such as iron, aluminium and copper are covered in great detail, whilst the less important metals receive much less coverage. Metallurgical aspects are covered in a most interesting and readable way.

US Geological Survey (USGS) (2003) *Minerals Information*. Available at http://minerals.usgs.gov/minerals/pubs/mes/.
An invaluable source of statistical and other data on the world's natural resources.

White, P.R., Franke, M. and Hindle, P. (1995) *Integrated Solid Waste Management: A Life Cycle Inventory*. Blackie Academic, London.
For those interested in waste management, an indispensable book! The life cycle approach is fully developed and methods of waste collection and disposal covered.

6 Risk and hazards

- An explanation of the terms 'risk' and 'hazard'
- How hazards are identified, with special reference to asbestos and cadmium
- Some of the properties, uses and dangers of cadmium and its compounds
- Types, uses and dangers of asbestos
- The dose–response relationship – ED_{50}, LD_{50}, TD_{50}, NOEL, TLV, PEL
- Ways of managing risk

Gifu Prefecture, Japan

The Jinzu River in Japan flows from the mountains of Gifu Prefecture to Toyama City on the coast of the Sea of Japan. Many of the people who live next to this river are very familiar with the term 'Itai–Itai' disease or 'Ouch–Ouch' disease. The expression describes a disease that leaves its victims suffering from constant and intense pain in their bones and joints. Their bones can become so fragile that even the gentlest of movements can cause fractures. Ultimately, the disease is fatal. The cause of the disease was cadmium, which had found its way into the Jinzu River from a mine being worked for its zinc and lead at Kamioka.

'One patient had a total of 76 fractures. . . . When my father held the wrist of a patient to feel her pulse, her wrist broke,' says Shigestsugo Hagino, the son of Dr Noboru Hagino who exposed the cause of the disease (quoted in the *Japanese Times* in December 1997).

Local farmers knew as early as 1910 that, because of its colour, the river water used to irrigate the rice paddies along its banks was contaminated by something from the mine. Continuing complaints resulted in some measures being taken in 1932 and 1940 to reduce the pollution but the build-up to the Second World War overtook concerns about pollution problems. In the 1940s and 1950s lead was a strategic 'war' material and mining became more intensive with the consequent increase in pollution and an increase in health problems. People were seen to suffer from a mysterious disease characterised by intense pain, brittle bones and liver and kidney damage. Dr Hagino started to investigate this disease in 1946 and in 1961 identified the culprit as cadmium (Hagino 1968). There was some considerable opposition to his findings, but such was the effect of his report that further investigations were carried out by the local government. They concluded in 1967 that it was difficult to confirm that cadmium was the only cause. However, in 1967 investigations carried out by a national government research team agreed with Dr Hagino's conclusions.

In 1968 the first claims against the mining company involved were started by Itai–Itai patients. In 1971 their claims proved successful and set the precedent for future compensation. Later in the same year, the Japanese Environmental Agency introduced environmental standards that were to limit the emission levels of cadmium and other toxic substances from industrial activities. From 1968, when the disease was officially recognised, to December 1997, a total of 181 people were diagnosed as suffering from Itai–Itai. A further 333 people were the subject of some considerable concern. Prior to 1968 it is not known how many people had died or suffered from cadmium-related diseases.

Could the above pain and disease have been avoided? Who, if anyone, was to blame?

The cities of Leeds and York, UK

In 1958 an asbestos factory in Armley in the City of Leeds was closed down. In the 16 years leading up to 1993, the deaths of some 180 men and women occurred, who had either worked at this factory or had relatives who had worked there. All of the dead had suffered from a form of malignant cancer known as mesothelioma, which is uniquely associated with asbestos.

The inhalation of even a minute amount of asbestos can kill, and for years the potentially deadly dust from the Armley factory had been a part of local life. Shortly before he died in 1988 at the age of 42, one victim of mesothelioma recalled how his mother came home from working at the factory covered with asbestos dust – he believed he had developed the cancer by coming into contact with his mother's coat as a small boy (Turton 1993). Yet another victim who died in 1990 at the age of 43 remembered how, as a boy, he had licked his fingers after writing his name in the asbestos dust on the pavements outside his home. Neither of these victims had worked at the factory!

During the early and mid-1990s many houses in the immediate vicinity of the closed factory were checked for asbestos pollution. Several hundred houses had to be evacuated and decontaminated. The factory itself was left in a badly contaminated state with asbestos dust and other debris lying about both inside the buildings and in the surrounding yard.

The factory at Armley started as a family business in 1870, making asbestos mattresses for lagging steam engines. In 1921 it was merged with another company to form the world's biggest and most powerful asbestos company. This new company ran 15 factories in the UK plus huge asbestos mines in Canada, South Africa and what was then Rhodesia. It also operated factories in India and the US. The technique which made this company particularly famous was developed from an Armley idea in 1931. This involved mixing raw asbestos with water and cement, which could then be sprayed onto a surface requiring insulation. The technique was portable, cheap and almost any surface could be sprayed to provide instant soundproofing and fireproofing. Asbestos thus became a component of thousands of public buildings, such as hospitals, schools and theatres, spread throughout the world.

In April 2000, the local York evening newspaper reported the seventy-fourth death caused again by the asbestos-related disease mesothelioma. This victim once worked for the now closed York railway carriage factory. Between 1954 and 1967 blue asbestos was used at the works as an insulator for locomotives and carriages. Workers came directly into contact with the asbestos in two ways. First, the process involved opening bags of dry powdered blue asbestos and pouring it into rotating drums, where it was dampened by water. This was then followed by the spraying of the asbestos on to the insides of exterior carriage panels coated with adhesive and on their undersides. White asbestos continued to be used until 1984 for insulating steam pipes. From 1975, the York site began stripping blue asbestos from rail carriages under the strictest of controls.

During the spraying days, workers recalled working in clouds of asbestos dust in the repair shops, some taking home the dust as a contaminant on their working clothes. They often ate their midday food in a similar atmosphere. Some described having 'snow-ball'-type fights with balls of asbestos, which exploded into fine clouds of dust upon making contact. Other workers state how they used brooms to sweep up the asbestos and how they then blew the asbestos dust outside using air hoses. All report how they daily came into contact with blue asbestos, but were never told about the dangers involved or equipped with protective clothing or face masks.

In March 2000, an ex-railway track employee who worked for British Rail in York in the 1970s and 1980s received substantial damages. He had developed mesothelioma as a result of being involved in the breaking up of concrete troughs containing asbestos used to protect signalling cable. He is probably the first person to be recognised as having acquired the disease whilst exposed to white asbestos when working outside in the open atmosphere.

What are the factors that link cadmium poisoning in Japan and asbestos deaths in Leeds and York in the UK in the 1990s? The answer to this question lies in an examination of the terms pollution, hazard, risk, risk assessment and risk management.

Risk assessment

It is clearly wrong to 'wait' for the results of pollution to manifest themselves in a group of people and then try to put things right. Risk assessment is about identifying the potential hazards and risks associated with a substance, process or activity and determining ways of managing those hazards and risks before adverse effects become evident. A **hazard** is that which has the potential to cause harm either to living organisms or to the physical environment. **Risk** is the likelihood or probability of suffering a harmful effect or effects resulting from exposure to some chemical, biological or physical agent, or of some other adverse effect occurring.

In the past, it has been in many cases relatively easy to identify a hazard and to devise ways of dealing with it based on common sense. For example, in every home there is always the risk of a fire hazard. The response to this risk has been based on previous experience in that it is known what commonly causes a fire, e.g. unswept chimneys, faulty electrical devices such as washing machines, smoking in bed, etc. Hence we try to avoid a fire in our home by having the chimney swept, ensuring that electrical safety codes are properly followed, and not smoking in bed! We try to lessen the effects of a fire by identifying escape routes, fitting smoke alarms, acquiring knowledge about the best ways of tackling a fire, and by establishing a fire brigade service. All of these responses have evolved over a period of time without resorting to elaborate calculations or analytical techniques. However, when a fire insurance policy is taken out with an insurance company, that company will have assessed the risk of a fire hazard. Such assessment will take into account how often house fires occur, how big claims tend to be, the age of the property, the district, previous fire accident claims, and so on. Thus, if the risk is to be properly assessed, many factors have to be taken into account, including the frequency of the event and how big the population is that is exposed. Risk depends on both the probability of an event occurring and the severity of the event should it occur.

The evolutionary approach to risk assessment becomes less useful the more complex a system becomes. Indeed, a more rigid and mathematically based approach to risk assessment is developing because many people-made systems are so complex that it is not possible for one single person to understand the whole system. Risk assessment attempts to quantify the probabilities and degrees of harm that result from a complex operation. It is not so easy to define risk assessment because of the wide range of situations in which it used. For example, the US Environmental Protection Agency (EPA) sees risk assessment as the process that clearly identifies the potential safety and health effects caused by the exposure of individuals and populations to hazardous materials and situations. It has also been defined as the identification of hazards and their causes, the estimation of probability that harm will result and the balancing of harm with any resulting benefit.

Risk assessment is done so that intelligent policy decisions can be made concerning the health and welfare of all living things; to understand and improve existing technology design; to educate people so they can make informed decisions; and to improve the economic and social welfare of people.

How is risk assessment done?

There are many ways to assess risk but most of them contain four essential elements: hazard identification; establishment of a dose–response (toxicity) relationship; determination of exposure levels and risk characterisation. These four steps are shown in Figure 6.1.

Risk assessment

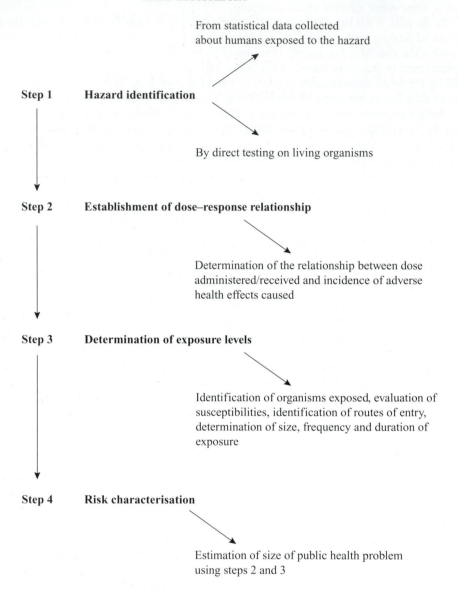

From statistical data collected
about humans exposed to the hazard

Step 1 **Hazard identification**

By direct testing on living organisms

Step 2 **Establishment of dose–response relationship**

Determination of the relationship between dose
administered/received and incidence of adverse
health effects caused

Step 3 **Determination of exposure levels**

Identification of organisms exposed, evaluation of
susceptibilities, identification of routes of entry,
determination of size, frequency and duration of
exposure

Step 4 **Risk characterisation**

Estimation of size of public health problem
using steps 2 and 3

Figure 6.1 The steps involved in risk assessment.

Hazard identification is the vital step because, if any hazards are missed out in the risk assessment, then it will have been carried out on the wrong basis. Hazard identification is largely a qualitative, complex process that often involves the use of a range of techniques to identify the hazard. Were the hazards correctly identified in the cases of cadmium and asbestos?

The presence of a hazard has often been identified by its effects. Until a hazard's effects have become noticeable, its very existence may not be appreciated, e.g. as in the

Table 6.1 *Dose and hazardous materials*

Effective dose-50 (ED_{50})	Dose resulting in 50% of animals tested responding in a specified way
Lethal dose-50 (LD_{50})	Dose resulting in the death of 50% of the organisms being tested
Toxic dose-50 (TD_{50})	Dose that produces signs of toxicity in 50% of the organisms tested

case of cadmium. Even if the effects are known, the identity and amount of the hazard may not be determined because the methods available may not be well developed enough at the time of testing.

The second step is the establishment of the dose–response relationship. The dose or amount of an agent received by an organism depends upon both the concentration and the time of exposure to that agent. The response is the named biological effect that the agent causes.

Acute effects are the easiest to deal with because of the short timescales involved and the large doses of agents used. The responses to acute effects are very quick, rarely reversible and often fatal. Acute effects are estimated by the use of studies on animals or plants.

Chronic effects are much more difficult to assess because they require much longer periods of exposure at much lower dose levels. These effects are often reversible but they can cause serious injury to the receiver. Sometimes it is possible to support these results with data available as a result of accidental exposure of humans to agents.

In the case of acute effects, the relationship between dose and response is established by testing the agent through bioassays. A bioassay involves the assessment of the strength and effect of an agent by testing it on living organisms and comparing the results with the known ones of another agent. There are a number of ways of expressing these strengths and effects (Table 6.1).

The ED_{50} depends upon the effect being measured. In general, the less severe the effect, the lower is the effective dose.

The LD_{50} is an average value covering many separate tests. This is usually expressed in milligrams per kilogram of body mass ($mg\ kg^{-1}$). A chemical with a small LD_{50}, e.g. $2\ mg\ kg^{-1}$, is considered to be highly toxic, whereas if it is $1,000+\ mg\ kg^{-1}$ it is considered to be for all practical purposes non-toxic. Note that the lethal dose gives no indication of any non-lethal effects. A chemical can still cause illness at a very low exposure level but have a very large LD_{50}. When viewing a lethal dose, it is necessary to know what the animals were that were tested – different animals have different responses to a chemical, e.g. for the dioxin TCDD the LD_{50} for a guinea pig is $1\ \mu g\ kg^{-1}$ whilst that for a hamster is $5,000\ \mu g\ kg^{-1}$. The other factor that is important is how the agent entered the animal. Some chemicals are extremely toxic if swallowed (oral ingestion) but not very toxic if splashed on to the skin (dermal exposure), e.g. a herbicide known as 2,4-D has an oral LD_{50} of 375–$666\ mg\ kg^{-1}$ for rats but one of $1,500\ mg\ kg^{-1}$ when administrated via the skin.

The larger the TD_{50} the more agent it takes to poison and produce symptoms.

Effects produced in organisms by hazards are expressed in a variety of ways (Table 6.2). In the case of the NOEL value, when dealing with food and drink the tolerance levels of poisons permitted are set between $1,000$–$10,000$ times less than the highest dose that produces a noticeable effect.

Table 6.2 *Effects produced by hazards*

No Observable Effect Level (NOEL)	The highest dose or exposure level that an organism can resist in that it produces no noticeable effect
Threshold Limit Value (TLV)	The airborne concentration (ppm) of a substance that produces no adverse effects on workers exposed for 8 hours per day for 5 days a week
Immediately Dangerous to Life and Health (IDLH)	The maximum level of a pollutant from which an individual could escape within 30 minutes without suffering temporary or irreversible health damage
Permissible Exposure Limit (PEL)	The concentration of a substance to which workers can be exposed without adverse effects, averaged over a normal 8-hour working day or a 40-hour working week
Time Weighted Average (TWA)	The concentration of a substance to which most workers can be exposed without adverse effects, over a period of time

In the case of the atmosphere it is the TLV that is measured. Limits are set to prevent minor toxic effects such as skin and eye irritations.

When applying data to occupational hazards, there are several ways of representing the hazard, i.e. the IDLH, PEL and the TLV.

For chronic cases, epidemiology is used. This is the study of disease in a human population, defining its incidence and prevalence, examining the role of external influences such as infection, diet or other toxic substances and examining appropriate preventative measures. Such information is not always available for every hazard and the study of human exposure is not always practical or ethical. The presence of external factors, in particular, makes it very difficult to establish a cause and effect link. In order to surmount these problems the effect of exposure of animals to hazards is used. The validity of data based on animal experimentation in order to establish human dose–response relationships is a subject of heated debate. This includes ethical considerations as well as the more scientifically based ones concerning the validity of extrapolation from high to low doses and extrapolation from animal data to humans.

The determination of **exposure levels** is a very important stage in risk assessment. It is about determining the rate at which an identified hazard reaches an organism.

Identifying and describing the population at risk often requires a long and in-depth study. This is because of the many factors that determine the extent to which a hazard will affect a particular population, e.g. occupation, living and working conditions, place of residence, state of health, gender and age.

Exposure assessment is done in three ways. The first is to determine the concentration of a pollutant at its source, e.g. air, soil or drinking water. Unfortunately, this only tells us that a particular pollutant is present and what its concentration is at the time of measurement. In an aquatic system, for example, the concentration of a pollutant may depend upon factors such as currents, tides, depth of sampling, amount of agricultural run-off received, time of the year, and so on. Such determinations can therefore be seen to be next to useless as retrospective or predictive indicators of a source pollutant.

The second way is to investigate some receiving body in the environment, which may give an indication of the level of pollution. For example, the analysis of samples taken from fish can give a good indication of the amount of a persistent chemical which is, or was, present in their aquatic environment. Past levels can be readily estimated using the size and age of the fish and determining how fast these animals accumulated the chemical.

The final way is via laboratory tests on the body fluids of an exposed or potentially exposed victim/victims. Such tests provide the most direct measurement of exposure levels but are not good estimates of the amount, duration or source of the exposure. For example, some chemicals build up to a maximum value in the human body so that no more will bio-accumulate, others are excreted in between exposures. The human race is now very mobile and therefore when testing a particular group it can be found that the members have experienced wide-ranging sources of exposure and exposure times. Habits can also influence the uptake of a pollutant, e.g. exposure to cadmium is greater in smokers than non-smokers. Thus, identifying and describing a population/group/individual at risk requires long and in-depth study. Many factors determine the extent to which a hazard will affect people, e.g. occupation, living and working conditions, place of residence, state of health, gender, age and habits.

Risk characterisation is determined by combining the information on the dose–response relationship and the exposure assessment. Any inherent uncertainties encompassed in the risk assessment should be specified when characterising the risk.

Cadmium – the hazard

The toxic nature of cadmium and its compounds led to the discovery of the element. In the early years of the nineteenth century, zinc carbonate (calamine) was used to produce zinc oxide for medicinal purposes. In Hanover, Germany, cases of poisoning were being reported resulting from the use of this zinc oxide. When tested with hydrogen sulphide a yellow precipitate was produced instead of the expected white precipitate of zinc sulphide. At first, the yellow colour was thought to be due to the presence of arsenic or iron but chemical analysis showed both elements to be absent. In 1817, the German chemist Stoymeyer at the University of Göttingen showed that the yellow colour was due to the presence of a new metallic element. Its similarities to zinc led Stoymeyer to call it cadmium (Gk: *cadmea* for calamine).

Over the next 70 years cadmium was mainly used in the form of the sulphide in paint pigments. However, cases of cadmium poisoning continued to be reported, e.g. servants were known to have fallen ill after using cadmium carbonate as a polishing compound.

In the early part of the twentieth century, cadmium was used in dental amalgams and other alloys. The First World War saw a shortage of tin and thus cadmium became a tin substitute in tin-based solders. The motor car industries used cadmium in electro-plating processes to produce shiny protective layers on headlights, bumpers, etc. For many years this was the main use of cadmium.

In the 1940s it was common for kitchen cooking pans to be protected by a layer of cadmium metal. Since the melting point of cadmium is so low (321 °C) this was done initially by dipping the utensil in the molten metal. This technique was later replaced by electroplating methods. Unfortunately, the use of organic acids such as citric acid, tartaric acid and ethanoic/acetic acid in food preparation caused the dissolving of cadmium from the protective layer. This caused several cases of cadmium poisoning. Once the cause had been identified this use of cadmium was stopped.

In 1956, cadmium propionate used for research purposes was being dried in an oven in a Leeds laboratory. The temperature of the oven became so hot that an explosion occurred, which caused the release of large amounts of cadmium oxide into the laboratory atmosphere. Four firemen who entered the laboratory to fight the fire were affected by the fumes and one died within six days of the exposure. In 1981, two youths in the UK gave cadmium chloride mixed in orange drinks to 11 children. None died but all suffered from cadmium poisoning symptoms.

In the case of the terrible events in Japan (see pp. 142–3), it is probable that up to the 1950s the mine operators were ignorant of the consequences of their operations. Although cadmium and its compounds were known to be hazards from the early nineteenth century, its presence in the mine effluents was not known. When the disease started to become more common and evidence accumulated against the mine waste, then appropriate remedial action began to take place – by then the damage had already been done, the action was retrospective and too late for a large number of victims. The Second World War and its political aftermath were no doubt instrumental in redirecting energies away from this point source pollution problem. The research was also hampered by lack of appropriate techniques of chemical analysis that could detect and measure the levels of cadmium accurately. Today, the toxic nature of cadmium in humans

Box 6.1

The toxicity of cadmium

Itai–Itai disease is not typical of the effects of cadmium poisoning. The disease particularly affected older Japanese women who suffered from malnutrition, calcium deficiency and low vitamin D intake. Thus there were extra factors that contributed to the effects of cadmium. However, the main target organs of cadmium are the liver and the kidneys. Indeed, cadmium gradually builds up in the cortex of the kidneys. When a cadmium level of about 20 μg g^{-1} (wet mass) in the kidneys is reached, then serious symptoms start to appear. Once cadmium enters the body, there is no medical treatment that can prevent its accumulation or remove it once it has been stored. The presence and amount of urinary b-2-microglobin excreted in the urine indicates the level of exposure to cadmium. Humans usually receive their cadmium intake from drinking water, foodstuffs and the air. High proportions of cadmium can be ingested particularly from seafood and meat such as kidneys. These can contain as much as 100 μg kg^{-1}. Most cadmium comes from wheat, rice, etc. The protein metallothionane, which contains large numbers of –S–H groups regulates zinc metabolism and also protects humans from low doses of cadmium. This protein forms a complex via its sulphydral groups with Cd^{2+}, which is then eliminated in urine. Cadmium is a cumulative poison, which stays for decades inside the human body. It appears currently to be a threat only in Japan and central Europe.

(Box 6.1) is well known, thanks to the studies made on Itai–Itai and other victims, and risk assessment has to be made to ensure its safe use.

Cadmium – the material

Cadmium is a soft, shiny bluish-white metallic element that occurs at a very low concentration (0.16 ppm) in the Earth's crust. It is found as the mineral greenockite, CdS, always in conjunction with the zinc sulphide ore sphalerite, ZnS. To be of economic value, the cadmium sulphide must make up some 0.3 per cent by mass of the zinc sulphide ore. Cadmium is thus a by-product in the extraction of zinc.

Cadmium forms the Cd^{2+} ion in ionic salts such as CdO, $CdCl_2$ and $CdSO_4$. Cadmium is not an essential element in metabolism and has no known biological function in humans. In 1998, the world production of cadmium was some 19,900 tonnes. Japan is the leading world producer at 2,400 tonnes followed by Canada at 2,300 tonnes. Cadmium is mainly used for the manufacture of nickel-cadmium batteries (as cadmium and cadmium oxide), pigments (as cadmium sulphide and cadmium selenide), stabilisers for PVC and other plastics (as cadmium stearate), electroplating metals for protection against corrosion (as cadmium sulphate) and non-ferrous alloys (as cadmium).

Table 6.3 shows how the use of cadmium has changed in the US since 1980.

Table 6.3 *Use of cadmium and its compounds in the US*

	Batteries (%)	Alloys and other (%)	Pigments (%)	Coatings and platings (%)	Plastics (%)
1980	16	8	27	34	15
1990	40	10	13	25	12
1998	69	3	13	8	7

Cadmium and risk assessment

Occupational exposure to cadmium and its compounds is the main way that humans encounter the element in quantities that are dangerous. Such exposure occurs in those industries that manufacture cadmium-based products or where the handling and treating of such products take place. Exposure via inhalation is much more dangerous than by ingestion. Hence, danger from cadmium usually results from the heating, incinerating, cutting, welding or soldering of cadmium which causes the formation of cadmium fumes or dust. Normal environmental levels have also been found to be much elevated in industrial areas, particularly around factories involved in the manufacture of cadmium metal and its alloys.

Non-occupational exposure can be greatly enhanced by anthropogenic effects that pollute the air, water or soil. Normally, non-occupational exposure levels are usually too small to add significantly to the body burden of those people who have been occupationally exposed. Humans will also be exposed to the larger quantities of cadmium that will be found in water that has been in contact with zinc mineral deposits, or with galvanised metal.

The major source of non-occupational exposure to cadmium is food. Kidneys and livers of adult farm animals, and certain sea foods such as mussels, oysters and crabs can all contain high amounts of cadmium. Grain may also contain significant amounts.

The use of sewage sludge or phosphate fertilisers can increase the cadmium content of soils, and thus increase the uptake of cadmium in food products.

Cadmium is found in small amounts in tobacco. Although it is in much lower quantities than in foods, cadmium is absorbed much more effectively through the lungs than the gut. For smokers not occupationally exposed to cadmium, then smoking is the main source of exposure – their body burden is twice that of a non-smoker!

Acute effects are usually associated with the workplace. Acute exposure by inhalation can cause a flu-like illness called metal fume fever. If enough cadmium has been inhaled, acute respiratory effects such as bronchial and pulmonary oedema may result. Even a single exposure to a high enough level of cadmium can cause long-lasting lung damage and even death. For example, Beton *et al.* (1966) reported that an exposure to cadmium fumes of concentration 9 mg m^{-3} for five hours is a lethal dose for humans. Fatalities have also occurred from exposure to concentrations of between 40 and 50 mg m^{-3} for one hour. For cadmium oxide fumes a lethal dose of not more than 85 mg m^{-3} for 30 minutes has been reported (Bulmer *et al.* 1938); 39 mg m^{-3} for 20 minutes has also produced fatal results.

Acute toxicity data have been derived from inhalation experiments performed on animals (see Tables 6.4 and 6.5).

Chronic effects (excluding cancer) in humans who have been exposed via inhalation or ingestion include kidney disorders, bronchiolitis and emphysema. For example,

Table 6.4 Some lethal concentrations of cadmium and cadmium oxide

Species	LC_{50} mg m^{-3}	Time of exposure/ minutes	Reference
Cd: rat	25	30	Yohikawa, H. and Homma, K. (1974)
Cd: rabbit	250	30	Friberg, L., Piscator, M., Nordberg, G.F., and Kjellstrom, T. (1974)
CdO: rat	780	10	Gates, M., Williams, J., and Zapp, J.A. (1946)
CdO: rat	500	10	Barrett, H.M., Irwin, D.A. and Semmons, E. (1947)
CdO: rabbit	2,500	10	Barrett, H.M., Irwin, D.A. and Semmons, E. (1947)
CdO: rabbit	3,000	15	Gates, M., Williams, J., and Zapp, J.A. (1946)
CdO: dog	400	10	Kotsonis, F.N. and Klasen, C.D. (1977)
CdO: dog	4,000	10	Barrett, H., Irwin, D.A. and Semmons, E. (1947)

Table 6.5 *Some lethal doses for cadmium, cadmium oxide and cadmium chloride*

Species	Route	LD_{50} mg kg^{-3}	Reference
Cd: rat	Oral	225	Kotsonis, F.N. and Klasen, C.D. (1977)
CdO: rat	Oral	72	National Institute for Occupational Safety and Health (1996)
CdCl$_2$: rat	Oral	60	Engstrom, B. (1981)

chronic cadmium exposure of 50 μg m^{-3} for ten years may cause kidney dysfunction (Jarup *et al.* 1988). Animal experiments have also shown that the kidney, liver, lung and many other organs are affected by cadmium.

The effects of chronic cadmium exposure on human reproductive and developmental processes are inconclusive. Animal experiments involving inhalation and oral administration show the development of low foetal masses, skeletal deformations, impaired neurological development and interference with foetal metabolism.

There is some inconclusive evidence for an excess risk of lung cancer in humans via inhalation. Animal inhalation studies have seen the development of several forms of cancer. The USEPA classifies cadmium and its compounds as probable human carcinogens. The EPA estimates that a person breathing air containing cadmium at 0.0006 mg m^{-3} over an entire lifetime has an increase in probability of 1 in 10^6 of developing cancer. The US Occupational Safety and Health Administration (OSHA) estimates 3 to 9 excess lung cancer deaths per 1,000 workers exposed to 5 μg m^{-3} of cadmium for 45 years. In 1992 OSHA set a limit of 0.005 mg m^{-3} for cadmium. Table 6.6 shows maximum inhalation exposure levels set by various US agencies.

In the UK the COSSH values used for cadmium exposure is the **MEL** or **M**aximum **E**xposure **L**imit. For cadmium and cadmium compounds (except CdS pigments) the MEL is 0.025 mg m^{-3} averaged over an 8-hour period. Cadmium sulphide MEL is 0.04 mg m^{-3} averaged over an 8-hour period. Cadmium oxide fume is set at a MEL value of 0.05 mg m^{-3} measured over a 15-minute period.

The Department of Energy (UK) give a guideline for urban atmospheric emissions of 5×10^{-3} to 1.1×10^{-2} mg m^{-3} for short-term exposure and a range of 5×10^{-6} to 5×10^{-5} mg m^{-3} for cadmium.

EU Directive 89/429/EEC has set a guideline of 0.2 mg m^{-3} for the total of cadmium and mercury in air.

In the case of drinking water, the WHO guideline for the recommended maximum tolerable weekly intake is 400–500 μg g^{-1}. The USEPA guideline is 0.0005 mg kg^{-1} per day. Scotland has set a regulatory maximum level for cadmium in water as 5μg dm^{-3}.

Table 6.6 *Some inhalation exposure level values for cadmium*

	Inhalation exposure level mg m^{-3}
National Institute of Occupational Safety and Health (cadmium dust or fumes) IDLH	40
Occupational Safety & Health Administration (cadmium dust) PEL	0.2
Ditto (cadmium fumes) PEL	0.1
American Conference of Governmental and Industrial Hygienists (cadmium dust and cadmium oxide)	0.01

Asbestos – the hazard

The problem for asbestos victims is the length of time it takes for the disease to become evident so that claims for compensation have proved to be difficult to bring to court. Indeed, many victims have died whilst waiting for their cases to be considered. This is because companies, such as the ones involved at York and Leeds, against which complaints have been made have claimed ignorance of the effects of asbestos and that much of the damage was caused before legislation and law controlling the use of asbestos was in place. In other words it is claimed that proper hazard identification, risk assessment procedures and control mechanisms did not exist. What is the truth behind this?

The first recognition of the dangers of asbestos dates back to a Factory Inspector's report of 1898, with the first recorded asbestos-related deaths of males all aged about 30. Hence, there appears to be strong evidence that something was not quite right about asbestos over 100 years ago. In 1929, a Leeds University pathologist made a study of the lungs of asbestos workers at Armley. He noted the presence of asbestos fibres not only in the workers themselves but in the lungs of a man who had never worked at the factory (Merewether 1930). This was perhaps the first recorded environmental exposure. In 1931, the Government of the UK passed the Asbestos Industry Regulations to control dust in the workplace. In 1935 the first cases of asbestos-related lung cancer were reported. In 1943, the company owning the factory at Armley, together with other asbestos companies, sponsored investigations of the effects of asbestos on white mice. This showed that 81 per cent of the mice that had inhaled long fibre asbestos developed lung cancer. In 1955 a further study showed that the lung cancer rate amongst this company's workers was ten times the national average. The real turning point came in 1960 when a South African physician published findings that showed an excess of deaths caused by cancer in a blue asbestos mining community in the Transvaal, and that mesothelioma tumours were killing both the miners and people who had never worked with asbestos. In the US a 1961 study of Second World War shipyard workers proved the connection between airborne asbestos and the development of disease 20 or more years after exposure, and indicated that it was probable that the asbestos industry knew of the dangers far earlier.

Asbestos – the material

Asbestos is the generic name given to certain inorganic mineral silicates that occur in fibrous form. They occur naturally in seams or veins generally between 1 and 20 mm thick, in many igneous and metamorphic rocks. These minerals belong to one of two large groups of rock-forming materials, the **serpentines** and the **amphiboles**. Such fibre can be processed into a variety of materials that are uniquely resistant to fire, heat and corrosion.

Chrysotile (white asbestos), the fibrous form of serpentine, is the most important source of asbestos and constitutes 95 per cent of world production. Its chemical formula can be represented as $Mg_3(Si_2O_5)(OH)_4$. Chrysotile is white, grey, green or yellowish with a silky lustre. It has a very high tensile strength and its fibres are soft and flexible. The material is soluble in acids. The largest producer of white asbestos is Russia (46 per cent of world production), but the main sources imported to the UK are from Canada (23 per cent world production) and South Africa (5 per cent world production). In 1997, the UK imported some 4,820 metric tonnes of this asbestos compared with 190,000 tonnes in 1973. White asbestos fibres can be spun and woven into fire-resistant textiles, matted into insulating materials, or used with other substances to make a wide variety of products, including brake linings and clutch pads, roofing and flooring materials, cement and insulation for electrical circuits. For many of these products, asbestos is almost irreplaceable. No other substance provides its stability, strength and heat resistance so cheaply and efficiently. Nevertheless, manufacturers are having to develop materials that can be substituted for asbestos.

The main two fibrous forms of the amphiboles, **crocidolite** or **blue asbestos** and **amosite** or **brown asbestos**, are now of far less importance as asbestos sources. Crocidolite has the idealised formula $Na_2Fe_3^{II}Fe_2^{III}Si_8O_{22}(OH)_2$ and is lavender or blue with a silky lustre. Amosite is ash grey, greenish or brown with a pearly to vitreous lustre. It has the formula $(Mg,Fe)_7Si_8O_{22}(OH)_2$. Blue and brown asbestos have properties, e.g. extremely high tensile strengths and resistance to acids, which lend themselves to specialised usage such as insulation mattresses, sprayed thermal and acoustic insulation, and where acid resistance is required.

Asbestos cement is familiar to many people in the form of corrugated shed roofing. When mixed with cement, white asbestos fibres gave the finished product good mechanical strength and resistance to corrosion. In 1996, approximately 55 per cent of the total asbestos used in the UK (3,000–4,000 tonnes) was in the form of asbestos cement products. Because of the now known health hazards associated with asbestos this kind of roofing material is no longer available.

In 1996, the use of white asbestos in friction products such as brakes and clutches constituted 26 per cent (1,850 tonnes) of its use in the UK. In the same year, 21 per cent (1,500 tonnes) of white asbestos was used for making gaskets and packaging materials. Textiles and composites in the form of woven tape, webbing, cloths and yarns were not manufactured in the UK but imported. These accounted for some 3.5 per cent (250 tonnes) of the total use of white asbestos.

Asbestos and risk assessment

One of the most difficult aspects of risk assessing asbestos is in identifying its presence! Even if a material is identified as being asbestos, it is very difficult to determine visually

Box 6.2

Diseases caused by asbestos

Asbestos causes several diseases, but the two that are causing considerable concern are **asbestosis** and **mesothelioma**. Asbestos is the only known cause of meso-thelioma, which is a malignant cancer of the pleura (the membrane that surrounds the lungs) or peritoneum (the membrane that lines the abdominal cavity and covers the abdominal organs) for which there is no treatment or cure. Although it can take between 10 and 50 years to develop, it is always fatal. Deaths normally occur at the age of 45 plus for asbestos workers themselves. If death occurs due to this disease below the age of 45, then it is probably due to indirect exposure. In the pleura, mesothelioma causes pain and breathlessness. Tumours in the peritoneum cause enlargement of the abdomen and obstruction of the intestines. Victims of this disease often show minimal fibrosis and no lung cancer. If the exposure has been to a high concentration of asbestos, the diseases of the lung are prominent. If exposure has been to a low concentration, then translocation to the pleura becomes more common and hence mesothelioma. Therefore, the more rigorously controlled asbestos has been, the more likely is the victim to die of the latter disease.

Asbestosis is caused by the replacement of the specialised tissues of the lungs and the pleura by scar tissue. It is similar to the coal-miners disease, pneumoconiosis, caused by coal dust, and to silicosis, caused by dust containing silica, which is asso-ciated with quartz mining, stone cutting, blasting and tunnelling. Asbestosis tends to progress even when the victim is no longer exposed to asbestos. The disease causes breathlessness and a dry cough, eventually leading to severe disability and death. The period from first exposure to the development of the disease is seldom less than 10 years and is usually much longer. Asbestosis increases the risk of developing lung cancer in smokers and non-smokers alike, but the combination of smoking and asbestosis leads to an even greater risk. Asbestosis is a prescribed disease, which entitles sufferers in the UK to industrial injury benefit.

How does asbestos cause such problems?

The cells lining the **bronchi**, and the larger bronchioles of the human respiratory system, have whip-like **cilia** at their free surfaces. These normally move dust parti-cles upwards towards the **trachea**, where they are coughed out. If the size of the dust particle falls below a certain size (4 μm or less), then they behave like gas particles. This means that such particles can enter the respiratory bronchioles and then the terminal air sacs (**alveoli**). Here there are no cilia and therefore the dust particles cannot be removed. In the case of fibrous particles, their ability to move through the air vessels and to reach the alveoli depends mainly on their thickness, e.g. a long but thin fibre is more likely to reach the alveoli than a long fat one. Once in the alveoli, macrophage cells engulf the asbestos particles. They pass through the alveoli walls and the walls of the larger bronchioles, blood vessel walls and through the pleura. If such a cell carries a dust particle through the wall of a bronchiole, it will reach cilia and be subsequently coughed up. If it moves through a blood vessel wall and into the blood, then the dust particle may be carried to another organ. It may, for example, move into the pleura where it can be transferred somewhere else and cause damage. Because of the process just described, asbestos workers tend to suffer from gastrointestinal cancer, biliary cancer and renal cancer.

the type, particularly if it is in the form of dust. Asbestos and particularly asbestos composites are subject to ageing with consequent colour changes that can be misleading. Microscopic analysis is usually necessary.

If mixed with, for example, cement, asbestos is less concentrated and therefore is less of a hazard than pure asbestos. However, workers who have regularly used composite asbestos floor tiles have died of mesothelioma. Roofing composite tiles have also been known to weather and produce asbestos fibres – victims are known to have developed mesothelioma as a result of brushing off mould from an asbestos cement roof on an old garden shed.

The main hazard is where asbestos has been sprayed or used as lagging, between 1935 and when it was banned in 1971. This asbestos (blue, brown and white) is now likely to be soft and very easily powdered. This can cause the contamination of a working area very easily and can, for example, be carried around a site via air conditioning ducting.

There are no studies yet available concerning the acute effects of asbestos on either humans or animals. Numerous epidemiological studies on humans have been made. Generally, these are consistent in reporting an increased incidence of asbestosis, lung cancer, mesothelioma and gastrointestinal cancer in occupationally exposed workers (Selikoff *et al.* 1979; Petro *et al.* 1977). These diseases are accompanied by an increase in mortality. Animal experiments on rats show similar findings for the inhalation of asbestos for lung cancer and mesothelioma (Reeves 1976; Wagner *et al.* 1974. Animal experiments to determine the effects of ingested asbestos are mainly inconclusive but male rats fed chrysotile fibres greater than 10 μm in length produced benign polyps. Studies of the connection between mesothelioma and smoking show no link, but smoking and the inhalation of asbestos do act synergically to produce lung cancer. Risks associated with low-level, non-occupational exposure are not known.

Asbestos fibres occur naturally in asbestos-bearing rocks and through erosion can find their way into the atmosphere, water and soil. In addition, airborne exposure may occur through pollution caused by asbestos-related industries and from the breaks and clutches of motor vehicles. Asbestos fibres are very light and aerodynamic so therefore there is a 'natural' very variable background level of asbestos to which everyone is exposed. Only when asbestos becomes available in higher concentrations for a long period of time does exposure result in a higher risk. As a consequence of the acceptance of asbestos as a serious health hazard, from 1969 until the present time protective regulations and laws have been passed in the UK. The use of asbestos and the modification of buildings and other structures involving asbestos have become so strictly controlled that only licensed holders can work with asbestos. The import, supply and use of the two most dangerous forms, blue and brown asbestos, together with any second-hand materials have been prohibited since 1973. In November 1999, the UK banned the import, supply and use of white asbestos other than for research purposes or where replacement cannot be immediately carried out. This ban included second-hand products. All aspects to do with asbestos involving its manufacture, processing, repairing, maintenance, construction, demolition, removal and disposal are now covered by legal requirements and rigorously enforced in the UK.

The US Environmental Protection Agency has also decided that there is no safe level of exposure to any forms of asbestos. The US Occupational Safety and Health Administration has set a permissible exposure limit (PEL) at 0.1 fibre cm^{-3} for an 8-hour time-weighted average. Chrysotile is now classified in the EC as a Category 1 carcinogen. Currently, in the UK, the control limits for aerial asbestos dust under the Control of Asbestos at Work (Amendment) Regulations of 1992 and 1998 are listed in Table 6.7.

Table 6.7 *Control of asbestos at work exposure limits (UK)*

Type of asbestos	Maximum exposure limit
Chrysotile	0.3 fibres cm^{-3} of air averaged over any continuous period of 4 hours
	0.9 fibres cm^{-3} of air averaged over any continuous 10-minute period
All other forms	0.2 fibres cm^{-3} of air averaged over any continuous period of 4 hours
	0.6 fibres cm^{-3} of air averaged over any continuous 10-minute period

Risk management

Risks cannot be avoided because it is neither practicable nor desirable. In order to make progress, a degree of risk has always been acceptable. What is required is to control and live with risks via the process of risk management.

Risk assessment is largely a scientific process, whilst risk management is a decision-making process that uses risk assessment. Risk management involves a wide range of expertise on the part of the practitioner involving political, economic and social considerations, together with interpersonal skills and a sound subject knowledge.

There are basically four ways of carrying out the management of risk. Having identified the hazard and associated risks, the *risk can be reduced*. This is achieved, for example, by designing more efficient incinerators and placing them in less well-populated areas, or by finding a less harmful substance as a substitute for the hazard material. In case of an accident there should be a system for *damage limitation* in place. This might take the form of ensuring that proper crisis planning has taken place within a process to minimise risk and prepare for accidents. A third way involves *accepting and sharing of risk*. Here, a waste site, which is in someone's backyard, is not endangering everyone equally. Therefore, part of the management process might entail the compensation of people who are at greater risk in case of any injury. Legislation in the form of insurance/taxation processes and higher prices of commodities may need to be managed to reflect what has been done by industry to pay for safety. Finally, *redressment for damage* will play its part and relates to the financial measures taken after an accident has occurred. The latter is currently of particular poignancy for asbestos workers.

Summary

- A hazard is anything that has the potential to harm living organisms or the physical environment.
- Risk is the probability of suffering adverse consequences as a result of exposure to chemical, biological or physical agents.
- Risk assessment involves four stages: (i) hazard identification, (ii) the establishment of the toxicity of the agent involved, (iii) determination of the level of exposure and (iv) risk characterisation.
- Acute effects of toxic agents are very quick, rarely reversible and often fatal.
- Acute effects are tested on animals and plants, the results then forming estimates for effects on humans.
- Acute effects are expressed as effective dose-50 (ED_{50}), lethal dose-50 (LD_{50}), toxic dose-50 (TD_{50}), or no observable effect level (NOEL).
- The threshold limit value (TLV) is the concentration of a substance that produces no adverse effects on workers exposed for 8-hour periods for 5 days a week.
- The immediately dangerous to life and health (IDLH) measure is the maximum level of a pollutant from which an individual can escape within 30 minutes without suffering temporary or irreversible damage.
- The permissible exposure level (PEL) is the concentration of a material to which workers can be exposed without adverse effects average over an 8-hour working day for 40 hours per week.
- Chronic effects manifest themselves over a long period of time, are often reversible but may cause serious harm to a receiving organism.
- Epidemiological studies are carried out to determine chronic effects.
- Exposure assessment is carried out in three ways, the determination of the concentration of a pollutant in air, water or soil, using animals or plants to determine the level of pollution, and the laboratory testing of body fluids such as blood and urine.
- Risk characterisation is the estimation of the size of a public health problem.
- The management of risk involves reducing the risk, damage limitation, the acceptance and sharing of risk and compensation for any damage caused.

Questions

1   Sprinkled throughout this chapter are the hazard symbols found on packaging such as chemical reagent bottles.
    (a)   What is the colour coding for these labels?
    (b)   What does each hazard symbol mean, and what additional details accompany them?
    (c)   What hazard symbols would you expect to be attached to containers of (i) cadmium(II) nitrate(V), (ii) white asbestos, (iii) a dilute solution of copper(II) sulphate(VI) and (iv) lead(II) sulphide?

2   Choose any element or one of its compounds and use it to distinguish clearly between the terms hazard and risk.

3   The preparation and properties of hydrogen sulphide gas was for many years an area covered in secondary school chemistry lessons. It is now 'frowned upon' since hydrogen sulphide is a hazardous material. Why is this gas hazardous?
    How hazardous to you think a 'one-off' demonstration experiment might be to the demonstrator and to the pupils?

4   What are the risks associated with chromium and its compounds?

5   In South America, gold has been recently extracted and purified using mercury. Find out what the real and potential effects of the extraction of gold have been on the environment.

6   The social and economic gains of transporting crude oil around the world far exceeds the hazards and risks involved. Discuss.

References

Barrett, H.M., Irwin, D.A. and Semmons, E. (1947) Studies on the toxicity of inhaled cadmium Part I: The acute toxicity of cadmium oxide by inhalation. *Journal of Industrial Hygiene and Toxicology*, **29**, 279–85.

Beton, D.C., Andrews, G.S., Davies, H.J., Howells, L. and Smith, G.E. (1966) Acute cadmium fume poisoning: five cases with one death from renal necrosis. *British Journal of Industrial Medicine*, **23**, 292–301.

Bulmer, F.M.R., Rothwell, N.F. and Frankish, E.R. (1938) Industrial cadmium poisoning: a report of fifteen cases, including two deaths. *Canadian Journal of Public Health*, **29**, 19.

Engstrom, B. (1981) Influence of chelating agents on toxicity and distribution of cadmium amongst proteins of mouse liver and kidney, following oral and subcutaceous exposure. *Acta Pharmacologica et Toxicologica*, **48**, 108–17.

Friberg, L., Piscator, M., Nordberg, G.F. and Kjellstrom, T. (1974) *Cadmium in the Environment*, 2nd edn. CRC Press, Cleveland, OH.

Gates, M., Williams, J. and Zapp, J.A. (1946) Arsenicals. In: *Summary Technical Report of Division 9, NDRC Vol. 1: Chemical Warfare Agents, and Related Chemical Problems Part 1*, pp. 173–78, Office of Scientific Research & Development, National Defense Research Committee, Washington, DC.

Hagino, N. (1968) *Fighting Against 'Itai–Itai' Disease* [in Japanese]. Asaki Shin Sha, Tokyo, Japan.

Jarup, L., Elinder, C.G. and Spang, G. (1988) Cumulative blood cadmium and tubular protein-uria. *International Archives of Occupational and Environmental Health*, **60**, 223.

Kotsonis, F.N. and Klasen, C.D. (1977) Toxicity and distribution of cadmium administered to rats at sub-lethal doses. *Toxicology and Applied Pharmacology*, **41**, 667–80.

Merewether, E.R.A. (1930) *Report on the Effects of Asbestos Dust on the Lungs and Dust Suppression in the Asbestos Industry*. Her Majesty's Stationery Office, London.

National Institute for Occupational Safety and Health (1996). *Documentation for Immediately Dangerous to Life or Health Concentrations (IDLHS): Cadmium compounds*. Available at: http://www.cdc.gov/niosh/idhl.

Petro, J., Doll, R., Howard, S.V., Kinlen, L.J. and Lewisohn, H.C. (1977) A mortality study among workers in an English asbestos factory. *British Journal of Industrial Medicine*, **34**, 169–72.

Reeves, A.L. (1976) The carcinogenic effect of inhaled asbestos fibres. *Annals of Clinical and Laboratory Science*, **6**, 459–66.

Selikoff, I.J., Hammond, E.C. and Siedman, H. (1979) Mortality experience of insulation workers in the United States and Canada 1943–76. *Annals of the New York Academy of Science*, **271**, 448–56.

Turton, J. (1993) Killer dust. *The Guardian Weekend* (8 May), London.

Wagner, J.C., Berry, G., Skidmore, J.W. and Timbrell, V. (1974) The effects of the inhalation of asbestos in rats. *British Journal of Cancer*, **29**, 252–69.

Yohikawa, H. and Homma, K. (1974) Toxicity of inhaled metallic cadmium fumes in rats [in Japanese]. *Japanese Journal of Industrial Health*, **16**, 212–15.

Further reading

Alloway, B.J. and Ayres, D.C. (1997) *Chemical Principles of Environmental Pollution*, 2nd edn. Blackie Academic, London.
    A comprehensive cover of a wide variety of pollutants and their effects on the environment. It includes the toxicological effects, methods of monitoring, analysing and disposal of

pollutants and elements of risk assessment. Almost encyclopaedic in content and requires some knowledge of basic chemistry.

Calow, P. (1998) *Controlling Environmental Risks from Chemicals*, John Wiley, Chichester.
A good account of ecological risk assessment covering many pollution control issues including cost-benefit analysis. Legislation is usefully covered but since this changes from month to month, it may already be a little dated.

Calow, P. (ed.) (1998) *Handbook of Environmental Risk Assessment and Management.* Blackwell Science, Oxford.
This is a very expensive book and should be referred to in a library. It is composed of many chapters written by international experts on risk assessment and the management of risk. It ranges from human health risks and chemicals to waste treatment and environmental risk assessment to genetically modified organisms. Well worth delving into!

Croner's Health & Safety at Work (1997) Croner Publications, Kingston-upon-Thames, Surrey.
All you need to know about the law and health and safety at work. Regularly updated.

Croner's Health & Safety Briefing Papers (2001) Croner Publications, Kingston-upon-Thames, Surrey.
A fortnightly newsletter for those responsible for health and safety. Regularly provides examples of case law, e.g. asbestos in the 27 January 1997 publication. You do not need to be a lawyer or a chemist to understand these!

Health & Safety Executive (1998) *Five Steps to Risk Assessment*. HSE Books, Sudbury, Suffolk.
This booklet is designed to show how to go about risk assessment in the workplace in a very straightforward and simple manner. An interesting read.

Health & Safety Executive (2001) *Asbestos Essentials*. HSE Books, Sudbury, Suffolk.
A very simple book which provides an interesting comprehensive guidance on working with asbestos and related products in the building maintenance and allied trades. It is not a chemistry book!

O'Riordan, T. (ed.) (1999) *Environmental Science for Environmental Management*. Longman, Harlow.
Covers key issues such as risk assessment, biodiversity, pollution, etc. written by experts in their respective fields. The sections on risk assessment are easy to read and understand. The level is aimed at first- and second-year undergraduates on environment-related degree programmes.

7 An introduction to the lithosphere

- The main rock forming minerals and the importance of the silicate structure
- Igneous rocks, their formation and composition
- Sedimentary rocks and the role of weathering
- Formation and types of metamorphic rocks
- Ore formation
- Plate tectonics and the formation and destruction of crustal material
- Ionic crystals, lattice energy and solubility
- The standard electrode potential
- The Nernst equation and its use
- The Pourbaix diagram
- Soil formation, cation exchange capacity and fertility

Introduction

Minerals are naturally occurring inorganic materials. They have crystalline structures which are not always externally evident, but their regular internal structure and constancy of interfacial angles can be shown using X-ray crystallography techniques. Some minerals show polymorphism, that is, they have the same chemical formula but different crystalline structures, e.g. silicon dioxide, SiO_2, can exist as quartz or cristobalite, and carbon can occur as graphite (Figure 7.1) and diamond. Minerals can be elements, e.g. gold, Au, but most are found as chemical compounds. Minerals have either a fixed chemical composition or chemical formulae which vary between limits, e.g. olivine varies from Fe_2SiO_4 via $(Fe,Mg)_2SiO_4$ to Mg_2SiO_4.

Rocks are a heterogeneous mixture of two or more minerals, e.g. the rock granite is a mixture of the minerals feldspar, mica and quartz (Figure 7.2). Some important rocks consist of a high proportion of a single mineral, e.g. limestone consists of more than 50 per cent calcium carbonate. Rocks cannot normally be represented by a simple chemical formula. The most important rock forming minerals are the silicates and the carbonates, but many others also contribute to the formation of rocks, e.g. galena, PbS and the chlorides such as halite, NaCl.

The main rock forming minerals

Table 7.1 shows the Earth's crustal abundance of the elements. Since oxygen and silicon are by far the most abundant elements, most rock forming minerals are made of silicates.

The silicate minerals are based on the tetrahedral silicate unit SiO_4^{4-}. This structure is often represented by a simple tetrahedron where the vertexes represent the positions

(a)

(b)

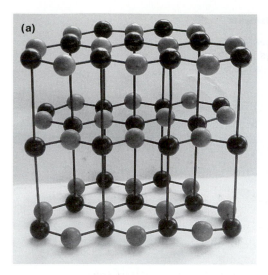

Figure 7.1 (a) The graphite structure (note the layers of carbon atoms). (b) Graphite – the pencil maker.

Figure 7.2 The components of granite: feldspar (top left), quartz (top right) and mica (middle bottom).

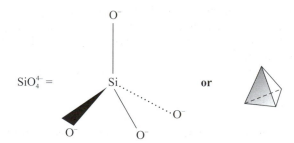

$SiO_4^{4-} =$

or

Figure 7.3 The representation of the tetrahedral silicate group.

Table 7.1 *Crustal abundance of the elements*

Element	Symbol	Percentage by mass (rounded to nearest %)
Oxygen	O	47
Silicon	Si	28
Aluminium	Al	8
Iron	Fe	5
Calcium	Ca	4
Sodium	Na	3
Potassium	K	3
Magnesium	Mg	2
Others		2

Table 7.2 *Common rock forming minerals*

Silicate mineral	Structure	Rock type occurrence	Common colour	Comments
Olivines	Independent SiO_4^{4-} units	Igneous, metamorphic, sedimentary	Pale green (Mg rich) to yellow/dark green (Fe rich)	Have high crystallisation points, melting points and densities Examples: Fosterite Mg_2SiO_4 and Fayalite Fe_2SiO_4
Pyroxenes	Single chains of indefinite length	Igneous	Dark green to black	Show concoidal fracture and glassy lustre Can be transparent or translucent. Augite is the most common
Amphiboles	Pairs of chains of indefinite length linked by Si—O—Si bonds	Igneous, metamorphic	Dark green to black	Very hard with prominent cleavage directions. Hornblende is the most common
Micas	Sheets of SiO_4^{4-} units	Igneous, sedimentary, metamorphic	Brown (Fe rich) for biotite mica to white (Fe poor) for Muscovite mica	Very pronounced single cleavage direction
Feldspars	Complex three-dimensional structures	Igneous, sedimentary, metamorphic	White Pink	Potassium orthoclase Calcium plagioclase
Quartz	Three-dimensional array of SiO_4^{4-} units	Igneous, sedimentary, metamorphic	Colourless	

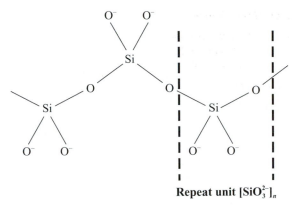

Repeat unit $[SiO_3^{2-}]_n$

Figure 7.4 The pyroxene chain of silicate tetrahedra.
(*Note*: Each silicon atom shares two oxygens with its neighbours.)

of the oxygen atoms (Figure 7.3). The charges on this silicate anion are balanced by appropriate cations. The silicon atom in this tetrahedron can be replaced by the cations Al^{3+}, Fe^{3+} or Mg^{2+}. The tetrahedra can be linked in chains, rings, multiple chains and ribbons, sheets or in more complex three-dimensional structures. This is because of the sharing of one or more oxygen atoms between silicon atoms to form strong Si–O single bonds (average bond enthalpy 368 kJ mol^{-1}). There are six main sub-groups of silicate minerals according to the number of joined silicate units: the olivines, pyroxenes, amphiboles, micas, feldspars and quartz. Some of their properties are shown in Table 7.2. As more and more silicate units combine, more open structures are formed and so larger cations such as K^+ and Ca^{2+} can fit within the structures. If a structure is open enough, the hydroxide ion OH$^-$ can fit into the gaps to give hydrated minerals.

The structure of the olivines is composed of independent tetrahedral units, SiO_4^{4-}. Much more commonly, silicate units are joined through the sharing of adjacent oxygen atoms. The group of minerals that form single chains of indefinite length by this method are the pyroxenes. Here one silicate unit shares two oxygen atoms (Figure 7.4) and thus the silicon to oxygen ratio is 1:3.

The pyroxenes are related to the third group of minerals called the amphiboles in that the chains are linked in pairs via Si—O—Si bonds. Thus some silicon atoms share two oxygen atoms whilst some share three. Their Si:O ratio is 4:11, forming a building block of $[Si_4O_{11}^{6-}]$ (Figure 7.5). The charges on these units are counter-balanced by Ca^{2+}, Na^+, Mg^{2+} or Fe^{2+} ions. Hydroxide ions OH$^-$ can fit into the structure because of the large gaps between the chains and the amphiboles are thus hydrated.

Micas are formed from sheets of silicate units where each silicon atom shares

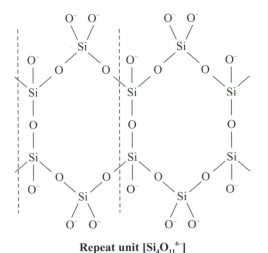

Repeat unit $[Si_4O_{11}^{6-}]$

Figure 7.5 The double chain structure of the amphiboles. (*Note*: Some silicon atoms share two oxygen atoms, some share three.)

Figure 7.6 Two-dimensional sheet structure – the basic structure of talc and the micas. (*Note*: Each silicon atom shares three oxygen atoms.)

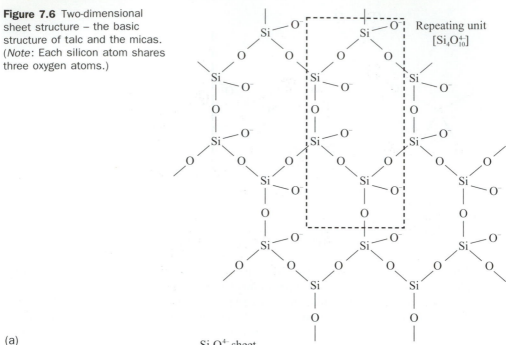

Repeating unit $[Si_4O_{10}^{4+}]$

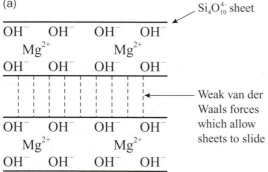

(a)

$Si_4O_{10}^{4+}$ sheet

Weak van der Waals forces which allow sheets to slide

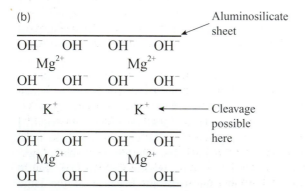

(b)

Aluminosilicate sheet

Cleavage possible here

In mica some Al^{3+} replaces Si^{4+}, the extra negative charge being counterbalanced by K^+

Figure 7.7 (a) Talc. (b) Mica.

three oxygen atoms to form rings (Figure 7.6). Biotite mica is a hydrated aluminium silicate containing K^+, Mg^{2+} and Fe^{2+} ions. In micas some Al^{3+} ions replace silicon atoms and therefore extra negative charge is counterbalanced by K^+ ions inserted between the sheets (Figure 7.7).

The feldspars are the most abundant rock forming minerals. These minerals have complex three-dimensional structures which will not be described here. In the silicate units from which they are made each oxygen atom is shared with two other tetrahedra. One in four silicon atoms are replaced by Al^{3+} ions and therefore the resultant negative charge is counter-balanced again by K^+, Na^+ or Ca^{2+} ions.

Quartz is formed from a three-dimensional array of silicate units in which each oxygen atom is shared with another silicon atom in an adjacent tetrahedron. Thus the chemical formula for quartz is SiO_2 (Figure 7.8).

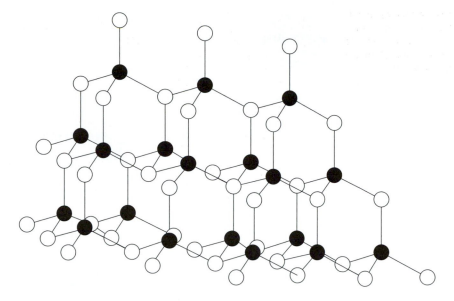

Figure 7.8 The silica structure.

Igneous rocks

Igneous rocks are of two main types, extrusive rocks and intrusive rocks. These rocks form about 95 per cent of the lithosphere and about 25 per cent of the rock that is exposed on the Earth's surface. All other rocks originate from igneous rocks (Figure 7.9).

Extrusive rocks are formed by the emission and cooling of lava (volcanic activity) on the Earth's surface. Crystallisation occurs quickly resulting in fine-grained rocks, which lie on top of older rocks. Cooling can occur so rapidly that a glassy material can be formed called obsidian (Figure 7.10). The appearance of extrusive rocks depends upon how they are formed, e.g. basalt resulting from a volcanic flow of lava is a dark, dense material, whilst pumice formed by lava being thrown into the air is a dark, very light material with many air bubbles (Figure 7.11). Basalt and pumice have very different physical properties but are of the same chemical composition. Pumice is termed a pyro-clastic rock after the manner in which it was formed. Other common extrusive rocks are andesite and rhyollite.

Intrusive rocks are formed by the cooling and solidification of molten magma which has penetrated into older rocks deep inside the Earth. An intrusion will cool slowly and remain hot for a very long time because it is well insulated by the surrounding colder rocks. At its centre a coarse-grained rock containing large crystals will be formed, whilst at its edges, because of faster cooling, a fine-grained rock will result. Intrusive rocks may have large crystals (called phenocrysts) embedded in a fine matrix (Figure 7.12). If there is a large number of phenocrysts then the rock is called a porphyry. Examples of intrusive igneous rocks are granite, gabbro and diorite.

The common igneous rocks can be identified by their mineral composition. The minerals used are quartz, feldspar (both potassium feldspar and plagioclase), pyroxene, olivine, biotite and muscovite micas, and the amphiboles (Table 7.2). For example, basalt contains mainly olivines and the plagioclase feldspars but no quartz. This rock is

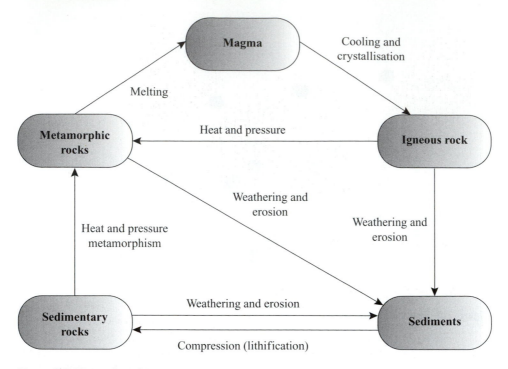

Figure 7.9 The rock cycle.

Figure 7.10 Obsidian.   **Figure 7.11** Pumice and basalt.

dominated by dark-coloured minerals due to their iron and magnesium content and therefore appears dark green or black.

The composition of rocks is often quoted in terms of the percentage amount by mass of silica, SiO_2, that they contain. This does not mean, for example, that, if a rock composition is given as 70 per cent SiO_2, then it is 70 per cent free quartz, because much of the SiO_2 is combined in the form of silicate salts. A rock containing more than 65 per cent silica is termed a felsic (feldspars and silica) rock whilst those which contain between 45 and 55 per cent are mafic (magnesium and ferric) rocks.

Figure 7.12 Various types of granite.

Sedimentary rocks and weathering processes

Sedimentary rocks are formed by the cementing together of particles which have been produced by (i) the weathering of existing rocks; (ii) precipitation from solution; or (iii) biological means. These rocks represent 70–75 per cent of the total surface of the Earth but only about 5 per cent of its crustal volume. The formation of sedimentary rocks is essentially the deposition of particles in horizontal layers (sediments) on the Earth's surface. Those particles that are at the bottom of the deposits become compacted and subject to increased temperatures and pressures. This eventually leads to the formation of sedimentary rocks.

All existing rocks whether igneous, sedimentary or metamorphic undergo mechanical/ physical and chemical weathering. Mechanical weathering is the breaking apart of rocks and involves the freezing and thawing of water trapped in cracks of rocks, and the action of plant roots when these enter a crack and exert pressure. Rocks formed from mineral particles (clasts) of pre-existing rocks that have been mechanically transported (by wind and water) to another place of deposition are called **clastic** rocks. The particle size can range from fractions of a micrometre to large boulders several metres in diameter. Because of their weight, larger particles are deposited close to their origin. Finer, lighter particles may be deposited at great distances from their source. Particles may also undergo many cycles of erosion and deposition before they ultimately become a sedimentary rock. The amounts of solids and solutes carried by moving water and ice are referred to as river loads and glacier loads respectively. In the course of transportation by a stream or river, solids become rounded and sorted into sizes and laid down in river beds, lakes or in the sea.

The three main types of clastic rocks are shale, sandstones and conglomerates (Figures 7.13, 7.14 and 7.15). Shale is a fine-grained rock that contains substantial amounts of clay minerals and silt- and clay-sized particles. Shales are formed from mud and silt deposited at the bottoms of lakes, the sea, flood plains and in low energy tidal flats. Shales formed in a marine environment are often mixed with other deposits such as calcium carbonate. Siltstones are of a coarser nature than shales and are formed mostly in flood plains and deltas. Sandstones contain predominantly well-rounded quartz particles (sand) that have been transported by wind and water over large distances. The deposits of sand that form do so in moderately energetic environments, and the layering that may be present within the sandstone that is subsequently formed often gives evidence of the type

Figure 7.13 Conglomerate: pudding stone.

Figure 7.14 Sandstones.

Figure 7.15 Shale.

of environment in which the deposition has occurred. Conglomerates are coarse-grained rocks (some resembling coarse concrete!) formed from rounded particles of gravel. Breccias which are very similar to conglomerates are formed from angular particles of gravel. Many of the particles in conglomerates and breccias exceed 2 mm in diameter and may include pebble, cobble and boulder-sized particles. Conglomerates and breccias are not always deposited in water environments. The sediments, which lead to the formation of conglomerates and breccias are deposited from high energy environments, which are able to carry large pebbles and other big objects.

If pyroclastic rocks are laid down in beds, they give rise to sedimentary rocks. Various-sized particles are emitted from a volcano, i.e. dust, ash, glassy material and pieces of rock, which may be spread out over a very large area. There are thus two types of pyroclastic sedimentary rocks, volcanic tuff, which contains small particles, and volcanic breccia, containing much larger particles.

Chemical weathering is more complex and can occur by three main processes. The first is by simple solution, e.g. both salt, NaCl, and silica, SiO_2, dissolve in water with the latter being very much less soluble than the former. The second process involves the reactions that can occur between various minerals and the acidic solution formed when carbon dioxide reacts with water,

i.e. $CO_2(g) + 2H_2O(l) \rightleftharpoons H_3O^+(aq) + HCO_3^-(aq)$

For example, this solution reacts with calcium carbonate to produce calcium hydrogen-carbonate which is 30 times more soluble in water,

$$CaCO_3(s) + CO_2(g) + 2H_2O(l) \longrightarrow Ca^{2+}(aq) + 2HCO_3^-(aq)$$

The third process is via oxidation, for example,

$$Fe_2SiO_4(s) \xrightarrow{O_2/H_2O} FeO(OH)(s) + SiO_2(s)$$
$$\text{olivine} \qquad\qquad \text{goethite}$$

The rate at which weathering occurs depends upon the type of mineral involved. Iron oxides, aluminium oxide and silica are the slowest to weather, followed by the clays, micas, potassium feldspars, sodium and calcium feldspars, hornblende, with augite and the olivines being the most rapid.

Rocks which result from chemical precipitation of minerals, the evaporation of water containing dissolved salts, or the conversion of organic matter to rock (coal) are sedimentary rocks.

If seawater becomes saturated with calcium carbonate, upon the cooling of the water it will precipitate out of solution and form deposits on the seabed. Deposition will vary according to how much calcium carbonate is present in the water and so beds of varying thickness will be formed. Warm, shallow seas in particular favour the formation of such chemical precipitates. Chemical precipitation can also occur in lakes and around hot springs. Carbonate rocks were formed not only by the chemical precipitation of calcium carbonate from water but also from the build-up of calcareous shells on the bottom of the seabed, and from the lithification of coral reefs.

Two types of carbonate-based sedimentary rocks can be found: the limestones and the dolostones. To be classified as a limestone, the sedimentary rock must contain at least 50 per cent calcium carbonate by mass. Limestone also contains clay, iron oxide, quartz and other rock fragments. Depending upon the impurities present, limestone takes on a variety of colours ranging from a light yellow to black. How the limestone has been formed will dictate its texture. For example, some limestone will contain fossils whilst others will have a very dense texture due to the close packing of microscopic sized crystals. Chalk is a particularly fine-textured limestone made up of the shells of micro-organisms.

Dolostone differs from limestone in that it is composed mainly of a double salt calcium magnesium carbonate, $CaCO_3.MgCO_3$. If seawater is saturated with Mg^{2+} as well as with Ca^{2+} ions, then magnesium will replace calcium to form the double salt.

Some sedimentary rocks contain a mineral that has grown during or after the deposition of the original rock, and are called authigenic rocks. Evaporites, phosphate rock, iron ore sediments and chert are all examples of authigenic rocks. Chert is formed from silica that was dissolved in fluids that found their way into the pores and spaces inside a sediment. Here the silica precipitated as chert nodules. If water containing dissolved materials is trapped in a lake or sea basin and has a restricted further supply of water, then as evaporation takes place minerals will be concentrated and precipitated from the water. The most important deposits formed in this way are the sulphates, gypsum and anhydrite, and the halides, particularly rock salt. Apatite is an example of a sedimentary phosphate rock that has been formed from the deposition of pieces of bone, shells, etc. Important sedimentary deposits exist, which contain iron compounds such as pyrite, siderite, hematite and limonite.

The final types of sedimentary rocks are those based on carbon. These were formed by the anaerobic decomposition of dead plants originating in freshwater swamps. Peat was initially formed but lithification processes led to the formation of coal. Different types of coal, e.g. lignite, bituminous coal and anthracite, were formed according to how long the material remained buried, the depth of burial, the temperature and the types of chemical reactions taking place.

Metamorphic rocks

Metamorphic rocks are those in which re-crystallisation has occurred as a result of increasing pressures and/or temperatures. The minerals contained in existing igneous, sedimentary or indeed metamorphic rocks can be formed into a new group of minerals that are more thermodynamically stable under the new prevailing conditions. Pressure and temperature changes can also change the shape, size and alignments of crystals. The most common types of metamorphic rocks are derived from sedimentary rocks such as limestone, shale and sandstones, and from various intrusive and extrusive igneous rocks. For example, limestone has been metamorphosed to marble, shale to slate, and sandstone to quartzite.

Metamorphic rocks can be recognised by the presence of interlocking coarse or fine grains. These rocks may show a preferred orientation of individual particles or parallel layers or bands of materials – this is termed foliation or banding. Rocks that show this kind of preferred orientation will tend to break along parallel lines. Unfoliated metamorphic rock examples are the marbles, quartzite, hornfels and anthracite. Examples of foliated metamorphic rocks are the slates, phyllites, schists and the gneisses. The size of the crystals present often gives an indication of the type of conditions under which the rock has been metamorphosed. Slate, for example, is fine grained and was formed under low pressure conditions, whilst schist is medium grained and was formed under moderately high temperatures and pressures. The mildest of conditions gives rise to a

Figure 7.16 Metamorphised granite.

Figure 7.17 Gneiss.

flaky rock with perceptible laminations; more extreme conditions are indicated by a coarser grained rock; and extreme conditions give rise to rocks such as gneiss that closely resemble the granites (Figures 7.16 and 7.17).

The formation, structure and properties of soil

Soil originates from the chemical and physical weathering of rocks. When, for example, the igneous rock granite is exposed to the elements, it will initially break down to form the parent material from which a soil might be formed. The loosely packed particles, or **regolith**, that are first formed may lie on the surface of the granite now referred to as the bedrock. This parent material may undergo further decomposition where it is, according to factors such as the prevailing weather conditions and the mineral content of the regolith. The regolith may also be transported and deposited elsewhere by erosion processes before it starts to undergo its transition to soil.

Mechanical weathering results in big rocks being broken down into smaller rocks with a corresponding increase in the surface area exposed to weathering. The smaller the particle size, the faster is the rate of any subsequent chemical reactions.

Chemical weathering is the result of chemical reactions of the minerals contained within a rock with water, with dissolved chemicals in water and with the gases of the air. Granite is a heterogeneous, hard material formed from interlocking crystals of feldspar, mica and quartz. The feldspar could be potassium aluminosilicate, orthoclase, $KAlSi_3O_8$, which will undergo chemical reaction with acidic waters. For example, rain-water reacts with the orthoclase to yield kaolinite clay, releasing potassium ion which may become available to plants,

$$2 KAlSi_3O_8(s) + 2H_3O^+(aq) + H_2O(l) \longrightarrow$$
$$Al_2Si_2O_5(OH)_4(s) + 4SiO_2(aq) + 2 K^+(aq)$$
$$\text{kaolite}$$

The formation of the clay weakens the interlocked structure of the granite, which thus starts its disintegration. The more acidic the water, then the faster the above reaction will occur.

Different minerals have different resistances towards weathering, e.g. quartz is very stable, whereas olivine readily undergoes chemical reaction. Calcium carbonate-based rocks such as limestone and dolomite readily react with acidic waters. For those common minerals not containing iron and magnesium, as the number of oxygen atoms *shared* with silicon atoms increases from zero to four, then there is a corresponding increase in stability towards weathering. In the case of the iron–magnesium minerals there is an increase in stability as the silicate structures change from single units, to single chains, to double chains, to sheets. As these silicates undergo weathering, insoluble iron oxides and iron hydroxides are produced which give the dark brown, reds and yellow colours to many soils.

Most minerals in parent material readily undergo chemical reactions; therefore organic activity and climate usually have the greatest effects on the ultimate nature of a soil. Since most weathering chemical reactions involve water, then the wetter the climate the faster will be the weathering processes. An increase in temperature also increases the rate of a chemical reaction and therefore a hot, humid climate will result in fast weathering.

Living organisms can have effects on both chemical and mechanical weathering. Plant roots can cause cracks in rocks to be widened and the rocks to break apart. Living organisms also produce chemicals that can react with minerals and help to dissolve and

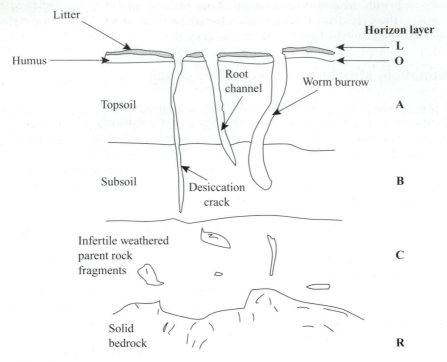

Figure 7.18 A soil profile.

break them down. Again, the warmer and wetter the climate, the more vigorous will be the effects of living organisms.

Mechanical weathering thus becomes dominant only in those areas that have a cold or a dry climate.

When the decaying remains of dead organic matter are added to the results of mechanical and chemical weathering, then the bedrock eventually becomes covered with soil. The decaying matter is called **humus** and the associated processes **humification**. Humus is the consequence of heterotrophic organisms decomposing the organic matter.

If a cross-section of a soil is taken down to its solid bedrock, then distinct layers of varying thickness, or the soil profile, will be seen. Figure 7.18 shows one possible soil profile. A soil profile contains a series of **horizons**, which are formed by the way in which water passes through the soil redistributing its minerals and organic matter. Thus each soil horizon has a different colour, chemical composition and physical characteristics. Profiles may show fewer and less distinct layers, or more sub-horizons, depending on how the soil developed and its age. Generally there are three basic horizons, A, B and C. The A horizon contains the most extensively weathered material by virtue of its proximity to the surface. It also contains most of the organic matter. As water percolates through the A horizon, it will dissolve soluble inorganic materials and carry them to the lower regions. This is called **leaching**, the extent of which depends upon the solubility of the minerals and the pH. **Cheluviation** can also occur, which is the dissolution of organic matter via the formation of chelated organic metal complexes. This again depends on the soil pH. Leaching and cheluviation can denude the top soil of nutrients leaving it both acidic and infertile.

The B horizon may accumulate much of the dissolved material, particularly in drier climates. **Precipitation** of matter may also occur lower down in this soil horizon, e.g. iron oxides and hydroxides can be precipitated which can lead to rock-hard layers called 'iron-pans'. There is only a small amount of organic matter to be found, which is not as well mixed as in the case of the A horizon.

The C horizon consists of little else than very coarse broken-up lumps of bedrock. The R horizon is the bedrock itself.

If the A horizon contains a topsoil, which is particularly rich in organic matter, then an O horizon may be identified, or if there is a thick litter layer an L horizon. Other horizons can also exist which show a gradation of properties between the A and B or the B and C horizons.

It is possible for water to return dissolved salts back to the surface via **capillary action**, where evaporation can cause the deposition of the salts as a hard crust.

Soil is a complex heterogeneous mixture of inorganic and organic matter. The particles found in soil range in size from colloidal (1×10^{-6} to 1×10^{-4} cm) to small pebbles (1–2 cm diameter). Soil also contains water, dissolved materials, gases and both living and dead organisms.

The three main solid inorganic components of soil are sand, silt and clay. The relative amounts of each of these components present in a soil dictates its texture.

Sand, silt and clay are differentiated by their particle sizes. Sand, which is mainly quartz, has a particle size ranging from 0.05 to 2.00 mm in diameter. This means that the spaces between particles are relatively large and thus sandy soils are usually lacking in structure, and are light and easily worked. However, such soils will allow the easy passage of water and therefore drainage. The leaching of valuable nutrients that can pollute ground and surface waters is also often a problem.

Silt contains particle sizes, which range from 0.002 to 0.05 mm. These particles are again mainly quartz, but silicates are also present which have been derived from both primary and secondary weathered materials. Silty soils lacking in organic matter have a weak structure. They are prone to water logging which restricts seedling growth and increases risk of erosion.

Clay contains particles that are less than 0.002 mm in diameter. Clay particles are mainly secondary silicates and aluminosilicates. Because of their small size the spaces between the particles are small. This causes a clay soil to exhibit poor drainage. If a dry clay soil is subjected to a sudden downpour the water will flow across the surface and flash flooding can be the result. In very dry conditions excessive shrinkage may occur which can lead to subsidence. Many clay soils have good buffering ability, and some are well structured.

The clays have layered structures and are labelled 1:1, 2:1 and 2:2 according to the arrangement of two different sheet structures. One of these sheets consists of tetrahedral 'SiO_4' units, where each Si atom shares *one* of its O atoms with the second sheet. The second sheet consists of aluminium in octahedral co-ordination represented as 'AlO_6'. A schematic diagram of the structure of a 1:1 clay is shown in Figure 7.19. The pairs of sheets are held by hydrogen bonding which stops the insertion of both water molecules and ions between the sheets. Thus a 1:1 clay will not expand when wet.

2:1 clays have three sheets, one 'AlO_6' sandwiched between two 'SiO_4' sheets (Figure 7.20). In muscovite mica, a quarter of the Si^{4+} are replaced by Al^{3+}, and in vermiculite some of the Al^{3+} are replaced by Mg^{2+} and Fe^{2+}. This leaves an overall negative charge on the layers. Hence, hydrated ions such as $K^+(aq)$, $Ca^{2+}(aq)$ and $Mg^{2+}(aq)$ can fit in between these layers helping to hold them together.

The 2:2 clays, e.g. chlorite, have a similar structure as the 2:1 clays but have an additional discrete layer of 'AlO_6'. In this latter layer, some of the Al^{3+} are replaced by Mg^{2+} and Fe^{2+}.

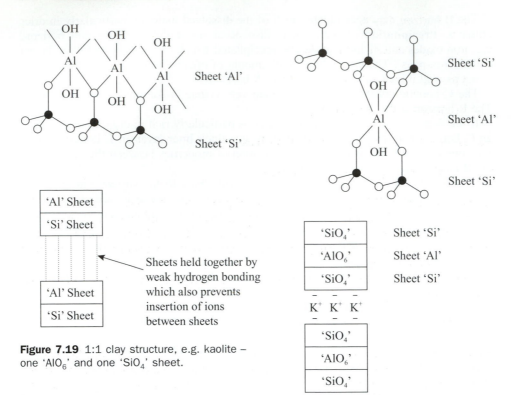

Figure 7.19 1:1 clay structure, e.g. kaolite – one 'AlO$_6$' and one 'SiO$_4$' sheet.

Figure 7.20 2:1 clay structure, e.g. illites – three sheets, one 'AlO$_6$' between two 'SiO$_4$'.

It is the presence of the negative charges on the large surface areas of the clays that gives them their excellent ability to retain and exchange ions with the surroundings, i.e. their **ion exchange capacity**. This ability is vitally important in supplying nutrients for plants and when pollution of the soil occurs. Organic matter is also a particularly good ion exchanger.

Both dead and living organisms also have important roles in soil development. Plants aid the passage of water through soil and help to retain mineral nutrients by taking them up through their roots. In temperate regions, small animals such as earthworms help to form an almost stone-free layer at the surface of soil. Humus is vital to the fertility of a soil in that it influences how much water a soil will hold, its ion exchange capacity and the retention of metal ions. During humification organic acids which act as cation exchangers, are released, along with other organic chemicals, e.g. proteins, that are capable of forming chelates with metal ions, thus preventing them from being leached out.

If Al^{3+} has replaced Si^{4+} in a clay, then it would be an cation exchanger since it carries an overall negative charge. At high pH, any OH groups present on the clays may lose their protons or be exchanged for metal ions. The formation of a chemical bond between a molecule, atom or ion and the surface of a material is termed **adsorption**. If an adsorbed species leaves the surface of a material, then it is termed **desorption**. Hence, cation exchange involves the adsorption and desorption of different chemical species.

If M represents a metal ion and C a clay, then the following series of equilibria show how clays and organic acids act as cation exchangers,

$$M_1^+ + C^- \rightleftharpoons C-M_1$$

$$M_1^+ + C-M_2 \rightleftharpoons C-M_1 + M_2^+$$

$$M^+ + C-OH \rightleftharpoons C-O-M + H^+$$

$$RCOOH + M^+ \rightleftharpoons RCOOM + H^+$$

For example, in the case of a nutrient such as K^+,

$$C-M + K^+(aq) \rightleftharpoons C-K + M^+(aq)$$

In this equilibrium, if $K^+(aq)$ was taken up by plants then the position of the equilibrium would move to the left and the soil would act as a source of potassium. This source is not endless, and so sooner or later a fertiliser would be needed to replenish the K^+. Additionally, a pollutant that suppresses or competes with the adsorption/desorption of a nutrient will be a problem.

Anion exchange is less important in soils than cation exchange. Typically, anion exchange can be represented by,

$$C-OH + X^- \rightleftharpoons C-X + OH^-$$

Nitrate (NO_3^-) is an anion that is weakly adsorbed by a clay and therefore can easily be washed out. Soil is not a good source of nitrate for plants. On the other hand, phosphate (PO_4^{3-}) can be so strongly adsorbed that it can become unavailable as a plant nutrient. The ions on the exchange sites at any time depend on their relative abundances and sizes. The order of attraction is known and represented in the Lyotropic Series that takes into account size and charge, i.e. $Al^{3+} > Ca^{2+} > Mg^{2+} > K^+ = Na^+ = H^+$.

Ion-exchange capacity is the sum of the exchangeable adsorbed ions expressed as milliequivalents per 100 g of soil of oven-dried soil at a specified pH. The SI unit is centimoles of charge per kg at a specified pH, but soil scientists much prefer the former units. The milliequivalent is 1 millimole of H^+ ions or the amount of any other ion that will displace it – it takes into account the charges on an ion, e.g. 1 milliequivalent = 1 millimole of H^+ = 1 millimole of K^+ = ½ millimole of Mg^{2+}. **Cation exchange capacity (CEC)** is important because cations are essential plant nutrients, e.g. Ca^{2+}, Mg^{2+}, K^+, NH_4^+, Fe^{2+}, etc. As indicated above, whilst they are held on the exchange sites, they cannot be leached from the soil profile. The CEC is a good guide to a soil's potential cation supplying powers and therefore a measure of soil fertility. However, because CEC is pH dependent this must be quoted when determining the CEC value, and when soils are compared it must be done at the same pH value. The values of CEC range from 5–40 meq/100 g dry soil for 'normal' mineral soils, though extremes of <1 and >100 have been recorded.

The type of clay present in a soil greatly affects the CEC value. For example, the 1:1 clay kaolinite can show relatively large changes in CEC values because of its dependency on pH, whereas many of the 2:1 clays have a permanent CEC value, which is independent of pH.

The CEC gives details on the capacity of the soil to hold useful cations. It does not, however, give information on how much of that capacity is actually being used to hold useful cations. This data is given by the percentage base saturation (%BS), which is defined as the percentage of exchange sites that are occupied by basic cations. These basic cations are principally Ca^{2+}, Mg^{2+}, K^+ and Na^+.

The pH of a soil influences that soil's fertility and is affected by what cations are adsorbed,

$$\text{e.g. } C\text{—}Na + H_2O(l) \rightleftharpoons C\text{—}H + Na^+(aq) + OH^-(aq)$$

In the above equilibrium the release of OH^- ions will cause the soil pH to rise.

When aluminium is adsorbed on a clay, then the following equilibrium will result,

$$C\text{—}Al + 4H_2O(l) \rightleftharpoons C\text{—}3H + Al(OH)_4^-(aq) + H^+(aq)$$

Here, the pH will decrease.

Soil pH normally ranges between 3 and 9. If large amounts of humus are present, then the pH of the soil can be quite low since humus formation releases carboxylic acids.

The availability of various cations depends upon the pH because different salts of the metals will be formed which have a different solubility in water. Calcium and magnesium will become unavailable at high pH values because of the formation of insoluble carbonates. Iron and aluminium form insoluble hydroxides at pH values of 6 and greater. Phosphorus is only available over a narrow range of pH from 6 to 7. Phosphorus is in its most soluble form as the $H_2PO_4^-$ ion in the pH range 2 to 7. Below pH 6.0 iron and aluminium form insoluble phosphates, whilst above pH 7.0 insoluble calcium phosphate is formed. Thus, although soil has a natural buffering effect, to ensure the availability of nutrients, the pH of a soil must be carefully regulated.

Soil pH can be lowered by the addition of a sulphate or sulphur since these will be oxidised by bacteria to sulphuric acid.

The acidity of a soil can be reduced by adding lime,

$$C\text{—}2H + CaCO_3(s) \rightleftharpoons C\text{—}Ca + H_2O(l) + CO_2(g)$$

Then, $C\text{—}Ca + 2 H_2O(l) \rightleftharpoons C\text{—}2H + Ca^{2+}(aq) + 2OH^-(aq)$

Ore formation

Ore formation can be divided into two main sets of processes: those that occur internally within the Earth and those that occur on its surface.

Internal processes can be further subdivided into magmatic segregation, contact mesomatic, pegmatite and hydrothermal processes. Magmatic segregation occurs when a hot magma penetrates or intrudes into a cooler rock. The magma, which is a very complex mixture, cools down and crystallisation of various minerals takes place. Which mineral solidifies first depends upon its solubility in the magma, and its melting point/crystallisation temperature. Denser crystals will sink and accumulate at the bottom of the intrusion. Chromite (Fe_2CrO_4), magnetite (Fe_3O_4) and ilmenite ($FeTiO_3$) are examples of minerals that have been formed in this way – they have high melting points and low solubilities in magma. The type of bonding can also influence the separation process. For example, sulphides tend to be covalent in nature, whereas silicates tend to be more ionic and therefore will not mix. Because of this, the denser sulphide minerals coalesce, sink and collect at the base of an intrusion, e.g. pyrite (FeS_2) and covellite (CuS).

Contact metasomatic processes occur when the magma in an intrusion bakes the cooler surrounding rock causing it to undergo change or metamorphism. Gases and fluids from magma may chemically react with the contact rock and thereby change its composition. If valuable elements are added to the contact rock, then an ore deposit may be formed.

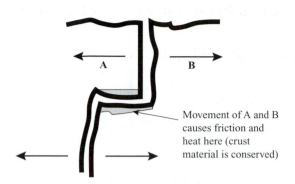

Figure 7.21 Transform fault boundary movement (causes earthquakes).

Movement of A and B causes friction and heat here (crust material is conserved)

If the intrusive magma contains water, then superheated water or steam may become available for reaction. Limestone in particular is readily dissolved by acidic fluids and often shows signs of contact metamorphism. Mineral deposits formed in this way are usually small in size but the ore is of high grade. The principal ores formed are mainly oxides of iron and some sulphides (copper, lead and zinc).

When the rock granite is formed from cooling magma, the main crystalline minerals produced are quartz (SiO_2) and feldspar ($KAlSi_3O_8$), together with some micas ($K_2O.3Al_2O_3.6SiO_2.2H_2O$). The acidic magma contained water, some of which helped to form the mica. As the granite forms, the remaining magma becomes richer in water, which can result in a separate water-rich silicate melt. This melt will also contain those elements whose ionic radius and charge prevented their inclusion in the crystals that make up granite. If this fluid enters cracks in the solidified granite or in surrounding rocks, it solidifies to form separate veins and pods called pegmatites. Pegmatites contain not only quartz, feldspar and mica but also concentrates of more valuable minerals. The elements lithium, tin and zirconium are concentrated in this way.

Hydrothermal processes result in the most common and important metallic ore deposits. These process involve the dissolving of material from rocks at great depths via mobile, high pressurised, superheated water and steam, plus heat from magma. This hot solution makes its way towards the Earth's surface and, in doing so, undergoes substantial heat and mass transfer to its surroundings. As a result, the deposition of relatively small volumes of concentrated material can occur in rock cavities and cracks with consequent precipitation and crystallisation. Chemical reactions may also occur between the surrounding rocks and the hot solution. Veins or thin sheets are often formed, many of which are barren. However, since crystallisation occurs from the contact rock surface inwards, cavities of well-formed crystals may be found. Sometimes, deposits can be very widely distributed and form no well-defined shape or pattern, making their exploitation difficult. Veins can show zoning. Wolframite ($FeWO_4$), cassiterite (SnO_2) and chalcopyrite ($CuFeS_2$) are minerals deposited at greater depths and higher temperatures. At intermediate depths and temperatures, chalcopyrite, pyrite and sphalerite (ZnS) are deposited. At shallower depths and lower temperatures minerals such as pyrite, argentite (Ag_2S) and stibnite (Sb_2S_3) are formed.

Ore formation at the surface of the Earth occurred under less extreme conditions. The surface processes involve sedimentary, residual and secondary enrichment processes. Sedimentary processes include **evaporite mineral deposits**, **chemical precipitates** and **placer deposition**. Evaporite mineral deposits are formed from shallow coastal lagoons, within inland lakes and within sediments formed upon gently shelving ground near to the sea in arid desert regions. As the water evaporates, all of its components become more concentrated. The least soluble components precipitate first, e.g. calcium carbonate in seawater will precipitate first, followed by calcium sulphate then sodium chloride, and finally by potassium and magnesium salts.

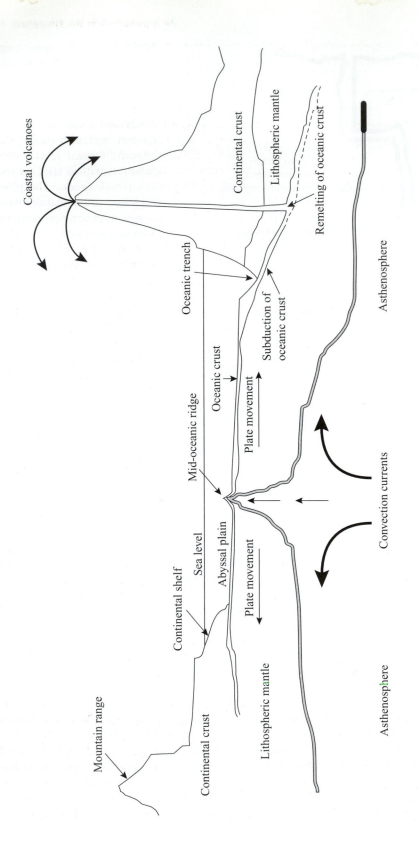

Figure 7.22 A cross-section through a hypothetical ocean. (*Notes*: 1 At the mid-oceanic ridge, new crust is formed and is referred to as a constructive zone. 2 Where the oceanic plate is subducted beneath a continental plate, it is called a destructive zone.)

Box 7.1

Plate tectonics

The lithosphere rides on top of the semi-molten part of the mantle known as the asthenosphere. It is segmented into rigid plates that move a few centimetres each year. The edges of these plates are characterised by seismic and volcanic activity. The movement of these plates is believed to be caused by convection currents. There are three kinds of boundaries between these plates, divergent boundaries where the plates are moving apart, convergent boundaries where plates are undergoing collision and transform fault boundaries where plates are sliding past each other. Figures 7.21 and 7.22 illustrate what can happen at these boundaries.

When plates diverge **rift** or crack-like valleys develop. For example, in the middle of the Atlantic Ocean, a mid-oceanic ridge or a sub-sea chain of mountains has formed. Thus, in these regions new crust material is formed. Where an oceanic plate meets a continental plate, a subduction zone is formed with deep-sea trenches being found offshore, e.g. Peru, South America. Here, the moving floor of the ocean dips down below the continental mass and the crust is reheated and melts. Thus, crust material is being 'destroyed'. In such a region granitic continental crust is formed as the highly fusible granite is distilled off and rises to the surface because it is a lighter material. This granitic material is then extruded as a lava, e.g. the volcanic activity of the Andes, which upon cooling gives andesite rock, or congeals into huge granitic batholiths, e.g. Sierra Nevada, California, US. At transform fault boundaries, crust material is conserved since the plates are sliding past each other, e.g. the San Andreas Fault in California.

Where plates diverge and magma is extruded, then basalt is formed. When plates collide and subduction occurs, melting causes heavier minerals to sink and lighter ones to rise. In this case granite is formed. If two continental masses collide, they stop and buckle giving rise to the metamorphism of rocks, e.g. the Himalayas, Northern India.

Chemical precipitation occurs because of reaction between anions and cations that produce insoluble chemicals. Chemical reaction depends on pH, the solubility of the chemical, oxidising conditions, etc. Most hydrated iron oxide deposits have formed by this method, e.g. the banded iron formation of Lake Superior in the US.

Placer deposits have been formed by mechanical weathering, i.e. they have been moved around by water or wind. The materials must be both hard and resistant to chemical reaction with water to survive weathering of this kind. Their densities are such that gravity enables a separation of lighter particles from heavier ones. Magnetite, chromite and cassiterite deposits can all be formed in this way.

Residual deposits are left over after the removal of soluble minerals by chemical and mechanical weathering. This process happens in hot climates with a moderate to very heavy rainfall. Bauxite ($Al(OH)_3$) and ($Fe(OH)_3$) are both formed as residual deposits.

If, for example, chalcopyrite is spread through a hydrothermal vein, which otherwise contains gangue, then once this ore is exposed to atmospheric weathering, the following oxidation reaction will take place,

$$2CuFeS_2(s) + 17[O] + 6H_2O(l) + CO_2(g) = 2Fe(OH)_3(s) + 2CuSO_4(aq) + 2H_2SO_4(aq) + H_2CO_3(aq)$$

The insoluble iron(III) hydroxide will form a deposit at the surface of the vein, forming a cap or gossan. The soluble copper(II) sulphate will be carried down the vein by percolating groundwater. This solution also contains K^+, Na^+, Ca^{2+} and Mg^{2+} ions. The bulk of the dissolved copper will remain in solution until it reaches the water table, where it meets more chalcopyrite that has not been exposed to an oxidising atmosphere. Here the reaction,

$$CuFeS_2(s) + Cu^{2+}(aq) = Cu_2S(s) + FeS(s)$$

takes place. The mineral chalcocite (Cu_2S) is thus deposited which contains a higher percentage of copper than chalcopyrite. The copper has been concentrated by a process known as secondary enrichment.

Energetics of ionic crystals

Many minerals form inorganic crystalline structures. One important aspect of this formation is the lattice energy of the structure. The internal energy change ΔU_0, when one mole of an ionic crystal is formed from its ions in the gaseous state at 0 K, is termed the **lattice energy** of that crystal. It is equivalent to the heat of formation of the ionic crystal at constant volume and has a negative value, i.e. is exothermic.

For approximate calculations ΔU_0 is replaced by ΔH_L^0, or the standard lattice enthalpy change. The Born–Haber cycle using experimental parameters is often used to calculate the lattice energy of a particular compound (Box 7.2),

$$\text{e.g. } Na^+(g) + Cl^-(g) \longrightarrow NaCl(s) \ \Delta H_L^0 = -787.3 \text{ kJ mol}^{-1}$$

The exothermic nature of the lattice energy is indicative of the energetic stability of the crystalline substance with respect to its gaseous ions. The more negative the value, the more energetically stable will be the crystalline structure.

A theoretical value of the lattice energy can be determined for a perfect crystal structure by assuming the ions are charged spheres and only electrostatic interactions exist between the ions (i.e. perfect ionic bonding exists). The theoretical equation obtained is of the form,

$$\Delta H_L^0 = \frac{-N_A A z_+ z_- e^2}{4\pi\varepsilon_0 r}$$

where
N_A = Avogadro constant
r = sum of ionic radii
z_+ = number of charges on the cation
z_- = number of charges on the anion
e = electron charge
A = Madelung constant
ε_0 = 8.8542×10^{-12} F m^{-1}

The Madelung constant is a number, which is derived by adding together all of the ions to form a particular lattice structure, i.e. it is the sum of a mathematically infinite series representing the effect of adding all the ions together to form the lattice. Its value depends upon the structure of the crystal. Values of this constant for various crystal structures are shown in Table 7.3.

Box 7.2

The Born–Haber Cycle for sodium chloride

The enthalpy of formation of a crystalline material like sodium chloride can be determined by examining the various steps that contribute in making that solid. For example, the reaction,

$$Na(s) + 1/2 \, Cl_2(g) \rightarrow NaCl(s)$$

can also take place in a series of steps:

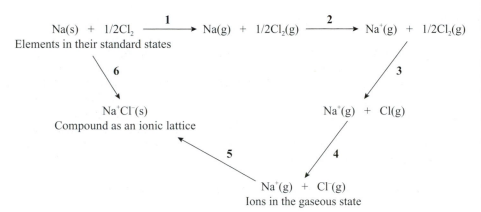

By the First Law of Thermodynamics, the sum of the energy change values in steps 1 to 5 must equal the energy change value associated with step 6. The energy changes for each step are:

Step 1   enthalpy of atomisation of sodium, $\Delta H^0_{sub}(Na) = +107.3 \, \text{kJ mol}^{-1}$

Step 2   first ionisation energy of sodium, $\Delta H^0_i(Na) = +498.3 \, \text{kJ mol}^{-1}$

Step 3   half the enthalpy of dissociation of chlorine, $1/2 \, \Delta H^0(Cl\text{–}Cl) = +121.7 \, \text{kJ mol}^{-1}$

Step 4   the electron affinity of chlorine, $\Delta H^0_{ea}(Cl) = -351.2 \, \text{kJ mol}^{-1}$

Step 5   lattice energy of sodium chloride, $\Delta H^0_L(NaCl) = $ to be determined

Step 6   standard enthalpy of formation of sodium chloride,
$\Delta H^0_f(NaCl) = -411.2 \, \text{kJ mol}^{-1}$.

Thus, if Step 6 = sum of the energy changes of Steps 1 to 5 inclusive, we have

$$-411.2 \, \text{kJ mol}^{-1} = +107.3 \, \text{kJ mol}^{-1} + +498.3 \, \text{kJ mol}^{-1} + +121.7 \, \text{kJ mol}^{-1} +$$
$$-351.2 \, \text{kJ mol}^{-1} + \Delta H^0_L(NaCl)$$

or $\Delta H^0_L(NaCl) = -787.3 \, \text{kJ mol}^{-1}$

Table 7.3 Madelung Constants for crystal structures

Co-ordination number	Structure	Madelung constant
8	Caesium chloride	1.763
6	Sodium chloride	1.748
4	Wurtzite	1.641
4	Zinc blende	1.638
8:4	Calcium fluoride	2.519

Table 7.4 Solubilities and enthalpy changes of solution of some simple inorganic materials

Compound	ΔH^0_{sol}/kJ mol^{-1}	Solubility (g of solute/100 g water @ 18 °C)
NaCl	+ 3.9	36.0
MgSO$_4$	−91.2	33.0
MgCO$_3$	−25.3	0.01
CaCO$_3$	−13.1	0.0014
CaSO$_4$	−18.0	0.21
KNO$_3$	+ 34.9	31.2

Thus using $r = 2.814 \times 10^{-10}$ m for the NaCl structure yields a lattice energy value of −863.0 kJ mol^{-1}. This value would appear to be slightly too large when compared with that obtained using the Born–Haber cycle. The above theoretical equation does indicate that the factors which favour the formation of an ionic solid are:

● a large ionic charge
● a small value for the sum of the ionic radii
● a large value of A
● a small value for the ionisation energy.

Corrections have been made to the theoretical equation, which take into account, for example, the repulsive forces that come into play due to the interaction between electron clouds as the ions get closer together. Such corrections have enabled very close agreement between theory and experiment. This close agreement would suggest that both the oxidation states of the elements involved and the nature of the bonding, i.e. that it is ionic, are correct.

When a solid dissolves in water, heat is either evolved or absorbed. If this heat is measured at 1 atmosphere pressure at 298.15 K for 1 mole of solid, then it is equivalent to the standard enthalpy change of solution, ΔH^0_{sol}. Many minerals are ionic in nature. Some of the more common minerals and their ΔH^0_{sol} values are shown in Table 7.4.

The solution process can be thought of as the breakdown of the ionic lattice into its constituent ions in the aqueous phase (which involves the negative of the value of the lattice energy since it is written in the reverse direction!) and the interaction of the ions with water to form the solution. The energy involved in the latter process is called the enthalpy change of hydration, ΔH^0_h. In terms of the energy changes involved, the solution can be viewed theoretically as shown in Figure 7.23. It can be noted from this diagram that the solution process will be exothermic or endothermic, depending upon the relative sizes of the lattice energy and the enthalpy change of hydration.

Thus for sodium chloride,

$$NaCl(s) \longrightarrow Na^+(g) + Cl^-(g) \quad \Delta H^0_L = -(-787.0 \text{ kJ mol}^{-1}) = +787.3 \text{ kJ mol}^{-1}$$

and $\quad Na^+(g) + Cl^-(g) \longrightarrow Na^+(aq) + Cl^-(aq) \quad \Delta H^0_h = -784.0 \text{ kJ mol}^{-1}$

Since $\quad -\Delta H^0_L = \Delta H^0_h + \Delta H^0_{sol}$

then $\quad \Delta H^0_{sol} = -\Delta H^0_L + \Delta H^0_h$

so that $\quad \Delta H^0_{sol} = (+787.3 \text{ kJ mol}^{-1} + -784.0 \text{ kJ mol}^{-1}) = +3.3 \text{ kJ mol}^{-1}$

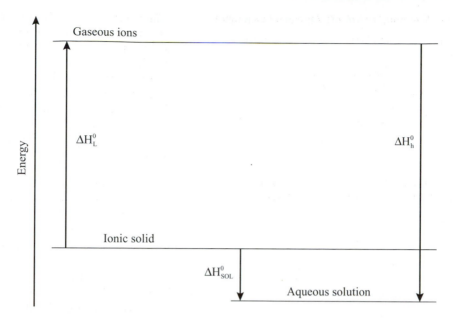

Figure 7.23 The relationship between the enthalpies of solution, hydration and lattice energy.

When the gaseous ions dissolve in the water (solvent), heat is liberated as the ions become hydrated (solvated) and attached to water molecules. Water is a polar molecule and therefore ion–dipole interactions occur between the cations and the negative end of the water molecule, whilst anions become hydrated via hydrogen bonding.

It is evident that the larger and more endothermic ΔH_L^0 is compared with the hydration energy, then the more endothermic will be the value of ΔH_{sol}^0. It would appear that, for a substance having an exothermic value of ΔH_{sol}^0, then ΔH_L^0 is more than compensated for by the ΔH_H^0 value leading to a higher degree of solubility. However, endothermic and exothermic values of ΔH_{sol}^0 are both common – a discussion of energy changes alone does not explain why!

Amongst the more soluble minerals are sodium chloride, found as halite deposits and in rocks generally, and calcium sulphate, found as gypsum and anhydrite. Silica is relatively much less soluble in water but in dissolving forms the equilibrium,

$$SiO_2(s) + 2H_2O(l) \rightleftharpoons H_4SiO_4(s) \qquad\qquad 7.1$$

This equilibrium contains silicic acid, H_4SiO_4, which is a weak acid ($pK_a = 9.8$ at 298.15 K),

$$H_4SiO_4(s) + H_2O(l) \rightleftharpoons H_3O^+(aq) + H_3SiO_4^-(aq) \qquad\qquad 7.2$$

Silicic acid, H_4SiO_4, exists in quantities to a greater extent than is suggested by the above equilibria because it is also formed by the hydrolysis of silicate minerals.

The solubility of a crystalline substance depends upon its lattice energy and on the hydration energy of its ions. Thus a solid with ions that have a low ionic charge and a large ionic radius will be the most soluble. Ions possessing a high charge and small ionic radius may be hydrolysed in water giving an insoluble hydroxide. The interaction of water with an ion of high surface charge density results in the weakening of the OH bond,

e.g.　$Al^{3+}(s)$ — $O\begin{smallmatrix} H \\ \\ H \end{smallmatrix}$ ⇆ $[Al(OH)]^{2+}(aq)$ + $H^+(aq)$ ⟶ $Al(OH)_3(s)$ + $3\,H^+(aq)$

The solubility of many minerals is also pH dependent, e.g. if the pH rises, then $H_3O^+(aq)$ ions in equilibrium equation 7.2 will be removed, causing a shift in the position of that equilibrium to the right and therefore increasing the solubility of silica. The crystalline form of the mineral may also influence its solubility because of a change in lattice energy. The solubilities of the common inorganic salts are shown in Box 7.3.

Standard electrode potentials

The relative abilities of metal and non-metal species to undergo oxidation or reduction can be explained via the concept of electrode potentials.

When a copper rod is placed in copper(II) sulphate solution containing a radioactive isotope of copper as a label in the form of Cu^{2+} ions, then eventually both the rod and the solution would become radioactive. This shows that some of the radioactive ions in the solution have been deposited or exchanged for some of the non-radioactive atoms in the rod. There is a tendency for copper atoms to form copper ions, releasing two electrons per atom, as well as a tendency for copper ions to accept two electrons and become atoms. An equilibrium is established, e.g.

$$Cu^{2+}(aq) + 2e^- \rightleftharpoons Cu(s)$$

The position of the equilibrium depends upon: (i) the concentration of the ions, (ii) the temperature, (iii) the tendency of the ions to gain electrons, and (iv) the tendency of the ions to hydrate.

Box 7.3

Solubility of inorganic compounds in water

1　All common ammonium, potassium and sodium salts are soluble.
2　All common nitrates are soluble.
3　All common metallic chlorides with the exception of those of lead(II), silver and mercury(I) are soluble.
4　All common sulphates with the exception of those of lead(II) and barium are soluble.
5　All carbonates with the exception of ammonium, potassium and sodium are soluble.

Applying the above rules to the reaction between soluble calcium chloride and soluble silver nitrate will lead to the prediction that soluble calcium nitrate and insoluble silver chloride will be formed.

Box 7.4

The meaning of the term activity

When solid silver chloride, AgCl, is stirred in a beaker containing distilled water, an equilibrium is established between undissolved silver chloride and its ions in solution,

$$AgCl(s) \rightleftharpoons Ag^+(aq) + Cl^-(aq)$$

The solubility product is thus,

$$K_{sp} = [Ag^+(aq)] \, [Cl^-(aq)] = 1.8 \times 10^{-10} \text{ mol}^2 \text{ dm}^{-6}$$

If solid sodium chloride, NaCl, is added, then upon dissolving it fully dissociates into its ions, providing the common ion $Cl^-(aq)$. This is the equivalent of adding excess chloride ions to the above equilibrium which moves to the left-hand side in order to maintain the value of K_{sp}. The solubility of AgCl thus decreases. This is termed the **common ion effect**. If, however, magnesium sulphate, $MgSO_4$, is added, then the solubility of AgCl is found to increase. In this case ions have been added that are not common to AgCl. Such an effect is termed the **uncommon ion effect**.

It is clear from the above observations that, whatever ions are present, they interact with each other. Therefore, ionic solutions do not behave in an 'expected' ideal way. For example, a 0.03 mol dm^{-3} solution of sodium phosphate, Na_3PO_4, does not act as if it were of that concentration. Ionic interactions are found to depend upon both the concentration of the ions and the charge that they carry. For a solute in solution, in order to indicate the level of deviation from ideal behaviour, the **activity coefficient**, γ, is used. By definition, the activity, a, is given by,

$$a = \gamma c / c^0$$

where c = concentration of the solute in mol dm^{-3}
c^0 = the standard concentration of 1 mol dm^{-3}
γ = the activity coefficient

The activity coefficient has a value less than or equal to 1. The closer it is to 1, the less the solution deviates from ideal behaviour, and the closer 'a' becomes in value to 'c'. Note that, by introducing the standard concentration, the activity is unitless. If the solution is an electrolyte, then the deviations can be attributed to the ions produced by the dissociation of the solute. It is not experimentally possible to determine individual ionic activity coefficients, γ_+ and γ_-, because it is not possible in solution to isolate the anions from the cations. The mean ionic activity coefficient γ has to be used. For an electrolyte M_pX_q, the mean ionic activity coefficient is defined as,

$$\gamma = (\gamma_+^p \gamma_-^q)^{1/n} \text{ where } n = p + q$$

For example, if the electrolyte were a solution of Na_3PO_4, then $p = 3$, $q = 1$ and $n = 4$. Therefore,

$$\gamma = (\gamma_+^p \gamma_-^q)^{1/4}$$

The **ionic strength**, I, of an electrolyte is defined by,

$$I = 1/2 \sum (c_i/c^0) z_i^2$$

where c_i = the concentration of the ith ion, and z_i = the number of charges on the ith ion. Thus I is a function of both the concentrations and charges of all the ions in the solution. Thus in the case of the 0.03 mol dm^{-3} Na$_3$PO$_4$ solution, assuming full dissociation, the ionic strength of the solution will be given by,

$$I = 1/2 \{3 \times 0.03 \times 1 + 1 \times 0.03 \times (3)^2\} = 0.18$$

It is possible, at concentrations of electrolyte below about 10^{-2} to 10^{-3} mol dm^{-3}, to express the relationship between γ and I at 298.15 K by the equation (Debye–Huckel theory limiting law),

$$\log \gamma = -0.51 |z_+ z_-| I^{1/2}$$

Since uncharged copper is in contact with charged copper ions, an electrical potential difference (pd) develops between the copper and its ions. Because of the dependency of the equilibrium position on temperature and concentration, if the pd is to be measured, then a set of **standard conditions** are required. These conditions are a temperature of 298 K and an ionic activity of 1 (see Box 7.4). The copper in contact with its ions in solution constitutes an electrode, a half-cell or half reaction.

The pd measured in volts and quoted under standard conditions is symbolised by E^0. Unfortunately, it is not possible to directly measure this pd. Another electrode is required against which the pd can be measured. The electrode chosen is the **standard hydrogen electrode**, which is given an internationally agreed arbitrary pd of zero volts, $E^0 = 0.000$ V, for a hydrogen gas pressure of 1 atmosphere, an activity of hydrogen ions of unity and a temperature of 298 K.

If the standard hydrogen electrode is coupled with the Cu^{2+}/Cu(s) electrode to form an electrochemical cell, then we can write this cell as,

$$(\text{Pt})\text{H}_2(g) \ (p = 1 \text{ atmosphere}) \mid \text{H}^+(\text{aq}) \ (a = 1) \parallel \text{Cu}^{2+} \ (a = 1) \mid \text{Cu(s)}$$

The single vertical line represents the interface between phases, and a double vertical line the interface between two half-cells. This latter interface is a porous pot or a device called a salt bridge which prevents the rapid mixing of the two solutions involved.

For this cell,

left-hand electrode:	$\text{H}_2(g) \rightleftharpoons 2\text{H}^+(\text{aq}) + 2e^-$
right-hand electrode:	$\text{Cu}^{2+}(\text{aq}) + 2e^- \rightleftharpoons \text{Cu(s)}$
overall cell reaction:	$\text{H}_2(g) + \text{Cu}^{2+}(\text{aq}) \rightleftharpoons 2\text{H}^+(\text{aq}) + \text{Cu(s)}$

If $E^0(\text{cell}) = E^0(\text{right-hand electrode}) - E^0(\text{left-hand electrode})$ then, if the emf of the cell is found to be $E^0(\text{cell}) = +0.34$ V and $E^0(\text{left-hand electrode}) = 0.000$ V, the standard electrode potential of the Cu^{2+}(aq)/Cu(s) electrode is $+0.34$ V. Thus the standard electrode potential of a half-cell is defined as the standard emf of a cell in which the standard hydrogen electrode is assumed to be on the left and the half-cell under investigation on the right. Note that, when writing the half equation associated with a standard electrode potential, the oxidised form is always written first, i.e. the half reaction is written as a reduction process,

e.g. $Cu^{2+}(aq) + 2e^- \rightleftharpoons Cu(s)$ $E^0 = +0.34$ V

$Zn^{2+}(aq) + 2e^- \rightleftharpoons Zn(s)$ $E^0 = -0.76$ V

These standard electrode potentials are referred to as **standard reduction potentials**. Standard electrode potentials for non-metal systems have also been determined,

e.g. $1/2\ I_2(s) + e^- \rightleftharpoons I^-(aq)$ $E^0 = +0.54$ V

$1/2\ Br_2(l) + e^- \rightleftharpoons Br^-(aq)$ $E^0 = +1.07$ V

$1/2\ Cl_2(g) + e^- \rightleftharpoons Cl^-(aq)$ $E^0 = +1.36$ V

Standard reduction potentials depend upon the oxidation state of the metal,

e.g. $Fe^{2+}(aq) + 2e^- \rightleftharpoons Fe(s)$ $E^0 = -0.44$ V

$Fe^{3+}(aq) + 3e^- \rightleftharpoons Fe(s)$ $E^0 = -0.04$ V

E^0 values can be determined for solutions containing two ions, e.g. $Fe^{3+}(aq)/Fe^{2+}(aq)$, $E^0 = +0.76$ V, using platinum wire as the inert electrode. The formation of complex ions also changes E^0 values, e.g. $[Fe(CN)_6]^{3-}(aq)/[Fe(CN)_6]^{4-}(aq)$, $E^0 = +0.48$ V.

In order to determine the emf of a cell, and therefore the E^0 value of the reaction, two electrodes are combined and the cell diagram written down,

e.g. $Zn(s) \mid Zn^{2+}(aq) \parallel Cu^{2+}(aq) \mid Cu(s)$

Using $E^0(cell) = E^0(\text{right-hand electrode}) - E^0(\text{left-hand electrode})$, then

$E^0(cell) = +0.34$ V $- -0.76$ V $= +1.10$ V

The sign of the standard cell emf gives the polarity of the right-hand electrode. In this case it is positive and therefore is the anode, i.e. if the cell is short circuited, electrons will flow from the left-hand electrode to the right (Figure 7.24). The positive sign of the combined E^0 value indicates that the reaction in which zinc metal displaces copper(II) ions from solution is thermodynamically possible.

So how does a knowledge of standard reduction potentials help in establishing what goes on in reactions involving oxidation and reduction? Generally, if two half reactions are combined to form a redox reaction then, if the E^0 value for the first is more positive than that of the second, the oxidised state in the first is thermodynamically capable of oxidising the reduced state in the second, under standard conditions.

The relationship between the molar Gibbs' Free Energy change and the standard reduction potential is given by,

$$\Delta G^0_m = -nE^0 F$$

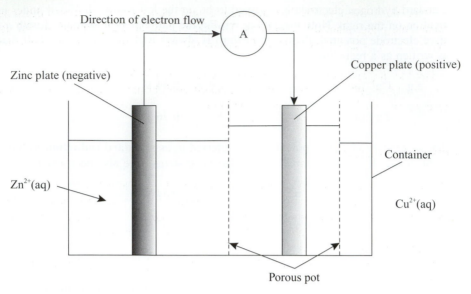

Figure 7.24 The zinc/copper cell.

where n = number of moles of electrons involved
         F = the Faraday constant, 96,485 C mol^{-1}

For the reaction, $H_2(g) + Fe^{2+}(aq) \rightleftharpoons 2H^+(aq) + Fe(s)$,

$$\Delta G_m^0 = -2 \times -0.44 \text{ V} \times 96{,}485 \text{ C mol}^{-1}$$

$$= +84.9 \text{ kJ mol}^{-1}$$

This positive ΔG_m^0 value indicates that the equilibrium lies well over to its left-hand side of the reaction, i.e. the reduction of Fe^{2+} to Fe is not thermodynamically favourable. Note that the half reaction involving hydrogen is not normally shown – it is assumed!

Now ΔG_m^0 for the $Fe^{3+}(aq)/Fe^{2+}(aq)$ half reaction, for which $E^0 = +0.76$ V, is -73.3 kJ mol^{-1}. This indicates an oxidising ability in that $Fe^{2+}(aq)$ is preferred to $Fe^{3+}(aq)$. For a particular half reaction, the more positive the E^0 value, the more strongly oxidising is that half reaction.

Consider the half reaction $Sn^{4+}(aq)/Sn^{2+}(aq)$, for which $E^0 = +0.15$ V and therefore $\Delta G_m^0 = -14.5$ kJ mol^{-1}. Will tin(II) reduce iron(III) to iron(II)?

i.e. $Sn^{2+}(aq) \rightleftharpoons Sn^{4+}(aq) + 2e^-$

$$\underline{2\ Fe^{3+}(aq) + 2e^- \rightleftharpoons 2\ Fe^{2+}(aq)}$$

$$Sn^{2+}(aq) + 2\ Fe^{3+}(aq) \rightleftharpoons Sn^{4+}(aq) + 2\ Fe^{2+}(aq)$$

Now, E^0(reaction) = +0.77 V − +0.15 V = +0.61 V.

Hence, $\Delta G_m^0 = -2 \times + 0.61$ V $\times$ F $= -117.7$ kJ mol^{-1}

Thus, the answer to the last question is yes! The reaction is thermodynamically favourable under standard conditions. The calculation of ΔG_m^0, though useful, is not necessary. We can simply subtract E^0 values without reference to the stoichiometry of the reaction.

For the tin(II)/iron(III) reaction, since $\Delta G_m^0 = -nRT \ln K^0$, so $\ln K^0 = 48.27$ and $K^0 = 9.15 \times 10^{20}$. This value of the standard equilibrium constant confirms that the reaction lies well over to the right-hand side.

The Nernst equation

Reduction potentials vary with both the temperature and the concentration of the ions taking part in a redox reaction. It is very rare to find reaction conditions in the environment to be near those under which standard electrode potentials are measured. Fortunately, the Nernst equation enables the reduction potential at any concentration and temperature to be calculated.

For the general reaction, $aA + bB + \ldots \rightleftharpoons pP + qQ + \ldots$, the Nernst equation can be written,

$$E = E^0 - \frac{RT \ln (a_P^p a_Q^q \ldots)}{nF \quad (a_A^a a_B^b \ldots)} = E^0 - RT \ln K^0$$

where
$E =$ equilibrium reduction potential compared with the standard hydrogen electrode
$E^0 =$ the standard reduction potential, at pH $= 0$, 298 K, 1 atmosphere pressure
$R =$ universal gas constant $= 8.314$ J K^{-1} mol^{-1}
$T =$ the absolute temperature, K
$n =$ number of moles of electrons being transferred
$F =$ the Faraday constant
$a =$ the activity of a particular species, taken to be 1.00 for a liquid or a solid, for a gas, $a = p/p^0$ where $p^0 = 1$ atmosphere pressure, for a solute in solution $a = \gamma c/c^0$ where $c^0 = 1$ mol dm^{-3}

Fortunately, it is found that calculations involving the Nernst equation are not necessary for many redox reactions. Often, the size and the sign of the standard reduction potential is sufficient to dominate the right-hand side of the equation. If this is so, the sign of E^0 may be taken as a sufficient indicator of a spontaneous change.

In the environment, redox reactions that involve the hydrogen ion or hydroxide ion are particularly important. In aqueous solutions the pH can vary by a factor of 14 and therefore the activity of the hydrogen ions can vary by about 10^{14}.

One possible reaction involving the oxidation of hydrogen ions is,

$$1/2 \, O_2(g) + 2H^+(aq) + 2e^- \rightleftharpoons 2H_2O(l) \quad E^0 = +1.23 \text{ V}$$

The Nernst equation for this reaction is,

$$E = E^0 - \frac{RT}{2F} \ln \left(\frac{a(H_2O)^2}{a(O_2)^{1/2} a(H^+)^2} \right)$$

Now if $a(H_2O) = 1$ and $p(O_2) = 1$ atmosphere then $a_{p(O_2)}/p^0 = 1$, therefore

$$E = E^0 - \frac{RT}{2F} \ln \{a(H^+)^2\} = E^0 + \frac{RT}{F} \ln \{a(H^+)\}$$

Since pH $= -\log \{a(H^+)\}$, then $E = 1.23 - 0.0592$ pH.

From this latter equation, we can obtain a measure of the oxidising ability of oxygen in the presence of water at any pH.

The mineral fayalite, Fe_2SiO_4, can undergo weathering according to the equation,

$$Fe_2SiO_4(s) + 4H_2O(l) + 4CO_2(g) \rightleftharpoons 2Fe^{2+}(aq) + 4HCO_3^-(aq) + H_4SiO_4(aq)$$

Once the cation Fe^{2+} is formed in aqueous solution, it can then be oxidised to the Fe^{3+} cation,

i.e. $Fe^{3+}(aq) + e^- \rightleftharpoons Fe^{2+}(aq)$ $E^0 = +0.77$ V

$1/2 \, O_2(g) + 2H^+(aq) + 2e^- \rightleftharpoons 2H_2O(l)$ $E^0 = +1.23$ V

Subtraction of the $Fe^{3+}(aq)/Fe^{2+}(aq)$ half reaction from the other yields,

$$Fe^{2+}(aq) + 1/2 \, O_2(g) + 2H^+(aq) \rightleftharpoons Fe^{3+}(aq) + 2H_2O(l)$$

Thus E^0 for the reaction is $+0.46$ V, i.e. Fe^{2+} will be oxidised to Fe^{3+} under standard conditions. The Fe^{3+} ion will then readily undergo hydrolysis to form insoluble iron(III) oxide, $Fe(OH)_3$, and iron(III) oxide, Fe_2O_3. This removes iron from the equilibrium shown in the last equation and therefore causes a shift in position to the left-hand side.

In the natural environment, aqueous solutions tend to contain low concentrations of ions. When trying to determine how an electrode potential varies with pH, it can be arbitrarily assumed that the concentration of a metal ion is 1×10^{-6} mol dm^{-3}. At such low concentrations, the activity coefficients of ions can be assumed to be unity, and their activities to have the same numerical value as their concentrations, i.e. if $c = 1 \times 10^{-6}$ mol dm^{-3}, $c^0 = 1.0$ mol dm^{-3}, then

$$a = \gamma c/c^0 = \frac{1.0 \times 1 \times 10^{-6} \text{ mol dm}^{-3}}{1.0 \text{ mol dm}^{-3}} = 1 \times 10^{-6}$$

The Pourbaix diagram

If iron and its redox reactions in water are to be understood, then the oxidation states that iron can normally exist in this environment must be known, and also which reactions depend on the pH. From these an E vs pH diagram or Pourbaix diagram can be constructed. The main features of this diagram are shown in Figure 7.25(a), and the three main types of E vs pH lines in Figure 7.25(b).

In the environment, the stability of water towards oxidation and reduction is a limiting factor. The oxidation of water is represented by,

$$2H_2O(l) \rightleftharpoons O_2(g) + 4H^+(aq) + 4e^-$$

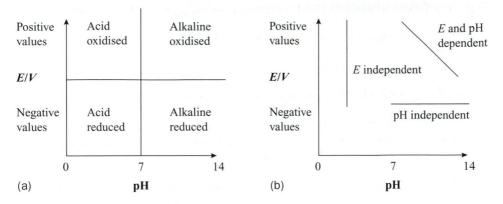

Figure 7.25 (a) Main areas of a Pourbaix diagram. (b) The three main types of E vs pH lines.

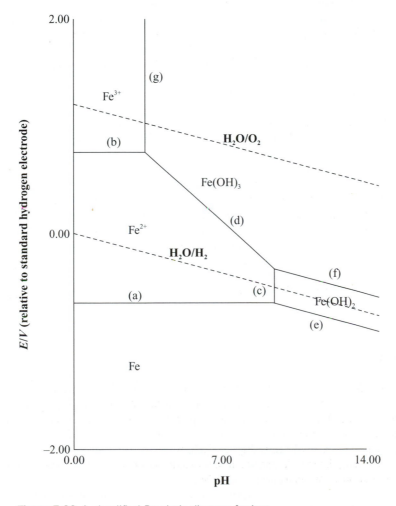

Figure 7.26 A simplified Pourbaix diagram for iron.

Rewriting this as a reduction process gives,

$$O_2(g) + 4H^+(aq) + 4e^- \rightleftharpoons 2H_2O(l) \quad E^0 = +1.23 \text{ V}$$

The reduction of water is given by,

$$H^+ + 2e^- \rightleftharpoons H_2(g) \quad E^0 = +0.00 \text{ V}$$

where the hydrogen ions originate from the self-ionisation of water.

The Pourbaix diagram for iron is shown in Figure 7.26. How such a diagram is constructed may be found elsewhere (Open University 1996).

What does the diagram tell us? First, it shows that elemental iron, Fe(s), is stable at large negative potentials with respect to the standard hydrogen electrode and will not corrode. Thus below lines (a) and (e) iron exists in an **immunity region**. Above line (a) iron is oxidised to iron(II) and therefore will corrode. Similarly above line (b), iron(III) will form. Hence the whole of the Fe^{2+} and Fe^{3+} regions can be described as **corrosion regions**. Iron thus does not occur naturally in the environment because its region of stability lies below the water reduction line. That is, water will undergo reduction before either $Fe^{2+}(aq)$ or $Fe(OH)_2(s)$. Iron will corrode (rust!), i.e. oxidise, in the natural environment. Second, $Fe^{3+}(aq)$ will occur only at low pH, i.e. less than or equal to about pH 3.0, and will therefore not be present in natural waters. Beyond lines (d), (f) and (g), solid $Fe(OH)_3$ is formed, which will coat iron and therefore protect it from further corrosion. Such a region is called a **passive region**.

The Pourbaix diagram is a description of the thermodynamic stability of iron in its various oxidation states in an aqueous system. Iron and its compounds can be similarly treated in iron minerals found in rocks and soil. Pourbaix diagrams also exist for many of the other elements and their compounds in a wide variety of situations.

Summary

- The silicate ion is the main building block of six main rock forming minerals: the olivines, pyroxenes, amphiboles, micas, feldspars and quartz.
- The three main types of rocks are igneous, sedimentary and metamorphic.
- Ores are formed within the Earth by magmatic segregation, contact mesomatic, pegmatite and hydrothermal processes.
- Ores are formed at the Earth's surface by sedimentation, residual and secondary enrichment processes.
- The lattice energy of a crystal is indicative of its energetic stability and is an exothermic quantity. The more negative its value the more energetically stable is the crystal.
- The Born–Haber cycle enables the lattice energy to be calculated from experimentally determined enthalpy changes. Thus comparisons can be made between experimental values and theoretically determined values enabling the nature of the bonding and the oxidation numbers of the elements involved to be identified.
- The relative sizes of the lattice energy and the enthalpy change of hydration will determine whether the enthalpy change of solution is exothermic or endothermic, and the degree of solubility of the crystalline solid.
- Standard electrode potentials are a way of expressing the relative thermodynamic stability of species in redox reactions, under standard conditions of temperature, pressure and an activity of 1.0 for the ions involved.
- Standard electrode potentials are measured relative to the hydrogen electrode and are normally quoted in the form of standard reduction potentials.

- The link between the standard electrode potential for a half reaction and the Gibbs' Free Energy change is given by,

$$\Delta G_m^0 = -nE^0 F$$

- The Nernst equation is used to calculate the EMF of a cell under non-standard conditions. For the general reaction, $aA + bB + \ldots \rightleftharpoons pP + qQ + \ldots$, the Nernst equation can be written,

$$E = E^0 - \frac{RT}{nF} \frac{\ln (a_P^p a_Q^q \ldots)}{(a_A^a a_B^b \ldots)} = E^0 - RT \ln K^0$$

- A Pourbaix diagram is a plot of E vs pH for various oxidation states of metals. From this diagram, the regions of stability with respect to the oxidation or reduction of water for various forms of a metal in different oxidation states can be established.
- Soil development and change depends upon (i) the nature of the underlying rock or sediments, (ii) activity of living organisms, (iii) climate, (iv) time, and (v) topology.
- The three main inorganic components of soil are sand, silt and clay. Clays and organic matter are very good ion exchangers.
- Soil fertility is affected by its pH and depends upon what cations are adsorbed.

Questions

1   What are the characteristics of an igneous rock, a sedimentary rock and a metamorphic rock?

2   Find out how the volcanoes of the Andes and those of Hawaii are formed. Why are the latter less dangerous than the former?

3   Draw a Born–Haber cycle for cadmium(II) bromide.
    Calculate the lattice energy of this compound from the following data:
    Enthalpy change of atomisation of cadmium, $\Delta H_{sub}^0(Cd) = +111.9 \text{ kJ mol}^{-1}$
    First ionisation energy of cadmium, $\Delta H_i^0(Cd) = +870 \text{ kJ mol}^{-1}$
    Second ionisation energy of cadmium, $\Delta H_i^0(Cd) = +1{,}600 \text{ kJ mol}^{-1}$
    Enthalpy of dissociation of dibromine, $\Delta H^0(Br–Br) = +193 \text{ kJ mol}^{-1}$
    Electron affinity of bromine, $\Delta H_{ea}^0(Br) = -342 \text{ kJ mol}^{-1}$
    Standard enthalpy change of formation of cadmium(II) bromide,
    $\Delta H_f^0(CdBr_2) = -314.6 \text{ kJ mol}^{-1}$.
    Assuming that cadmium(II) bromide forms a perfect crystal with a sodium chloride structure ($A = 1.763$), calculate its lattice energy. The ionic radius of the Cd^{2+} ion is 0.097 nm and that of the Br^- ion 0.196 nm. Comment on your answer.

4   Using the lattice energy of $CdBr_2$ determined from the Born–Haber cycle in Question 3 above, ΔH_h^0 for Cd^{2+} of $-1774.8 \text{ kJ mol}^{-1}$ and that for Br^- of $-351.0 \text{ kJ mol}^{-1}$, calculate the value of ΔH_{sol}^0 for cadmium(II) oxide.

5   (a)   Consider the two half reactions below,
    $$H_2O(l) + e^- \rightleftharpoons OH^-(aq) + 1/2 \, H_2(g) E^0 = -0.83 \text{ V}$$
    $$Al^{3+}(aq) + 3e^- \rightleftharpoons Al(s) E^0 = -1.66V$$
    From your knowledge of the chemistry of aluminium, does it react with water at 25 °C?
    Is your answer consistent with the given data?

(b) Draw a cell diagram representing a cell made from the two half-cell reactions,

$$Ni^{2+}(aq) + 2e^- \rightleftharpoons Ni(s) \quad E^0 = -0.23 \text{ V}$$
$$Sn^{2+}(aq) + 2e^- \rightleftharpoons Sn(s) \quad E^0 = -0.14 \text{ V}$$

Which is the negative pole of your cell?

What is the EMF of this cell?

6   A simplified Pourbaix diagram for zinc is shown in Figure 7.27.

What will be the redox conditions and the most stable species of zinc in:

(a) seawater at a pH of 8.0;

(b) a lake with a pH of 5;

(c) an industrial effluent with a pH of 13?

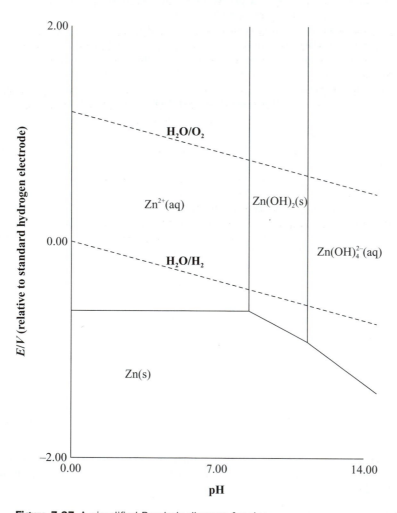

Figure 7.27 A simplified Pourbaix diagram for zinc.

References

The Open University (1996) S342 Physical Chemistry, Principles of Chemical Change, Block 7: *Equilibrium Electrochemistry*. Open University, Milton Keynes.

Further reading

Atkins, P.W. (2001) *The Elements of Physical Chemistry*, 3rd edn. Oxford University Press, Oxford.

Atkins, P.W. and Beran, J.A. (1992) *General Chemistry*, 2nd edn. Scientific American Books, W.H. Freeman, New York.

Compton, R.G. and Sanders, G.H.W. (1996) *Electrode Potentials*. Oxford Chemistry Primers, Oxford Science Publications, Oxford.
This is, like many of the books in this series, a neat and readable summary of the origin of electrode potentials and what they are used for. The maths required is minimal and where it is needed reasonable attempts are made to explain. This book is intended for first-year chemistry undergraduates.

Gribble, C.D. (ed.) (1988) *Rutley's Elements of Mineralogy*, 27th edn. Unwin Hyman, London.
This is a mine of information! It covers the chemistry and physical properties of minerals, elementary crystallography and the occurrence and classification of minerals. Individual minerals are extensively described.

Hammond, C. (2001) *The Basics of Crystallography and Diffraction*, 2nd edn. Oxford University Press, Oxford.
A very good, well illustrated introduction in particular to crystals and crystal structures. It is clearly written and covers topics such as crystal symmetry, the reciprocal lattice and point groups. This book is an excellent start to crystallography.

Pellant, C. (1990) *Rocks, Minerals and Fossils of the World*. Pan Books, London.
Contains hundreds of field photographs in colour. The text is also very readable and informative. It is relatively easy to identify rocks, minerals and fossils using this book.

8 Heavy metals and pollution of the lithosphere

- Importance of complex formation with metal ions – the co-ordinate bond, ligands and chelating agents. Naming of complexes
- The thermodynamic stability of complexes in aqueous solutions – the overall stability constant and successive stability constant
- Lead and mercury as examples of heavy metal pollutants – their effects on human health
- The occurrence of lead and mercury in soil, and some of their consequences
- Lead and mercury in solid wastes
- Solid waste disposal methods

The Romans, Beethoven and the Mad Hatter

The Romans used large amounts of lead for their bathing and drinking water distribution system. Roman baths were lined with thick sheets of cast lead; the pipes were made by bending thin sheets of lead around a cylindrical object and soldering the seam. In addition, the Romans added lead salts to their wine to sweeten the taste. Analysis of the skeletons of Romans have shown the lead content of the bones to be 100 times greater than that of modern-day Europeans. Some historians have suggested that the fall of the Roman Empire can be, in some part, attributed to plumbism or lead poisoning.

For most of his rather short life, Beethoven (1770–1827) was dogged by ill-health and personality disorders. Until his early twenties, his health appears to have been normal. Thereafter, he complained regularly about abdominal pain and of generally feeling unwell. He is reported to have died of 'dropsy', a term that often hid the real cause of death. His chronic symptoms correlate well to those found in cases of lead poisoning. Recent measurements (Russell 2000) carried out on samples of Beethoven's hair have shown the presence of lead at a concentration of some 60 ppm. This is 100 larger than would be expected for normal hair. Thus Beethoven had been exposed to lead and probably suffered from lead poisoning. Although possible, it is thought that his deafness was not caused by lead. It could, though, have contributed to his early demise. Where the lead came from is unknown.

Mercury and its compounds have also long been known to be dangerous! Some 300 years BC, the Romans used slaves to mine mercury in Spain. To be sent to the mercury mines usually meant a death sentence, since the life expectancy was said to be less than three years. The slaves died of then mysterious illnesses!

In the nineteenth century, mercury(II) nitrate was used to treat the fur used in the manufacture of felt hats. As a consequence, the workers were exposed to high levels of mercury over long periods of time, which resulted in mercury poisoning. Lewis Carroll in his book *Alice in Wonderland* introduces the character known as the 'Mad Hatter', who exhibits the signs of mental disorder associated with mercury poisoning. To be described as being as 'mad as a hatter' meant that you were suffering the symptoms seen in hatters such as a peculiar gait and erratic behaviour.

The use of mercury to treat syphilis in the nineteenth century was also known to cause severe side effects and even death.

In more modern times, point source pollution episodes have resulted in dire consequences. Industrial waste containing mercury(II) chloride was released into the bays of Minimata (1953) and Niigata (1960) in Japan. The mercury eventually found its way into the food chain. When Japanese people ate contaminated fish from the bays, many became victims of mercury poisoning. Over 80 people died, and many adults and children were badly affected by degenerative neurological disorders (Tamashiro *et al*. 1985).

Again, in 1972, seed grain treated with methyl mercury fungicide and intended for planting was distributed to villages in Iraq. Despite warnings, the grain was used to make bread. Some 6,530 people were hospitalised. At least 500 died (Bakir *et al*. 1973).

In 1970, in Ontario, Canada, fishing was banned in the St Clair River, Lake St Clair and the Detroit River because of contamination by mercury from industrial sources. The analysis of fish flesh in 1935 showed concentrations between 0 and 0.01 ppm, but by 1970 it had risen to a mean of 0.5 ppm (in some cases the concentration had risen to as high as 7.0 ppm). The levels of mercury found in Minimata fish were as high as 100 ppm – compare this with a present USEPA fish limit of 0.5 ppm.

Currently, there are considerable fears for the environment in Brazil's Amazon region, where mercury is used in the extraction of gold. Mercury is amalgamated with gold and the amalgam is heated in open containers to boil off the mercury. Not only is this directly dangerous to its handlers but has led to the illegal dumping of tonnes of mercury each year.

Heavy metals and soil

The metals mercury and lead are classified as **heavy metals**. The term 'heavy metal' is usually taken to mean those metallic elements that have a density equal to or greater than 6.0 g cm^{-3}. Such metals also include copper, cadmium, chromium, nickel and zinc (Table 8.1).

Heavy metals occur naturally in soils, sedimentary deposits and water bodies, and so normal background concentrations of these metals exist. Such metals are considered to be pollutants where their concentrations, relative to background concentrations, have risen to such an extent that they present a real or potential risk to living organisms. As described on p. 198, heavy metal pollution has made its presence felt in numerous cases of point source pollution – Itai–Itai disease caused by cadmium has also been described in Chapter 6. The main way in which heavy metals have found their way into soils and sediments has been via airborne particulate matter in the form of dust, smokes and aerosols (Figure 8.1), i.e. by anthropogenic processes. Heavy metals have also been added to agricultural land by the spreading of industrial wastes, sewage sludges and dredgings from rivers. Until recently, the spreading of sewage sludge has occurred on only about 1 per cent of agricultural land in the UK. However, it is feared that this may increase because the dumping of wastes in the sea was stopped in 1998. The over-application of inorganic fertilisers and animal manure can contribute to the soil's burden of heavy metals. Those pesticides that contained heavy metals have now been withdrawn and are no longer a source of this kind of pollution.

The quality of soil can be badly affected by the presence of metal pollutants. This is because the colloidal-sized clay particles carry negative electrical charges, which are

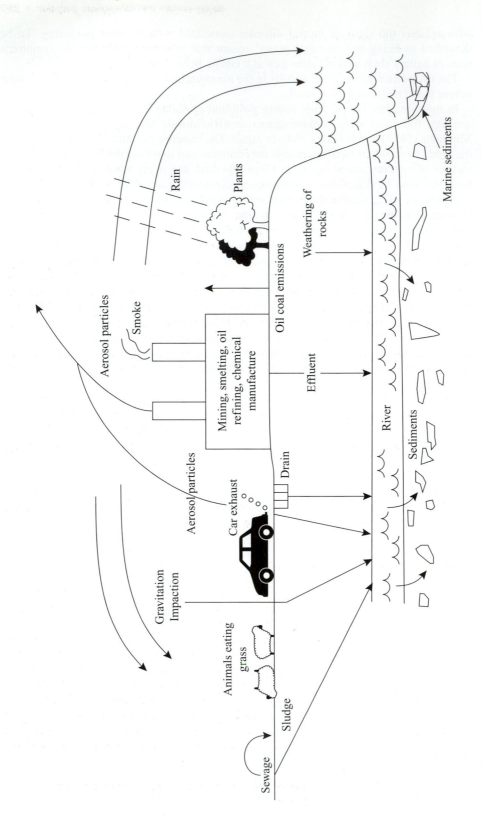

Figure 8.1 The sources of heavy metal pollutants.

Table 8.1 *Properties of some heavy metals*

Metal	Density g cm^{-3}	Atomic number	Relative atomic mass	Number of natural isotopes	Melting point/ °C	Boiling point/ °C
Cadmium	8.65	48	112.41	8	320.8	765
Chromium	7.14	24	51.00	4	1,900	2,690
Copper	8.95	29	63.55	2	1,083	2,570
Lead	11.34	82	207.2	4	328	1,751
Mercury	13.53	80	200.59	7	−38.9	357
Nickel	8.91	28	58.69	5	1,455	2,920
Zinc	7.14	30	65.38	5	419.5	907

surrounded by an atmosphere of cations such as Ca^{2+}, Mg^{2+} and H^+. These cations undergo cation exchange, depending upon the ions that are present in the water that surrounds the clay particles. In addition, the humus in soil contains lignin, which can undergo oxidation that results in the formation of humic acids and fulvic acids. Although the formulae of these acids are complex and wide ranging, they do contain carboxylic acid groups (−COOH) and phenolic groups (OH). It is the presence of these groups that enables these acids to form complexes with cations. Water-insoluble complexes tend to be formed with the humic acids, whilst water-soluble complexes are formed with fulvic acids.

Thus, when heavy metal ions enter the soil, they can be retained in three main ways. First, they can exchange with other cations on the surface of the colloidal clays and become adsorbed. Second, they can form complexes with humic and fulvic acid groups. Third, they can react with anions present and be precipitated out of solution. Generally, once heavy metals have entered the soil they have very long lifetimes and their removal can be very expensive.

To understand how heavy metals enter the environment in a chemical form that makes them become mobile and hazardous, and how heavy metal poisoning can be treated, then an understanding of complex formation is necessary.

The formation of metal complexes

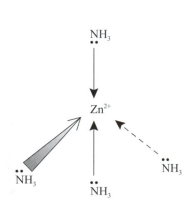

A metal complex can be viewed as being formed by the donation of a lone pair of electrons from a negative ion, e.g. CN^-, or molecule, e.g. H_2O, to that metal. The bonding is assumed to be covalent and thus does not take into account any ionic bonding character. Such an approach is called the valence-bond approach. The species donating the electron pair is referred to as a **ligand**. The bond that is formed is termed a co-ordinate chemical bond. The number of ligands involved in the bonding process is termed the co-ordination number. For example,

Figure 8.2 The structure of the complex molecule $[Zn(NH_3)_4]^{2+}$.

the zinc(II) ion forms a tetrahedral shaped complex with ammonia via the lone pairs on the nitrogen (Figure 8.2).

Here ammonia is the ligand and the co-ordination number is 4. The outer electron shells of the zinc atom can be represented by,

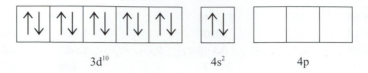

$3d^{10}$ $\quad\quad\quad$ $4s^2$ $\quad\quad\quad$ $4p$

To form the Zn^{2+} ion two electrons are lost,

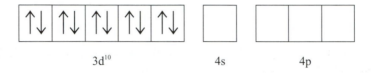

$3d^{10}$ $\quad\quad\quad$ $4s$ $\quad\quad\quad$ $4p$

The empty 4s and 4p atomic orbitals 'mix' or 'hybridise' and produce four sp^3-hybridised atomic orbitals which take up a tetrahedral geometry. Four lone pairs (represented by X) of electrons are donated to these atomic orbitals to form the complex,

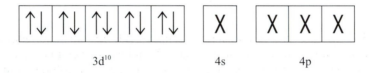

$3d^{10}$ $\quad\quad\quad$ $4s$ $\quad\quad\quad$ $4p$

Similarly, iron(II) forms a six co-ordinate octahedral complex, with water as the ligand, i.e. $[Fe(H_2O)_6]^{2+}$. Its structure is,

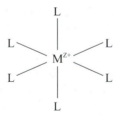

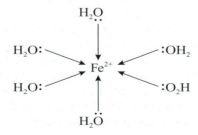

The outer electronic configuration of the iron atom is,

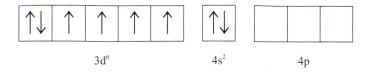

The outer electronic configuration of the iron(II) ion is therefore,

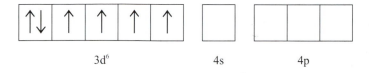

When the complex $[Fe(H_2O)_6]^{2+}$ is formed then again a lone pair of electrons is donated by each of six water molecules,

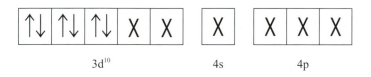

Here, two 3d atomic orbitals, one 4s and three 4p hybridise to form six d^2sp^3 atomic orbitals each accepting a lone pair of electrons. To minimise electron repulsion, the complex has an octahedral shape. Many heavy metal ions exist in solution in the form of octahedral aqua complexes.

There is a wide range of ligands that can be involved in the formation of complexes. Table 8.2 lists some of the more common ligands that donate only one pair of electrons to the central metal ion. These are termed **monodentate** (meaning 'one-toothed') ligands.

Some monodentate ligands are capable of co-ordination through more than one of their atoms, e.g. thiocyanate, CNS^-, can co-ordinate through either its sulphur atom or its nitrogen atom.

According to the groups they contain, ligands can have the ability to co-ordinate more than once to a metal. For example, the ligand is said to be bidentate if it co-ordinates twice, tridentate if three times and so on. Ligands capable of co-ordinating more than once are termed **polydentate ligands**, and are referred to as **chelating agents**. Two common chelating agents are ethylenediamine (abbreviated to en), a bidentate ligand, and ethylenediaminetetraacetic acid (abbreviated to EDTA), which provides a hexadentate ligand (Figure 8.3). Three ethylenediamine molecules can thus occupy the six octahedral sites around a central metal ion. One molecule of EDTA can occupy all six octahedral positions. How complexes are named is shown in Box 8.1.

The valence bond approach is qualitatively useful over a wide range of metal oxidation states and shapes of complexes. However, whilst it can explain what happens, it is not able to explain why the bonding occurs in the way that it does. The magnetic properties of complexes can be explained to a limited extent, but the valence bond approach cannot explain why the complexes are coloured (their spectral properties!).

Table 8.2 *Some common monodentate ligands*

Anionic ligands		Neutral ligands	
Name	Formula	Name	Formula
Fluoro	F^-	Aqua	H_2O
Chloro	Cl^-	Ammine	NH_3
Bromo	Br^-	Nitrosyl	NO
Iodo	I^-	Carbonyl	CO
Hydroxo	OH^-		
Cyano (bonded to metal via C atom)	CN^-		
Isocyano (bonded to metal via N atom)	CN^-		
Nitrito (bonded to metal via O atom)	NO_2^-		
Nitro (bonded to metal via N atom)	NO_2^-		
Thiocyanato (bonded to metal via S atom)	SCN^-		
Isothiocyanato (bonded to metal via N atom)	SCN^-		
Oxo	O^{2-}		

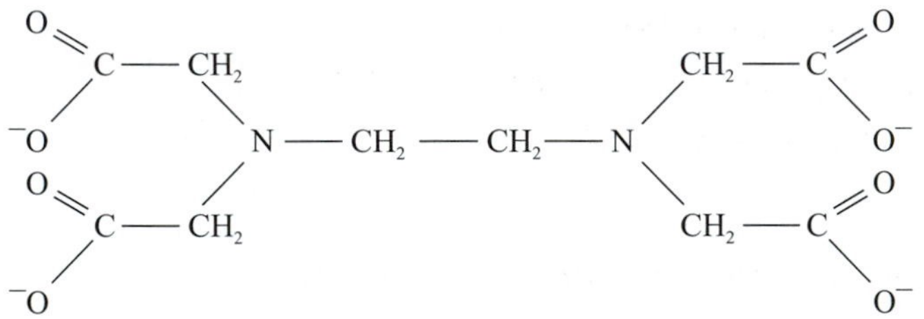

Ethylenediaminetetraaceto anion (EDTA)

EDTA is a hexadentate chelating ligand,
co-ordinating to the central metal ion via
the two N-atoms and four O^- atoms

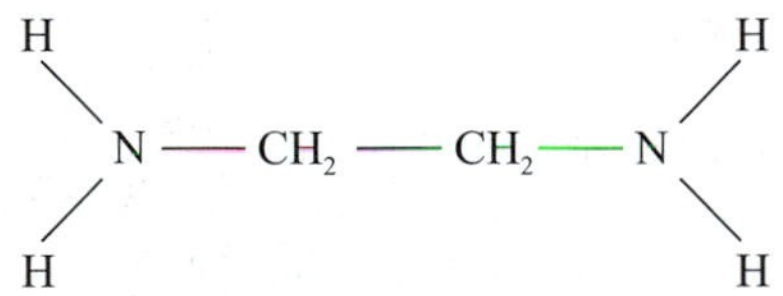

Ethylenediamine (en)

en is a bidentate chelating ligand,
co-ordinating through its two N-atoms

Figure 8.3 Ethylenediamine and the ethylenediaminetetraaceto anion (EDTA).

Box 8.1

The naming of complexes

Please refer to Table 8.2.

The rules are:

1   All anionic ligands ending in -ide such as cyanide, chloride and bromide change the ending to -o. Those ending in -ite change the ending to -ito. Those ending in -ate change the ending to -ato.
2   The three most common neutral ligands are water, ammonia and carbon monoxide. Water is referred to as 'aqua', ammonia as 'ammine' and carbon monoxide as 'carbonyl'.
3   Metals are quoted with their oxidation states in brackets.
4   When different ligands are present, then they are named in alphabetical order.
5   The number of the same kind of ligand is shown by a prefix di-, tri-, tetra, etc.
6   Where there may be a confliction between a prefix and a ligand that already contains a prefix, e.g. ethylenediamine, then the numbers of that type of ligand are referred to by bis- for two and tris- for three.
7   If the complex is an anion, then the suffix -ate is added to the metal's name.
8   The overall name of the complex is completed using the same method for naming simple compounds.

Example 1

   $[Co(NH_3)_6]^{3+}$ hexamminecobalt(III) ion

   $[Co(NH_2.CH_2.CH_2.NH_2)_3]^{3+}$ tris(ethylenediamine) cobalt(III) ion

   $[Co(NH_2.(CH_2)_2.NH.(CH_2)_2NH_2)_2]^{3+}$ bis(diethylenetriamine) cobalt(III) ion

Example 2

   $K_3[Fe(CN)_6]$ potassium hexacyanoferrate(II)

   $Ni(CO)_4$ tetracarbonyl nickel(0)

   $[Al(OH)(H_2O)_6]^{2+}$ hydroxopentaaquoaluminium(III) ion

Example 3

   $[Co(NH_3)_3(H_2O)_4]Br_3$

First, deal with the complex cation part, i.e. $[Co(NH_3)_3(H_2O)_4]^{3+}$.
This is the triamminetriaquocobalt(III) ion.
Now the whole compound!
The full name is therefore triamminetriaquocobalt(III) bromide.

Example 4

   $[Co(NH_3)_3(H_2O)_4][Fe(CN)_6]$

The anion part is either $[Fe(CN)_6]^{4-}$ or $[Fe(CN)_6]^{3-}$ depending upon whether the oxidation state of iron is II or III. If it were the former, then it would require cobalt to be in an oxidation state of IV, which is highly unlikely. Hence we may assume it is the latter anion.
   Thus the name of the complex is triamminetriaquocobalt(III) hexacyanoferrate(III).

For more advanced theories of how bonding occurs in complexes and accompanying explanation of their properties, the crystal field, ligand field and molecular orbital theories, the reader is referred to more advanced chemistry texts.

Thermodynamic and kinetic stability of complexes in aqueous solution

There are two ways of expressing the thermodynamic stability of metal complexes in terms of equilibrium constants (note that all quoted equilibrium values are for standard conditions throughout this section). Metal ions in aqueous solution often form six co-ordinated complexes with water, e.g. $[Co(H_2O)_6]^{2+}$. Each water molecule may be replaced successively by a monodentate ligand L, e.g. the ammonia molecule. If M represents any central metal ion and *assuming* that six water molecules are initially co-ordinated to the central metal ion, then the successive replacement of these six water molecules by L can be written,

$$M + L \overset{K_1}{\rightleftharpoons} ML \qquad \text{for which } K_1 = [ML]/[M][L]$$

e.g. $[Co(H_2O)_6]^{2+} + NH_3 \rightleftharpoons [Co(H_2O)_5NH_3]^{2+}$

$$ML + L \overset{K_2}{\rightleftharpoons} ML_2 \qquad \text{where } K_2 = [ML_2]/[ML][L]$$

$$ML_2 + L \overset{K_3}{\rightleftharpoons} ML_3 \qquad \text{where } K_3 = [ML_3]/[ML_2][L]$$

And so on down to,

$$ML_5 + L \overset{K_6}{\rightleftharpoons} ML_6 \text{ where } K_6 = [ML_5]/[ML_4][L]$$

The constants $K_1, K_2, K_3, \ldots K_6$ are called the **successive stability constants** or **stepwise formation constants**.

A second way of forming a complex can be written,

$$M + L \rightleftharpoons ML \quad \text{for which } \beta_1 = [ML]/[M][L]$$

$$M + 2L \rightleftharpoons ML_2 \quad \text{for which } \beta_2 = [ML_2]/[M][L]^2$$

$$M + 3L \rightleftharpoons ML_3 \quad \text{for which } \beta_3 = [ML_3]/[M][L]^3$$

This process continues until $M + 6L \rightleftharpoons ML_6$ for which $\beta_6 = [ML_6]/[M][L]^5$
The expression for β_3 can be written,

$$\beta_3 = [ML_3]/[M][L]^3$$
$$= [ML_3]/[M][L]^3 \times [ML]/[ML] \times [ML_2]/[ML_2]$$
$$= [ML]/[M][L] \times [ML_2]/[ML][L] \times [ML_3]/[ML_2][L]$$
$$= K_1 \times K_2 \times K_3$$

Table 8.3 *Some stability constants for complex metal ions in aqueous solution*

Ion	Ligand	$\log k_1$	$\log k_2$	$\log k_3$	$\log k_4$	$\log k_5$	$\log \beta_n$	n
Co^{2+}	NH_3	1.99	1.51	0.93	0.64	0.06	4.39	5
	en	6.0	4.8	3.1			13.9	3
	EDTA						16.3	6
Cd^{2+}	NH_3	2.51	1.96	1.30	0.79	–	6.65	4
	CN^-	5.18	4.42	4.32	3.19	–	17.11	4
	EDTA						16.5	6
Hg^{2+}	NH_3	8.8	8.7	1.0	0.8	–	19.3	4
Pb^{2+}	EDTA						18.0	6
Cu^{2+}	NH_3	4.25	3.61	2.98	2.24	−0.52	13.08	4
	EDTA						18.8	6

Generally, $\beta_n = K_1 K_2 K_3 \ldots K_n$ where β_n is the **overall formation/stability constant** for the appropriate number of steps.

Example successive stability constants and overall stability constants for some aquated metal ions are listed in Table 8.3.

Table 8.3 shows that, as the number of ligands increases, the values of the successive stability constants decrease. This is because there is (i) a reduction in the number of sites available for substitution to take place, i.e. statistical reasons, (ii) increased steric hindrance if the replacing ligands are bulkier than those being replaced, and (iii) electrostatic repulsion if the ligands are negatively charged.

As a rough guide, if the value of β_n is greater than 10^8, the complex can be considered to be thermodynamically stable.

If the ligands already present in a complex can be rapidly replaced by other ligands, then that complex is described as being kinetically labile. If the rate is slow, then it is kinetically inert.

Thus, the complex ion $[Co(NH_3)_6]^{3+}$ exists for days in acid solution, though it is thermodynamically very unstable,

$$[Co(NH_3)_6]^{3+}(aq) + 6H_3O^+(aq) \rightleftharpoons [Co(H_2O)_6]^{3+}(aq) + 6NH_4^+(aq)$$

$$\log \beta_6 \sim 25$$

Generally, complexes are greatly stabilised when polydentate species are present, i.e. chelation has taken place. For example,

$$[Ni(H_2O)_6]^{2+}(aq) + 6NH_3(aq) \rightleftharpoons [Ni(NH_3)_6]^{2+}(aq) + 6H_2O(l)$$

$$\log \beta_6 = 8.61$$

$$[Ni(H_2O)_6]^{2+}(aq) + 3en(aq) \rightleftharpoons [Ni(en)_3]^{2+}(aq) + 6H_2O(l)$$

$$\log \beta_3 = 18.3$$

Thus the latter complex is approximately 10^{10} times more thermodynamically stable than the former complex. The structure of the $[Ni(en)_3]^{2+}$ ion is shown schematically below.

The 'ends' of the ethylenediamine molecule have now encompassed the central nickel(II) ion rather like a claw – the term chelate is derived from the Greek word for 'claw'!

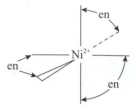

Chelating agents are very common in the environment. For example, lichens and mosses use chelating agents to form stable complexes with some of the trace metals they require from the rocks they live on. In humans, chelating agents are used in the treatment of poisoning by heavy metals such as copper, lead, mercury and arsenic. Penicillamine (not a penicillin drug!) is a commonly used chelating agent, which forms particularly stable complexes with these metals. This reduces the availability of these metals for metabolism and therefore their poisonous nature. The complexes may then be excreted via the urinary system.

A study of lead and some of its compounds

Lead occurs naturally in four stable isotopic forms, ^{206}Pb, ^{207}Pb, ^{208}Pb and ^{204}Pb. The first three isotopes are the end products of the uranium, actinium and thorium decay series, respectively, whilst the fourth has no natural radioactive precursor. Lead is a relatively poor conductor of electricity, and is a dense, soft, malleable, ductile metal with a low tensile strength.

Lead does not occur naturally as the free element, and does not occur naturally in air. It may well have been the first metal to be smelted. It is mentioned in the Old Testament, and lead artefacts dating from about 6500 BC have been found at Catul Huyuk in South Turkey. When freshly cut, lead has a bright shiny lustre, which quickly tarnishes to a blue–grey colour upon exposure to air. This is due to the formation of a thin layer of lead(II) hydroxide and lead(II) carbonate which protects the metal from further attack. Lead is attacked by concentrated nitric acid to form lead(II) nitrate and oxides of nitrogen. It is little affected by concentrated sulphuric or hydrochloric acid at ordinary temperatures. Lead reacts slowly with water in the presence of air to form lead(II) hydroxide which is slightly soluble in water,

$$2Pb(s) + 2H_2O(l) + O_2(g) \longrightarrow 2Pb(OH)_2(s)$$

If carbonate or sulphate ions are present in the water, i.e. it is hard water, then an insoluble, protective layer of lead(II) carbonate or lead(II) sulphate layer is formed. For many years, lead was used in making pipes and cisterns for drinking water systems. Thus, such pipes are safest in hard water regions because of this protective layer which coats the insides of the pipes thus protecting them from further corrosion. If fruit juices, wines or spirits are placed in containers made of pottery containing lead-based glazes or if they are of leaded glass, then the slightly increased acidic conditions can lead to soluble lead compounds being formed,

$$2Pb(s) + O_2(g) + 4H_3O^+(aq) \longrightarrow 2Pb^{2+}(aq) + 6H_2O(l)$$

The main ore of lead is galena, PbS, from which lead is obtained by roasting it with air in a reverberatory furnace. The reactions taking place at lower temperatures are,

$$PbS(s) + 2O_2(g) \longrightarrow PbSO_4(s)$$

$$2PbS(s) + 3O_2(g) \longrightarrow 2PbO(s) + 2SO_2(g)$$

When the temperature is raised and the air supply terminated, the following reactions occur,

$$PbSO_4(s) + PbS(s) \longrightarrow 2Pb(l) + 2SO_2(g)$$

$$2PbO(s) + PbS(s) \longrightarrow 3Pb(l) + SO_2(g)$$

Note here the potential for atmospheric pollution by SO_2, and water, soil and air pollution by lead and its compounds.

The lead produced initially from the roasting process is called 'pig lead' and contains small amounts of zinc, silver and gold. If zinc is added to this molten lead, silver and gold will form an alloy with the zinc. This alloy floats on the top of the molten lead and is skimmed off. Very pure lead is obtained by electrolysis.

Elemental lead is used in the manufacture of lead acid accumulators; sheathing for electrical cables; lining of pipes and tanks where chemical resistance is required; shielding for X-ray equipment and when using radioactive materials; the making of solder, type-metal and other alloys; and the manufacture of lead compounds. There is no evidence that lead is an essential element to man.

Lead(II) oxide, PbO, is produced industrially by blowing air over molten lead. It is used in the manufacture of lead glass crystal, as a drier in paints and varnishes, and in various insecticides. It dissolves slowly in nitric acid or ethanoic (acetic) acid to give the colourless $Pb^{2+}(aq)$ ion,

$$PbO(s) + 2H_3O^+(aq) \longrightarrow Pb^{2+}(aq) + 2H_2O(l)$$

'Red lead', Pb_3O_4, acts as if it is a mixed oxide of formula $2PbO.PbO_2$. It is made by heating lead(II) oxide in air at elevated temperatures. It is used as a pigment in paint, as a rust inhibitor for structures made of iron and steel, and in lead acid accumulators.

The white solid basic lead carbonate, $PbCO_3.Pb(OH)_2$, 'white lead', is made from sheets of lead, acetic acid and air, which are left in a confined space enriched with carbon dioxide. White lead was used for hundreds of years as a white pigment in paint and in ceramic glazes. As a paint, it is a major source of lead contamination in a number of older buildings.

Lead(II) chromate, $PbCrO_4$, is a yellow solid widely used as a pigment known as 'chrome yellow'. Because of its insoluble nature, it is used for road markings.

Lead acetate, $Pb(CH_3COO)_2$, is soluble in water and is one of the very few lead compounds that will give a high concentration of $Pb^{2+}(aq)$ ions. It is a white sweet-tasting solid known as 'sugar of lead'. It is used in dyeing, as a drier in varnishes and for making other lead compounds.

All of the above materials are inorganic lead compounds, which are much more common than organic lead compounds. Inorganic lead pollutes most of the environment because it has been used for several thousands of years. It cannot be chemically destroyed nor does not it become sufficiently concentrated to reform ores that can be mined. It is thus a permanent pollutant. Ice cores drilled out in Greenland have shown a 400-fold increase in lead deposition between 800 BC and 1965. By far the biggest increase occurs after the Industrial Revolution.

Inorganic lead compounds are hazardous because the metabolism of lead is very similar to that of calcium and lead inhibits many enzymes. Since the Ca^{2+} ion is very important at a cellular level, the Pb^{2+} ion is a broad-spectrum poison.

In the case of organic lead compounds two are of major interest. Tetraethyl lead, $(C_2H_5)_4Pb$, is a colourless liquid that has been used extensively as an anti-knocking agent in petrol engines. This compound, together with tetramethyl lead $(CH_3)_4Pb$, have been responsible for a great deal of the increase in lead pollution. Both of these organic lead compounds dissolve in petrol and are also fat soluble. Tetraethyl lead and tetramethyl lead are particularly hazardous because they undergo dealkylation in the liver to form trialkyl lead,

$$\text{e.g. } (C_2H_5)_4Pb \longrightarrow (C_2H_5)_3Pb^{2+}$$

Trialkyl lead can also be formed in exhausts, i.e.

$$(C_2H_5)_4Pb + HBr \longrightarrow (C_2H_5)_3PbBr + C_2H_6$$

Trialkyl lead compounds are respiratory irritants, strongly mutagenic, can cross blood barriers more easily than inorganic lead compounds, and cause personality and behavioural problems.

The threat to humans from tetraethyl lead was appreciated when it was first made in 1923. Thirteen workers died and 126 became mentally disturbed. This resulted in the stopping of production of the compound until appropriate safety precautions were introduced (Box 8.2).

A study of mercury and some of its compounds

Although mercury has occasionally been found as the free element in rocks, the main source of mercury is the mineral cinnabar mercury(II) sulphide, HgS. Mercury is a dense, silvery-white metal. At ordinary room temperatures, it is a very mobile liquid with a vapour pressure of 0.002 mm Hg at 25 °C. Its mobility is thought to have earned it the nickname 'quicksilver'. Mercury has such a low viscosity that when its droplets collide they rapidly coalesce into one. It has a high surface tension and therefore forms spherical droplets very readily. Mercury has an exceptionally high electrical resistivity and is a relatively poor conductor of electricity – it is used as an international standard for electrical resistance.

Because of its high boiling point, ability to conduct heat and uniform expansion, mercury is often used in thermometers. It is used as a coolant because of its good heat conducting ability. Since mercury does not wet glass and has a high density, it is also used in barometers. In electrical devices, despite its conductivity, mercury is used where a liquid conductor is required, e.g. in switches. Some electrical discharge tubes use mercury vapour as the source of light for street and road lighting.

Mercury does not react with oxygen or air at room temperatures, but just below its boiling point will form red mercury(II) oxide. This oxide decomposes upon further heating to 500 °C. It does not react with cold or hot water. Mercury will react with hot concentrated sulphuric acid to give sulphur dioxide and a sulphate of mercury. Nitric acid reacts in a similar way, but yields oxides of nitrogen and nitrates of mercury. The metal does not react with hydrochloric acid or with alkalis.

Box 8.2

The effects of lead and its compounds on human health

Lead enters the human system as a result of inhaling lead and lead salt fumes, and by swallowing lead salts. It accumulates in the bones, teeth, muscle and skin. Once stored, the body excretes lead very slowly taking months and even years to rid itself of its lead burden. When it enters the bloodstream as Pb^{2+}, lead becomes biologically active and therefore potentially at its most dangerous.

In cases of acute exposure (now rare in the UK) lead can cause paralysis of the limbs, neurological disorders, coma and even death (ATSDR 1993a,b). Blood levels in excess of 125 µg per 100 cm^3 will cause death in children. Brain and kidney damage is known to occur at blood levels greater than 100 µg per 100 cm^3 in adults and 80 µg per 100 cm^3 in children, while gastrointestinal problems occur at blood levels of 60 mg per 100 cm^3.

Chronic exposure can cause anaemia in adults at blood levels between 50 and 80 µg per 100 cm^3 and in children between 40 and 70 µg per 100 cm^3. Neurological symptoms appear in adults at blood levels between 40 and 60 µg per 100 cm^3 and in children between 10 and 30 µg per 100 cm^3 or lower. In the case of children, lead has also caused growth problems (ATSDR 1993a,b). Other chronic effects include increased blood pressure, kidney malfunction and interference with Vitamin D metabolism. Lead(II) acetate and lead(II) phosphate have been identified as carcinogens based on animal experiments. In humans there is less evidence of their carcinogenic nature, but the EPA classifies lead as a carcinogen.

Exposure to lead is more dangerous for unborn and young children (Pocock et al. 1994; Tong et al. 1996). Lead will cross the placenta and cause premature births, smaller babies and decreased mental ability in the infant. The most worrying aspect of lead poisoning is that lead levels in children may cause mental impairment without any obvious physical symptoms being evident. Above blood levels of 10 µg per 100 cm^3 it has been reported that lead has affected the brain development of children and hence their IQ (ATSDR 1993b). Although it has proved difficult to dissociate all the other factors that may affect IQ, it is evident that the more lead present in a child's blood, the lower their average IQ. Very young children, in particular, have suffered because of their regular contact with dirt and dust whilst playing, and the habit of licking dirty fingers and placing objects in their mouths. Strong links between old flaking lead-based paint is the main contributor to soil lead levels. Children aged two or three may be at risk of exposure to lead-containing soils because pica tendencies (eating soil) are most common at this age (Pocock et al. 1994; Tong et al. 1996).

In the US, for example, it was reported (ATSDR 1988) in 1988 that some 12 million children under the age of seven were at risk from lead-based paint. Some 5.9–11.7 million children were potentially exposed to lead in dust and soil.

Mercury forms amalgams with many metals, for example, when zinc, lead, tin, cadmium, silver or gold are rubbed with mercury. Dental amalgam is a mixture of tin, silver and mercury.

When mercury is extracted from cinnabar, the ore is crushed and concentrated by froth flotation. The ore is then roasted in air,

$$HgS(s) + O_2(g) \longrightarrow Hg(g) + SO_2(g)$$

The mercury vapour is distilled off and condensed. It is purified by repeated distillation.

Mercury forms two series of compounds, mercury(II) and dimercury(I) (also known as mercuric and mercurous, respectively). Mercury(II) compounds appear to contain the Hg^{2+} ion and mercury(I) the Hg_2^{2+} ion.

Mercury(II) chloride, $HgCl_2$, is a white, very poisonous solid readily soluble in hot water but much less soluble in cold. It is the most important inorganic compound and is manufactured by heating mercury in a current of dry chlorine. When dissolved in water, the solution does not conduct electricity well, thus indicating that few ions are present. This, together with its solubility in ethoxyethane and its volatility, suggest the bonding is covalent. Mercury(II) chloride is used to make organo-metallic compounds, as a wood preservative, and as an antiseptic for sterilising surgical instruments. Mercury(II) sulphide is used in the red pigment, vermilion.

Dimercury(I) sulphate, Hg_2SO_4, is made by precipitating it from a nitric acid solution of dimercury(I) nitrate by the addition of dilute sulphuric acid. This compound is used in electrochemical cells.

Organic compounds of mercury are much more dangerous to humans than the element or its inorganic salts. Perhaps the two forms of organic mercury that have caused the most problems to date are methyl mercury, CH_3Hg^+, and dimethylmercury, $(CH_3)_2Hg$. Mercury(0), mercury(I) and mercury(II) inorganic compounds can be changed to dimethylmercury by the action of microbes, particularly under anaerobic conditions in sediments of lakes and rivers, and in wet soils. The process is referred to as **methylation**. The amount of methylation taking place depends upon the amount of dioxygen present, the pH of water and the presence of particles of clay or organic matter. In particular, small amounts of dioxygen favour the formation of methylmercury and low pH increases methylation. Dimethylmercury is volatile and can evaporate relatively quickly both from soils and water bodies.

Methylmercury is much more soluble in water than is dimethylmercury. The latter compound is readily converted to the former under acid conditions. Methylmercury is the form in which mercury is most commonly found in the tissues of animals. Methylmercury exists as CH_3HgX where $X = Cl^-$, Br^-, I^-, OH^-, etc. It is the predominant species in acidic or neutral aqueous solutions. Methylmercury is the more potent toxin because its salts are soluble in the fatty tissues of animals where it accumulates. Ingestion leads to the formation of the CH_3Hg^+ group attaching itself to the sulphur atom of an amino acid. The new species is soluble in tissues and crosses both the blood–brain and placental barriers.

In order of toxicity, methylmercury is more toxic than elemental mercury, which is more toxic than dimethylmercury (Box 8.3). The mercury(II) ion is not readily transported across membranes. Mercury(I) ion is not very toxic when ingested because it forms insoluble Hg_2Cl_2 in the acidic environment of the stomach.

Mercury in humans is in the methylmercury form and is usually ingested by eating contaminated fish. In fish this compound forms at least 80 per cent of the mercury present and is distributed throughout the body of the animal. It is absorbed through the gills and by eating contaminated food. Predatory fish contain an even higher amount of

Box 8.3

The effects of mercury and its compounds on human health

In the UK and other western countries, the use of elemental mercury and its compounds in industrial processes has almost gone. Therefore, the general population is now not significantly exposed to mercury. Exposure is limited to dentistry activities and low-level exposure from fillings, from equipment containing mercury such as barometers and thermometers, some pigments, some fungicides and insecticides, and some primary batteries.

Inhalation at the workplace is the major way mercury enters the human system, some 90–100 per cent being absorbed via the lungs. For example, since a large number of deceased people are now being cremated in the UK, there is evidence that volatilised mercury from their fillings is increasing the body burden of crematoria workers. Eating contaminated food, particularly fish, provides a prominent source of methylmercury.

Acute exposure to high levels of mercury affects the central nervous system causing hallucinations, delirium and suicidal tendencies. Gastrointestinal and respiratory effects are also present, i.e. chest pains, cough, lung impairment and inflammation of the intestines (ATSDR 1999). Acute oral effects include a metallic taste, nausea, vomiting and severe abdominal pains. The acute lethal dose for most inorganic mercury compounds is between 14 and 47 $mg\,kg^{-1}$ for a person of mass 70 kg (ATSDR 1992). Methylmercury causes blindness, deafness, an impaired level of consciousness and death. The minimal lethal dose in this case is between 20 and 60 $mg\,kg^{-1}$ for a 70 kg person (WHO 1990).

Chronic effects caused by elemental mercury are those associated with attack on the central nervous system, e.g. increased excitability, irritability, excessive shyness, insomnia, severe salivation tremors and kidney problems. Inorganic mercury causes kidney damage and methylmercury central nervous system damage. Higher doses of the latter can cause deafness, difficulties in speaking and restriction of the field of vision (WHO 1991).

Reproductive and developmental effects in the case of elemental mercury is uncertain, but there is some evidence to suggest enhanced abortions and birth defects (ATSDR 1999). There are currently no data available concerning inorganic mercury. Methylmercury has been found to affect the central nervous system of infants, causing mental retardation, lack of co-ordination, deafness, blindness and cerebral palsy. Lower doses have led to developmental problems and abnormal reflex action.

As a carcinogen, the evidence against elemental mercury is inconclusive (WHO 1991). Inorganic mercury has been observed to cause increases in renal, stomach and thyroid cancers in mice (National Toxicological Program, 1991). Data for methylmercury are not available. The EPA has classified inorganic mercury and methylmercury as possible human carcinogens, whilst elemental mercury is not included.

methylmercury than non-carnivorous fish. Since the half-life of CH_3Hg^+ in humans is 70 days, much longer than inorganic mercury, it can readily bio-accumulate (Box 8.3).

Lead and mercury in soil

Lead compounds are released naturally into the environment by weathering processes and volcanism. Lead has become more widely dispersed in the environment as a consequence of (i) wastage during the mining and smelting of its ores; (ii) wastage when lead containing products have been manufactured and used; (iii) the disposal of unwanted lead-containing products; and (iv) the burning of fossil fuels.

Emissions of lead into the atmosphere in the UK originate from a number of sources, e.g.

- power station combustion
- commercial and domestic combustion
- industrial combustion plants and processes
- production of coke
- production of iron and steel
- chemical processes
- road transport
- waste treatment and disposal.

In 1970, total emissions from all sources amounted to some 6,456 tonnes of lead. By 1998 this had fallen to 1,033 tonnes.

In 1995, lead pollution in the UK was caused mainly by particulate emissions in the form of inorganic compounds ($PbBr_2$, $PbClBr$) from petrol-driven internal combustion engines. Lead originating from the combustion of petrol amounted to some 1,067 tonnes or 73 per cent of the total. In order to combat this source of pollution, the maximum amount of permitted lead allowed to be present in UK petrol was reduced from 0.40 g dm^{-3} to 0.15 g dm^{-3} in 1986. This reduction has also been accompanied by a marked increase in the use of unleaded petrols. Hence, lead levels caused by emissions from petrol engines have steadily reduced in the UK – in 1970 the emissions caused by the petrol engine were 6,456 tonnes of lead. As from 2000 the sale of leaded petrol in the UK has been banned and so pollution by lead from this source should be essentially zero. Industrial processes (3 per cent), combustion processes (8 per cent), the production of non-ferrous metals (9 per cent) and waste treatment and disposal (7 per cent) are also current major sources of lead pollution. Although airborne lead compounds can be inhaled, lead also enters the human system via ingestion of contaminated food, water and dust.

Natural levels of lead found in surface soils are usually below 50 ppm. However, soil has been contaminated by lead as a consequence of flaking lead-based paints, from incinerators, and particularly from motor vehicles that have used leaded petrol. Hence, soils in remote and rural areas contain levels of lead in keeping with normal geological processes, whilst those in urban and industrial areas contain much higher levels of lead. In lead-mining areas, lead has been deposited in the local environment, generally in the form of galena (PbS).

The build-up of lead in soil depends mainly on how fast it is deposited from the atmosphere. What happens after deposition depends upon whether it is adsorbed on to clay particles and other surfaces, stays in solution, is precipitated from solution or is able to form complexes. Lead appears to enter the soil as lead(II) sulphate ($PbSO_4$)

or is rapidly converted to this compound on the surface of the soil. Lead sulphate is relatively soluble and thus would leach down through the soil unless it was changed into something less soluble. In soils of pH equal to or greater than 5.0 containing at least 5 per cent by mass of organic matter, lead tends to remain in the top 2–5 cm of undisturbed soil (USEPA 1986). The leaching of lead is very slow under most natural conditions (Burgess 1977).

Leaching is particularly favourable when there is so much lead in the soil that it exceeds the soil's sorption capacity, if chelating agents are present that enable soluble lead complexes to be formed, and if there is an increase in acidity of the leaching solution. Lead may also leave the soil in the form of dissolved and undissolved compounds in run-off caused by heavy rainfall. This can cause the pollution of surface waters. Soil particles that are contaminated by lead may also be transported elsewhere by the action of wind.

Unless people are working in a lead-based industry, most lead taken in by humans is via food found mainly in vegetables, beverages and milk. When airborne lead is involved, leafy crops and cereals in particular would benefit from washing to reduce the amount of lead that may be ingested. The uptake of lead from soil by plants can also occur, the amount being dependent on the species of plant and how much is available. When sewage sludge has been applied to agricultural land as a fertiliser, this has also increased the lead content of soil.

Because of its high volatility, mercury is being continually emitted to the atmosphere from rocks and soils. How much mercury finds its way into the atmosphere depends upon how much mercury is present in the soil and rocks, how strongly the mercury is adsorbed on to the soil, and the prevailing temperatures. A geochemical mercury belt of soils containing high levels of mercury and mercury deposits passes through the southern parts of Europe. Mercury also enters the environment from volcanic activity, for example in 1999 it was estimated that Mount Etna contributed some 0.47 tonnes of mercury per year, whilst Stromboli some 0.07 tonnes (Ferrara et al. 2000). The geothermal province of Iceland contributed some 0.08 tonnes. Worldwide the estimated total amount of naturally emitted mercury is between 26×10^3 and 150×10^3 tonnes per year (Edner and Svanberg 1991).

Anthropogenic emissions of mercury to the atmosphere from the UK in 1990 have been estimated at 31.4 tonnes. Since then the emissions have steadily dropped to 12.4 tonnes in 1998 (NAEI 1970–98). Also in 1998, the EPA reported annual anthropogenic emissions for mercury to be about 158 tonnes in the US, the majority of which (about 87 per cent) coming from incinerators and coal-burning utility boilers. Overall, worldwide anthropogenic emissions amount to between 16×10 and 20×10^3 tonnes per year. Elemental and inorganic mercury enters the soil by being 'washed' out of the air by rainfall and snowfall, as well as by dry fallout from the atmosphere. The application of sewage sludges on agricultural land and the agricultural use of pesticides containing mercury also contribute to the soil burden. Improper dumping/disposal methods and the placing of mercury-containing materials in landfill sites can also be contributing factors. Clay minerals adsorb mercury in acidic soils – the optimum acidic conditions are pH 6.0. Clay minerals and iron oxides will also adsorb mercury in neutral soils. In acid soils, mercury is mostly adsorbed by organic matter. When organic matter is not present, mercury becomes much more mobile in acid soils, leading to its eventual evaporation to the atmosphere or leaching to groundwater. The flooding and waterlogging of vegetated areas can also release soluble methylmercury formed from soil-bound mercury(II) ions.

Lead and mercury as solid wastes

Lead, mercury and cadmium are all to be found as solid waste in municipal waste dumps. The major source of mercury in such dumps is from dry primary cells. For example, in 1989 in the US, household batteries accounted for 86 per cent of dumped mercury. In the UK the domestic waste stream contains between 20,000 and 40,000 tonnes of batteries. Of these batteries, the small 'button' type in particular contains mercury(II) oxide as one of their components. In 1997, 53 million 'button' cells were sold in the UK (Walmer Bulletin 2000). The disposal of such batteries in UK landfill sites is acceptable providing that the sites are appropriately controlled.

Secondary cells include the very common lead–acid type used on vehicles of all kinds, and the smaller nickel–cadmium type used in power tools, mobile telephones and portable computers. Some 5.4 million of both kinds were sold in the UK in 1997 (Walmer Bulletin 2000). Some lead–acid batteries do end their lives in landfill sites but many are recycled. Recycling reduces the demand on non-renewable virgin lead (also cadmium!) and the energy used in its production. In addition, lower emission levels are achieved from landfills and incinerators. In 1997, the UK processed 95,000 tonnes of lead–acid batteries. However, a slump in the price of lead and the costs of tighter emission controls have affected the profitability of those companies involved in lead recycling.

In 2000, the cost of recycling mercury(II) oxide was quoted as being £2,000 per tonne (Walmer Bulletin 2000). If mercury were to be eliminated from batteries, then the cost of recycling would reduce to £125 per tonne. Companies who are members of the European Portable Battery Association (EPBA) now sell mercury-free batteries, but imports continue to be used which contain up to 0.025 per cent by mass of mercury. For a battery to be economically recycled, mercury content must be less than 5 ppm.

In Switzerland, the use of the Batrec process enables most metal components in batteries to be extracted without generating toxic emissions. Batteries are heated in the absence of air to between 300 and 700 °C. This causes the mercury to evaporate so that it can be condensed and collected. Approximately 1.5 kg of mercury is recovered from 1,000 kg of spent batteries.

Table 8.4 *The components of solid waste*

Type	Origin/composition of type
Municipal solid waste (rubbish)	Composed of household waste, commercial waste and institutional waste; contains decomposable wastes from food and combustible materials, e.g. paper, wood; non-combustible materials, e.g. glass, metal
Commercial waste	Solid waste from offices, shops, restaurants, etc. may be included within municipal waste
Ashes	Residues from combustible solid fuels
Mining and quarrying waste	Residues from mines, e.g. slag heaps, rock piles, coal refuse piles
Large-sized waste	Construction and demolition debris, mainly inert mineral or wood waste
Sewage treatment solids	Organic waste matter from sewage treatment screens, settled solids, biomass sludge; may result from industrial or domestic waste water treatment
Agricultural waste	Mainly farm animal manure and crop remains
Industrial waste	Chemicals, sands, plastics, etc.; may contain substances dangerous to life, i.e. hazardous waste, and therefore may require special treatment

Table 8.5 *Typical annual municipal waste figures for some European countries*

Country	Annual domestic waste/ 10^3 tonnes	Equivalent per person/ kg
Austria	4,783	325
Belgium	3,410	343
Denmark	2,430	475
Finland	3,100	624
France	20,320	328
Germany	27,958	350
Italy	20,033	348
Netherlands	7,430	497
Norway	2,000	472
Spain	12,546	322
Sweden	3,200	374
Switzerland	3,000	441
UK	20,000	348
United States	173,520	720
Japan	4,977	410

(*Note*: Adapted from *Integrated Solid Waste Management* by P.R. White, M. Franke and P. Hindle (1995), by kind permission of the authors.)

Another source of mercury is from the disposal of fluorescent tubing, which contains about 0.01 per cent by mass of mercury. Most of the 1.6 tonnes of mercury resulting from the disposal of some 80 million tubes per year in the UK finds its way into land-fill sites.

The major components of solid waste are given in Table 8.4. Domestic waste can thus be seen to contain essentially rubbish and ashes. Although relatively small in overall terms, vast amounts of domestic waste are produced each year. Typically, in Great Britain 20,000,000 tonnes of domestic waste are produced annually, which is equivalent to about 348 kilograms per person (see Table 8.5 for comparisons with other EU countries, the US and Japan). The general public have the most contact with this type of waste and thus it has great political and social significance. Household waste is one of the most difficult sources of waste to manage effectively because it is a complex mixture of a wide variety of materials, each present in relatively small proportions. Its composition depends on a number of factors which can include seasonal variations (winter waste in the UK dustbin for example differs significantly from summer waste), the geographical position of a country, and whether it has been derived from rural or urban communities. In contrast, commercial, industrial and other waste materials tend to be much larger in bulk and more homogeneous in nature. Thus it can be argued that, if an effective way can be developed to manage household waste, then it should be possible to transfer the skills and resources used to the treatment of other sources of waste.

Solid waste disposal methods

A variety of methods for the disposal of solid waste have been developed and include recycling, composting, biogasification, heat treatment and landfilling. In most of the developed and emerging nations, the disposal of solid waste in landfill sites or by dumping is by far the most common method. Thermal or heat treatment (burning) accounts for most of the waste not disposed of in these ways. Composting of waste accounts for a relatively insignificant amount. The disposal method adopted depends usually upon cost which itself depends very significantly on local circumstances.

Landfill is probably the cheapest and simplest means of disposing of solid waste if done in a sanitary fashion, and is the only method which deals immediately with all of the waste (Figure 8.4). All other methods of waste treatment produce waste, which in turn has to be added to landfill. Landfill sites will always be necessary for the treatment of solid waste. Any suitable land must be situated within reasonable travelling distance of the waste source to reduce both costs and the effects of transport on the environment. The UK and most other European countries currently use landfilling as the main waste disposal method. However, environmental pressures and the increasing cost of land in the UK are having an effect on the availability of land for new sites.

Land filling may involve the creation of a dedicated hole in the ground. However, in many countries, solid waste has been used to fill exhausted quarries and clay pits and therefore could be described as contributing to land reclamation, i.e. a form of conservation. In the UK above-ground structures, such as slag and rubbish heaps built up by past generations have joined the natural contours of the surrounding district via landscaping techniques. Other countries like Japan and Hong Kong have used solid wastes to 'sea-fill' thus producing artificial islands in Tokyo Bay and extensions to runway facilities in Hong Kong harbour.

In modern landfill methodology, the solid waste is spread in thin layers, each of which is compacted by the use of heavy vehicles before the next layer is added. When about 3 metres thickness of waste has been laid down, it is covered by a thin layer of clean soil, which is then compacted (Figure 8.4). Great care has to be taken in the construction and placing of landfill sites so that local pollution does not occur.

Pollution of surface and groundwater by leaching is minimised by lining the hole and contouring the landfill site; compacting and planting the cover; selecting proper soil; diverting upland drainage; and placing waste in sites not subject to flooding or high groundwater levels. Gases can be generated in landfills via anaerobic decomposition of any organic solid waste which can result in problems unless proper ventilation

Figure 8.4 The construction of a surface landfill site – Harewood Whin Landfill Site, near York, UK. (*Note*: The caterpillar tractors spread and compact the waste prior to covering with soil.)

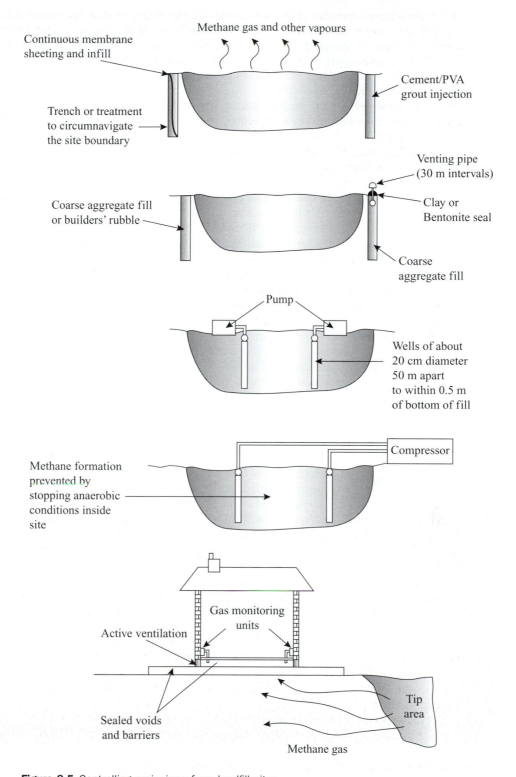

Figure 8.5 Controlling emissions from landfill sites.

is installed. Very expensive methods of controlling landfill gas have been based either on the prevention of gas emissions or on the prevention of the gas from entering into critical or confined spaces. Figure 8.5 shows how the lateral migration of gas can be controlled. In some parts of the world, landfill sites are operated as large 'bioreactors' in which the water content of the landfill is carefully controlled to maximise the production of methane gas.

The methane gas is then used in heating applications or for power generation, and thus can contribute to recovery of the costs involved in land acquisition, site construction, site operation, site closure and its long-term monitoring and aftercare.

Appropriate waste can undergo **thermal (heat) treatment**. This can occur via one of three processes. Perhaps the most well known is the mass burning, or incineration, of domestic waste in large incinerator plants. This waste may be burned at a municipal incinerator to produce an ash and energy. It may be treated using two 'select burn' processes where combustible fractions from solid waste are burned as fuels. These fuels (paper, wood, etc.) can be separated from the mixed domestic waste mechanically to form what is known as Refuse-derived Fuel (RDF) at the municipal site. Alternatively, it could be collected from people's homes and local collection points where it has already been separated. This separated paper can then be burned as a fuel and thus can be considered as a waste-to-energy technique which improves the value of the waste. Alternatively, it can be viewed as a useful pre-treatment prior to final disposal. All of the thermal methods used are very similar in terms of the underlying physical processes and issues involved.

In most incinerators the waste is burned on moving grates in refractory-lined chambers; the combustible gases and any solids they carry are then burned in secondary chambers. Combustion is normally 85–95 per cent complete for the materials that can be burned. In addition to heat, the products of incineration can include carbon dioxide, water, oxides of sulphur, oxides of nitrogen, other gases and smoke. Many of these products are potentially harmful environmental pollutants. The non-gaseous products are ash and unburned solid residues. The emission of particulate matter is often controlled by wet scrubbers, electrostatic precipitators and filters.

The burning of solid materials has four basic objectives. The first is that both its volume and mass are considerably reduced. Typically, the volume reduction can be as much as 90 per cent and the mass reduction 70 per cent. This has major environmental and economic impacts because ash will fill a smaller space in a landfill site and there will be less need for the use of vehicles to transport the waste to the site. The second objective is to reduce the amount of organic matter in the waste thus rendering it less subject to landfill gas formation and leaching. The third is to recover useful heat energy and in doing so reduce the demand on fossil fuel use. Energy derived from waste could provide 3–4 per cent of the UK energy needs. The fourth objective is to sanitise the waste by destroying potentially dangerous disease-causing organisms before the waste is placed in a landfill site. In addition to the destruction of pathogens, efficient heat treatment can destroy toxic chemicals such as dioxins. Heavy metals such as lead and cadmium are also concentrated in the ashes and clinker in quantities, which could be harmful to the biosphere. These should be found as oxides which effectively reduces the availability of the metals via leaching because these oxides are not very soluble in water.

Thermal treatment of solid waste has made a number of impacts upon the environment. First, fuel is needed to start up and continue the process of incineration. Hence, efficient as well as effective incinerators are required to lower the demands on non-renewable energy resources. The second impact is the emission of pollutants to the atmosphere via the flue gases. Whilst the emission of pollutants is now physically and legally carefully controlled in all developed countries, in the past there have been some serious cases of the continual expulsion to the atmosphere of hazardous by-products of burning, e.g. the toxin dioxin.

The **biological treatment** of solid waste uses naturally occurring micro-organisms to decompose its biodegradable components derived from plant and animal materials. There are two basic ways of decomposing organic material. One involves aerobic treatment and the other anaerobic treatment. Aerobic treatment is usually called **composting**, whilst anaerobic treatment is often referred to as anaerobic fermentation/digestion or **biogasification**. Composting is very familiar to the gardener, but on the large scale biodegradable material must be collected at centralised biological treatment plants. The type of initial treatment that takes place depends upon the origin of the organic matter. Household waste, for example, would require extensive separation techniques ranging from hand sorting to the use of electromagnets to remove ferrous-based metals. If the organic matter has been separated at source, it is called biowaste, VGF (vegetable, fruit and garden) and green wastes. This latter material would require no expensive separation at the plant before treatment. Under aerobic conditions, the products are carbon dioxide, water vapour and compost. Heat is also produced, i.e. the process of decomposition is exothermic. In the case of anaerobic conditions, methane gas, carbon dioxide, water vapour and an organic residue are produced. The latter may be useful as compost.

What are the objectives of such biological treatment and the environmental consequences? When used as a pre-treatment, it first helps to reduce the volume of the waste via loss of material in the form of carbon dioxide, water vapour and methane gas. The loss of water in particular is useful because it can help to lessen the amount of leachate should the reduced material be placed in a landfill site. Second, pre-treatment helps to stabilise the waste because it would be less subject to biological decomposition and therefore a better landfill material. Third, in the case of composting in particular, it is an effective way of sanitising the waste because the aerobic temperatures developed destroy the majority of pathogens and seeds that may be present. However, the composting process consumes more energy than it produces. The energy produced is not in a useful form and thus this process causes a net loss of useful energy from the environment.

Anaerobic conditions are mildly exothermic but the waste can be warmed by the application of external heat to a temperature that sterilises the waste. A major aim of biogasification is the production of compost and a fuel known as biogas which contains a high percentage of methane. The biogas can be stored easily, burned on site to provide electricity or sold and delivered elsewhere. Biogasification consumes energy during the process but the production of a fuel means that there is a net gain in useful energy. The compost produced by both processes can be usefully used to improve the quality of soil and act as a fertiliser. The major drawback is that the origin of the solid waste giving rise to the compost needs to be carefully monitored so that the compost is not sold containing a dangerous level of hazardous materials.

Whatever process is chosen to treat solid waste, recyclable materials can be removed before treatment is started, during its operation and at its end. As we have seen in the case of lead and mercury, the drive to recycle is often based on economic criteria and market demands. The cost of energy consumption in the transport of recyclable material to a treatment centre and in its processing needs to be balanced against the same costs involved in using virgin material. Recycling may appear to lessen environmental impact but it does involve a very careful balancing of costs vs resources and energy used.

Summary

- A heavy metal has a density equal to, or greater than, 6.0 g cm^{-3}, and includes all of the transition elements.
- Heavy metals have been deposited in soils and sediments in the form of dusts, smokes and aerosols.
- Metal pollutants can adversely affect the cation exchange capacity of a soil.
- Metal complexes are formed by the donation of a lone pair(s) of electrons to empty metal atomic orbitals. The bond formed is known as a co-ordination bond or dative covalency.
- Polydentate ligands are also called chelating agents.
- The thermodynamic stability of a complex is described by the successive stability constants/ stepwise formation constants and the overall formation/stability constant.
- When a ligand attached to a central metal ion/atom can be rapidly replaced by another ligand, then the complex is described as being kinetically labile. If the reverse is true, the complex is kinetically inert.
- Chelation stabilises a complex. Chelating agents can therefore be used to remove metals from a biological system.
- The Pb^{2+} ion is the most biologically active form of lead and is therefore the most dangerous.
- Inorganic mercury(II) ion does not readily pass through cell membranes and mercury(I) is not normally dangerous. Organic compounds of mercury are particularly dangerous to humans. Methylmercury and dimethylmercury have caused most problems.
- Lead enters the environment naturally via weathering, volcanism and plant metabolism.
- Emissions of lead to the atmosphere in the UK come from a variety of sources.
- Lead(II) sulphate appears to be the main form of entry to the soil. Its solubility in water depends upon pH.
- Mercury metal has a high volatility; therefore, it is continually being emitted to the atmosphere from soil and rocks. Volcanic activity is also a major contributor.
- Anthropogenic sources of mercury are coal-burning fires, incinerators, sewage sludges and some pesticides.
- In waste dumps, the main source of mercury contamination arises from batteries and from lead secondary cells. Both metals are extensively recycled.
- Solid waste is ultimately disposed of at landfill sites.

Questions

1 Arsenic is often classified as a 'heavy metal'.
   (a) What is the element and its compounds used for?
   (b) Find out what the natural and anthropogenic sources of arsenic are in the environment.
   (c) What are the acute and chronic effects associated with arsenic and its compounds?

2 Summarise the UK, WHO, EC and US regulations concerning the limits on arsenic in soil and sediments.

3 Mercury(II) forms the complex ion $[Hg(NH_3)_4]^{2+}$ with ammonia. Using the valence bond approach, describe how this complex may be formed and suggest a possible shape.

4 Identify the oxidation number of each of the transition metals in the following compounds and give their full names.
   (a) $[Co(NH_3)_6] Cl_2$ (b) $K_4[Fe(CN)_6]$ (c) $[Cu(H_2O)_4]SO_4.2H_2O$

5 Write down the molecular formula for each of the following compounds.
   (a) Tetraamminediaquanickel(II) chloride.
   (b) Pentaamminenitrocobalt(III) bromide

    (c)   Potassium tetrachloplatinate(III)

    (d)   Potassium tris(oxalato)rhodium(III) chlorohydroxybis(oxalato)rhodium(III) octahydrate.

6   Before building houses on some derelict land, investigations were carried out to identify its previous uses. About 50 m^2 of this land was found to be heavily contaminated with chromium salts as a result of its use as a storage area for treated cowhides. What action could be taken to make this land safe enough to build houses on or use as a garden area?

References

Agency for Toxic Substances and Disease Registry (ATSDR) (1988) *The Nature and Extent of Lead Poisoning in Children in the United States: A Report to Congress*. US Public Health Service, US Department of Health and Human Services, Atlanta, GA.

Agency for Toxic Substances and Disease Registry (ATSDR) (1992) *Case Studies in Environmental Medicine: Mercury Toxicity*. US Public Health Service, US Department of Health and Human Services, Atlanta, GA.

Agency for Toxic Substances and Disease Registry (ATSDR) (1993a) *Case Studies in Environmental Medicine: Lead Toxicity*. US Public Health Service, US Department of Health and Human Services, Atlanta, GA.

Agency for Toxic Substances and Disease Registry (ATSDR) (1993b) *Toxicological Profile for Lead*. US Public Health Service, US Department of Health and Human Services, Atlanta, GA.

Agency for Toxic Substances and Disease Registry (ATSDR) (1999) *Toxicological Profile of Mercury*. US Public Health Service, US Department of Health and Human Services, Atlanta, GA.

Bakir, F., Damliyi, S. and Amin-Zaki, L. (1973) Methyl mercury poisoning in Iraq. *Science*, **181**, 230–41.

Burgess, W.R. (ed.) National Science Foundation (1977) *Lead in the Environment*. NSF/RA-770214, NSF, Washington, DC.

Edner, H. and Svanberg, S. (1991) Lidar measurements of atmospheric mercury. *Water, Air and Soil Pollution*, **56**, 131–40.

Ferrara, R., Mazzolai, B., Lanzillota, E. *et al.* (2000) Volcanoes as emission sources of atmospheric mercury in the Mediterranean basin. *The Science of the Total Environment*, **259**, 115–22.

National Atmosheric Emissions Inventory (NAEI) (1970–98) *Annual Report: UK Emissions of Air Pollutants: Heavu Metals, Mercury*. Available at http://www.aeat.co.uk/netcen/airqual/naei/annreport/annrep98/naei98.html/

National Toxicological Program (1991) *Toxicology and Carcinogenesis of Mercuric Chloride in Rats and Mice*. US Public Health Service, US Department of Health and Human Services, Bethesda, MA.

Pocock, S.T., Smith, M. and Baghurst, P. (1994) Environmental lead and children's intelligence: a systematic survey of the epidemiological evidence. *British Medical Journal* **309**, 1189–97.

Russell, M. (2000) *Beethoven's Hair*. Bloomsbury, London.

Tamashiro, H., Arakiri, M., Akagai, H., Futatsuka, M. and Roht, L.M. (1985) Mortality and survival for Minamato disease. *International Journal of Epidemiology*, **14**, 582–8.

Tong, S., Baghurst, P. and McMichael, A. (1996) Lifetime exposure to environmental lead and children's intelligence at 11–13 years: the Port Pirie Study. *British Medical Journal* **312**, 1569–75.

USEPA (1986) *Air Quality Criteria for Lead*, June; *Addendum*, Sept. USEPA 600/8–83–018F, Research Triangle Park, NC.

USEPA (1998) *Mercury Study Report to Congress, Vol II: An Inventory of Anthropogenic Mercury Emissions in the United States,* Dec. 1997. USEPA–452/R–97–004 USEPA, Research triangle Park, NC.

Walmer Bulletin (2000) Battery recycling. *Journal of Sustainable Waste Management*, **70**, January.

White, P.R., Franke, M. and Hindle, P. (1995) *Integrated Solid Waste Management: A Life Cycle Inventory*. Blackie Academic, London.

World Health Organisation (WHO) (1990) *Methyl Mercury, Vol 101*. Distribution and Sales Service, International Program on Chemical Safety, Geneva, Switzerland.

World Health Organisation (WHO) (1991) *Inorganic Mercury, Vol 118*. Distribution and Sales Service, International Program on Chemical Safety, Geneva, Switzerland.

Further reading

Earnshaw, A. and Greenwood, N.N. (1997) *Chemistry of the Elements*, 2nd edn. Butterworth-Heinemann, Oxford.

Good chapter on co-ordination compounds. The transition metals are extensively treated in their respective groups, e.g. zinc, cadmium and mercury.

Mingos, D.M.P. (1995) *Essentials of Inorganic Chemistry*. Oxford Chemistry Primers, Oxford University Press, Oxford.

This is an alphabetical list of some of the important concepts in inorganic chemistry. It is better than a dictionary! First-year undergraduates will find it useful as a quick reference, as well as those studying ancillary chemistry.

The Walmer Bulletin. *Journal for Sustainable Waste Management*.

This international journal is published six times per year. It contains wide ranging articles on waste management, e.g. battery recycling, scrap tyre management, life cycle assessment, anaerobic digestion, etc. Very easy to read with useful references.

White, P.R., Franke, M. and Hindle, P. (1995) *Integrated Solid Waste Management: A Life Cycle Inventory*. Blackie Academic, London.

Chapters 5, 8, 9, 10 and 11 are particularly relevant to the generation and treatment of solid waste.

9 The chemistry of the atmosphere

- Current structure and composition of the Earth's atmosphere – number density, atmospheric regions
- The ideal gas. Universal gas constant. standard temperature and pressure
- The gas laws – the laws of Boyle, Charles, Avogadro, Dalton, Graham and Henry
- The absolute temperature scale
- Photosynthesis
- Geochemical cycles – the carbon, nitrogen and sulphur cycles
- Free radicals and complex reactions
- Initiation, propagation, branching and termination steps in free radical mechanisms
- Ozone in the atmosphere. Chapman oxygen-only mechanism. Catalytic families, reservoir molecules

The current structure and composition of the atmosphere

Table 9.1 lists some of the most common components currently found in the Earth's atmosphere. Note that not all substances are listed, e.g. neon, helium, hydrogen, particulate matter or pollutants. The concentrations are expressed as mixing ratios, or in ppm, and are average figures for the troposphere.

The atmosphere extends to an altitude of approximately 800 km above the Earth's surface. It becomes progressively less dense the higher the altitude and so the atmospheric pressure declines. This variation is shown schematically in Figure 9.1. A convenient measure of pressure is the **number density**,

$$\text{number density} = \frac{\text{total number of molecules/atoms present}}{\text{volume of air}} = \frac{N}{V}$$

The **mixing ratio** by volume of an atmospheric component 'A' then becomes,

$$\text{mixing ratio} = \frac{\text{number density of component } A}{\text{total number density of air}} = \frac{N(A)/V}{N/V} = \frac{N(A)}{N}$$

The mixing ratio can thus be viewed as the **fractional abundance** of each component. Dinitrogen and dioxygen are the main or bulk components of the troposphere up to an altitude of about 80 km. These two gases are well mixed in a ratio of about 4:1.

Table 9.1 *The most common components of the Earth's atmosphere*

Gas	Concentration
Dinitrogen, N_2	Mixing ratio 0.781
Dioxygen, O_2	Mixing ratio 0.209
Argon, Ar	93,000 ppm
Water, H_2O	3,000 ppm
Carbon dioxide, CO_2	353 ppm
Ozone, O_3	0.01–0.1 ppm
Methane, CH_4	1.72 ppm
Nitrous oxide, N_2O	0.31

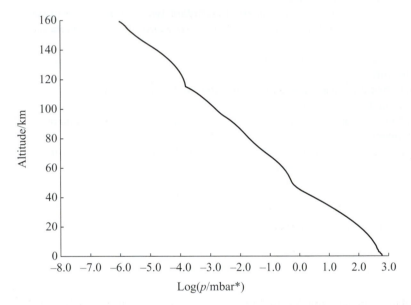

Figure 9.1 The variation of atmospheric pressure with altitude. (*Note*: *mbar is the pressure unit used by meteorologists and scientists who study the atmosphere.)

For the trace components, however, their mixing ratios vary considerably with longitude, latitude and sometimes with the season and time of day.

The atmosphere is separated into four main regions (Figure 9.2). The lowest region, the **troposphere**, contains about 90 per cent of all of the matter found in the atmosphere. The troposphere extends to a height of between 8 and 15 km depending upon the latitude – for example, it is higher and colder at the equator than at the poles. The temperature of the troposphere falls with increasing altitude to a minimum of between 200 and 220 K ($-73\ °C$ to $-53\ °C$) at the boundary between this layer and the next (the **tropopause**). Mixing ratios can show very distinct and abrupt changes at the tropopause.

The next region, the **stratosphere**, extends to a height of about 50 km. The temperature within this region rises to approximately 290 K (17 °C) until the upper boundary of the stratosphere is reached (the **stratopause**).

The Sun's rays warm the Earth's surface, which then warms the air immediately in contact with it. Warmer air is less dense than colder air and so rises causing convection

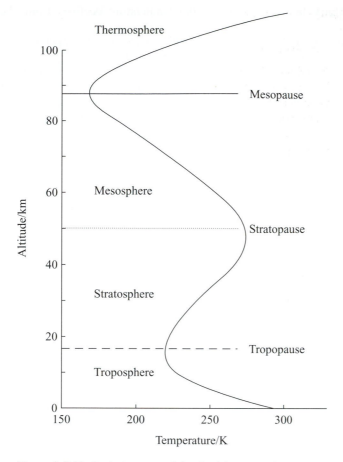

Figure 9.2 Vertical structure of the Earth's atmosphere.

currents that warm the base of the troposphere. Thus the troposphere is warmed from *below*. The troposphere is therefore characterised by strong vertical mixing, which gives rise to weather patterns and the rapid mixing of its material content. It is within the troposphere that many important chemical reactions occur such as photosynthesis and nitrogen fixation.

Chemical reactions occurring in the troposphere ensure that it is warmer than the troposphere. This change from a decreasing temperature at the top of the troposphere to an increasing temperature is called a **temperature inversion**. The region where this starts to happen is the tropopause. The temperature inversion acts like a porous lid that separates the turbulent troposphere from the calmer stratosphere. The vertical movement and mixing through the stratosphere is much slower than in the case of the tropopause (taking years as opposed to days).

Rapidly rising air from the troposphere can penetrate the tropopausal lid and so transfer matter to the stratosphere. Violent tropical storms in particular can contribute to such mixing. Additionally, stratospheric material can be transferred through the tropopause, particularly in the middle latitudes. There is thus continual, slow exchange of matter between the two regions. Above the stratosphere, the **mesosphere** extends from an altitude of 50 km to about 90 km. This is followed by the **thermosphere**

(90 km to about 800 km). In the mesosphere the temperature declines from about 280 to 200 K.

A second temperature inversion occurs at about 90 km where the temperature rises from 200 K to 300 K. Again, chemical reactions are responsible for this change in temperature. The study of the movement or 'general circulation' horizontally of air in the lower atmosphere is carried out by meteorologists and will not be discussed in any detail here. However, there are patterns to be observed such as the occurrence of prevailing winds and weather conditions, which are important in what happens to any matter that enters the atmosphere. The general circulation of air is caused by the temperature difference between the warm low and cold high latitudes. This leads to strong zonal wind movement, i.e. east to west. Wind moves and helps to mix materials in the atmosphere. Zonal mixing is much faster than north–south (medicinal) mixing. The movement of a pollutant, for example, takes approximately one month to uniformly mix along a line of latitude, whilst it takes six months to move and mix throughout a hemisphere. To cover the globe takes between one and two years.

The gas laws

An examination of the behaviour of a gas often involves practical calculations using equations, which relate the pressure 'p', volume 'V', the number of moles 'n' and the absolute temperature 'T' of that gas.

An **ideal gas** is a hypothetical one that is said to obey Boyle's, Avogadro's and Charles' laws at all temperatures and pressures.

The **ideal gas equation** linking these three laws is,

$$pV = nRT$$

It is an approximate equation which is obeyed by gases only at low pressures.

Great care must be taken in using this equation to ensure that the units of each component are compatible. The SI unit for pressure is the Newton per square metre ($N\,m^{-2}$) otherwise called the Pascal (Pa), that for volume the m^3, and for temperature Kelvin (K). The mole is abbreviated as **mol**. The constant R is termed the **universal gas constant** and has the value of 8.3l4 $J\,K^{-1}\,mol^{-1}$ in SI units.

Both the numerical value of R and its units depend on the units of p, V, n and T. For example, for 1.00 mole of a gas at a pressure of 1.00 atmosphere and a temperature of 273.15 K that occupies a volume of 22.4 litres, the value and units of R would be calculated as follows:

From $pV = nRT$, then $R = pV/nT$, so that

$$R = \frac{1.00 \text{ atm} \times 22.4\, l}{1.00 \text{ mol} \times 273.15 \text{ K}} = 0.08206 \text{ atm}\, l\,K^{-1}\,mol^{-1}$$

When p is in torr and V in litres, $R = 62.37$ torr $l\,K^{-1}\,mol^{-1}$

Special attention must be paid to the units of pressure, volume and temperature when calculating any properties of a gas (Tables 9.2 and 9.3). *Always use SI units where possible.*

The standard state of a gas is the **standard temperature and pressure (STP)**. The standard temperature is 0 °C and the pressure 1 atmosphere.

Table 9.2 *Pressure units and conversion factors*

SI unit of pressure	Pascal (Pa)	$1 \text{ Pa} = 1 \text{ N m}^{-2}$
Conventional units	Bar	$1 \text{ bar} = 100 \text{ kPa}$
		$= 1 \times 10^5 \text{ Pa}$
		$= 1 \times 10^5 \text{ N m}^{-2}$
		$1 \text{ mbar} = 1 \times 10^2 \text{ Pa}$
	Atmosphere	$1 \text{ atmosphere} = 101.325 \text{ kPa}$
		$= 1.013 \times 10^5 \text{ N m}^{-2}$
		$1 \text{ atmosphere} = 760 \text{ torr}$
		$1 \text{ atmosphere} = 760 \text{ mmHg}$
		$= 14.70 \text{ lb in}^{-2}$
	Torr	$1 \text{ torr} = 133.3 \text{ Pa}$
	Mm of Hg	$1 \text{ mm Hg} = 133.3 \text{ Pa}$

Table 9.3 *Volume units and conversion factors*

SI Unit	m^3
Conversion factors	$1 \text{ m}^3 = 1 \times 10^6 \text{ cm}^3$
	$1 \text{ l (litre)} = 1 \text{ dm}^3$
	$= 1 \times 10^3 \text{ cm}^3$
	$1 \text{ dm}^3 = 1 \times 10^{-3} \text{ m}^3$

(*Note*: The ml (millilitre) normally taken to mean 1 cm^3 is still in common use but is very dangerous to use in calculations!)

The value of the ideal gas equation can be seen by the examples in Boxes 9.1 and 9.2.

Boyle's Law is an empirical law, i.e. one determined by experiment, and relates the decrease in volume of a gas as the pressure on it is increased. For a fixed amount (n) of gas at a constant temperature (T), a plot of volume (V) vs pressure (p) would give,

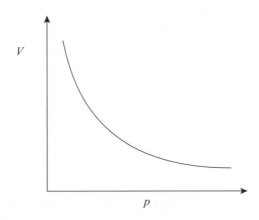

That is, as pressure increases the volume decreases.

Box 9.1

The ideal gas equation (1)

Question
A 10-litre cylinder containing 500 g of dichlorine gas is to be used for the sterilisation of water. What will be the pressure inside the cylinder at a temperature of 20 °C?

Answer
Step 1   Convert the volume to SI units.

$$V = 10 \text{ dm}^3 = 10 \times 10^{-3} \text{ m}^3$$

Step 2   Convert Celsius to Kelvin.

$$T = (273 + 20)\text{K} = 293 \text{ K}$$

Step 3   Calculate the number of moles of dichlorine.

Dichlorine has a relative molecular mass of 71.00. Hence its molar mass is 71.0 g mol^{-1}. Thus the number of moles 'n' of dichlorine in the cylinder is given by,

$$n = 500 \text{ g}/71.0 \text{ g mol}^{-1} = 7.04 \text{ mol}$$

Step 4   Calculate the pressure 'p' using the ideal gas equation.

$$p = nRT/V = 7.04 \times 8.314 \times 293/10 \times 10^{-3} \text{ Pa}$$

$$p = 1.715 \times 10^6 \text{ Pa} = 17 \text{ atmospheres}$$

In order to give a straight line relationship, a plot of V vs $1/p$ is plotted, i.e. as V increases, p decreases so $1/p$ increases,

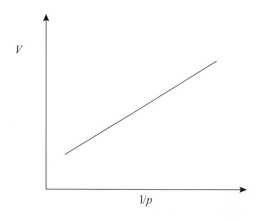

Box 9.2

The ideal gas equation (2)

Question
A gas at 390 °C and a pressure of 4 atmospheres occupies a volume of 10 m^3. What volume would it occupy at STP?

Answer
Because the units involved here are mixed, pressure, temperature and volume must all be in SI units.

STP is now given by $p = 1.01 \times 10^5$ Pa and $T = 273$ K.

From the ideal gas equation, $pV/T = nR$. The amount of gas has not changed and therefore pV/T = a constant. So, for conditions p_1, V_1 and T_1, p_1V_1/T_1 = a constant and for conditions p_2, V_2, T_2, p_2V_2/T_2 must equal the same constant. Hence,

$$p_1V_1/T_1 = p_2V_2/T_2$$

Thus if $p_1 = 1.01 \times 10^5$ Pa, $T_1 = 273$ K, we need to calculate V_1.

Let $p_2 = 4 \times 1.01 \times 10^5$ Pa and $T_2 = (273 + 390)$K $= 663$ K and $V_2 = 10$ m^3.

The unknown in this equation is V_1, thus rearranging for V_1 gives,

$$V_1 = \frac{p_2V_2T_1}{p_1T_2} = \frac{4 \times 1.01 \times 10^6 \times 10 \times 273}{1.01 \times 10^6 \times 663}$$

$$V_1 = 16.5 \text{ m}^3$$

Boyle's law thus states that 'the volume occupied by a fixed amount of gas at a constant temperature is inversely proportional to its pressure', i.e. $V \propto 1/p$ or $V = k/p$ and therefore $Vp = k$, where k is a constant. Gases obey this equation at, and below, about 1 atmosphere pressure.

Charles' Law is also an empirical law which relates the volume and temperature of a gas. For a fixed amount of gas at a constant pressure, a plot of its volume vs temperature in Celsius (C) would give,

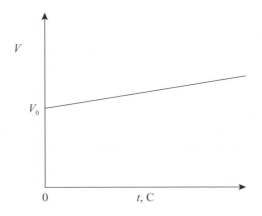

If the slope of the line were to be measured, then it would be found that the volume of the gas expands by approximately 1/273 of its volume at 0 °C for every degree rise in temperature. So one way of stating Charles' Law is 'fixed amounts of gases expand by 1/273 of their volume at 0 °C per degree rise in temperature, providing the pressure is kept constant'.

If the volume of a gas at 0 °C is V_0, then for 1 °C rise in temperature, the new volume V will be given by $V = V_0 + V_0/273$. Hence for a t °C rise,

$$V_t = V_0 + tV_0/273 = (273V_0 + tV_0)/273 = V_0 (273 + t)/273$$

Now, $V_0/273 =$ a constant $= k^1$, and hence $V_t = k^1(273 + t)$

When $t = -273$ °C, then $V_t = 0$, so at a temperature of -273 °C a gas will theoretically have a zero volume. This is equivalent to extrapolating the V vs t graph to the $V = 0$ point, e.g.

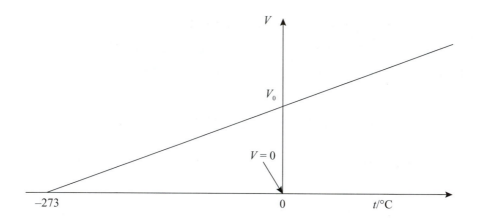

Accurate work shows the theoretical temperature at which an ideal gas has zero volume to be -273.15 °C.

The Kelvin temperature scale has its origin at 0 K which corresponds to -273.15 °C. Hence to convert the Centigrade scale to the Kelvin scale, one adds on 273.15. Temperatures expressed in this way are **absolute temperatures.**

In the equation

$$V_t = k_1(273.15 + t)$$

$$273.15 + t = T\,K$$

or $V = k_1 T$

where T is the absolute temperature measured in Kelvin.

Charles' Law now states that the volume of a fixed amount of gas at constant pressure is directly proportional to its absolute temperature. An example application of this law is given in Box 9.3.

Box 9.3

Charles' Law

Question
An industrial tank with a floating top (similar to the North Sea gas tanks at domestic supply centres) contains 10,000 m³. During the night, the temperature of the gas is 1 °C, which warms up to 27 °C during the daytime. Assuming that no gas is used and the pressure inside the tank remains constant, what volume will the tank have at 27 °C?

Answer

$$\text{Let} \quad T_1 = (273 + 10) \text{ K} = 283 \text{ K, and } V_1 = 10.000 \text{ m}^3$$

$$T_2 = (273 + 27) \text{ K} = 300 \text{ K, and } V_2 = ?$$

To solve this problem, we can use Charles' Law,

$$T_1 V_1 = T_2 V_2$$

Thus,

$$V_2 = \frac{T_2 V_2}{T_1} \quad \text{or} \quad V_2 = \frac{300 \times 10,000}{273} = 10,989 \text{ m}^3$$

Avogadro's Law states that, at a constant temperature and pressure, the volume of a gas is proportional to the amount (number of moles) of gas present,

i.e. $V \propto n$

or $V = k_2 n$ where k_2 is a constant

The ideal gas law is a combination of the preceding laws.

We have $pV = k$, $V = k_1 T$, and $V = k_2 n$

i.e. $pV \propto nT$

or $pV = RnT$

where R is the constant of proportionality called the universal gas constant.

The latter equation is known as the ideal gas equation and is usually written as,

$pV = nRT$

Box 9.4

Partial pressures

Question
Dinitrogen is approximately 78 per cent of the atmosphere by volume and dioxygen 22 per cent. At a prevailing atmospheric pressure of 1 atmosphere, what would be the partial pressures of these two gases?

Answer
Since 1 atmosphere is equal to 101 kPa, then the partial pressures of the two gases will be given by,

partial pressure of N_2 = 78/100 × 101 kPa = 78.8 kPa

partial pressure of O_2 = 22/100 × 101 kPa = 21.2 kPa

(*Note*: If CO_2 is 0.035 per cent by volume of the atmosphere, then its partial pressure would be, 0.035/100 × 101 kPa = 0.035 kPa.)

Dalton's Law applies to mixtures of gases and their partial pressures. The **partial pressure** of a gas in a mixture is the pressure that the gas would exert if it occupied the container by itself (Box 9.4). Hence in a mixture of gases, assuming that there are no intermolecular forces, Dalton's Law states that the total pressure exerted by that mixture will be the sum of the partial pressures of the components A, B, C, etc.

Hence, the total pressure p_T in the container is,

$$P_T = p_A + p_B + p_C + \dots$$

Graham's Law of Diffusion/Effusion is based on the observation that gases occupy all available space. Effusion is the passage of a gas through a small orifice under the influence of a pressure gradient. It is often seen in the form of leaks from high pressure and vacuum apparatus. The law states that the rate of diffusion/effusion of a gas is inversely proportional to the square root of its molar mass. This means that the time required for diffusion is directly proportional to the square root of its molar mass, i.e. $t \propto \sqrt{M}$.

If two different gases A and B are compared at the same temperature and pressure then,

$$t_A/t_B = \sqrt{(M_A/M_B)}$$

The solubility of a gas depends upon its pressure – the higher the pressure the more soluble it is. How the solubility depends on pressure is expressed by **Henry's Law**, i.e. the solubility of a gas in a liquid is directly proportional to its partial pressure, or

$$S = K_H \times p$$

Box 9.5

Henry's Law

Question
Assuming that air contains 0.035 per cent by volume of CO_2, what is the solubility of this gas in water at 20 °C? (K_H for CO_2 is 1,720 mg dm^{-3} atm^{-1} at 20 °C.)

Answer
Using Henry's Law, $S = 1,720 \times 0.035/100 = 0.6$ mg dm^{-3}

---

where S = equilibrium solubility, p = partial pressure and K_H is called Henry's constant. This latter constant depends upon the gas, the solvent and the temperature. Its units depend upon those used for solubility and pressure. Box 9.5 illustrates the use of Henry's Law.

Photosynthesis

Photosynthesis is a complex redox process by which plants, algae and certain bacteria, using the energy of sunlight, convert carbon dioxide and water into carbohydrates (sugars) and dioxygen. The catalysts used in plants are the chlorophyll pigments, carotenoids and the phycoblins. These pigments absorb light and help to convert it into chemical energy via the formation of new chemical bonds. The overall reaction can be simply represented by,

$$CO_2\,(g) + H_2O(l) \xrightarrow[\text{sunlight energy}]{\text{chlorophyll}} [CH_2O](aq) + O_2(g)$$

Photosynthesis is the main way that foodstuffs are produced for the higher animals, atmospheric dioxygen is replenished and energy obtained from the Sun is stored. Plants that can photosynthesise are therefore referred to as the primary producers in the food chain. All other organisms that feed on plants in order to use their organic compounds in respiration and as an energy source are called consumers.

The above reaction is the overall result of two processes. The first process involves the light-induced oxidation of water, referred to as the **light reaction**. This reaction can be written,

$$2H_2O \longrightarrow O_2 + 4H^+ + 4e^-$$

This oxidation is accompanied by a reduction reaction which results in a compound called NADPH (reduced nicotinamide adenine dinucleotide phosphate),

$$\text{i.e. } NADP^+ + H_2O \longrightarrow NADPH + H^+ + 1/2\,O_2$$

A second reaction is coupled with the above to form a highly energetic compound called ATP (adenosine-5-triphosphate). Here ADP (adenosine-5-diphosphate) adds on an additional phosphate group PO_4^{3-} during the light reaction. This reaction is termed **photophosphorylation**,

Box 9.6

More about photosynthesis

In the cells of plants there are structures called chloroplasts, which are organelles, containing liquid **stroma** and sacs called **thykaloids**. The thykaloids contain the light-capturing pigments that are necessary for the light reaction. The dark reaction occurs in the stroma.

Chlorophyll a (Figure 9.3) and **chlorophyll b** are the primary pigments used in photosynthesis. Other **accessory pigments**, including some chlorophyll molecules, help to serve certain chlorophyll a molecules with energy. In Figure 9.3, the central Mg^{2+} ion is surrounded by a **porphyrin ring**. This ring is found in many biologically important molecules such as haemoglobin and vitamin B_{12}. The porphyrin ring can be seen to contain a number of alternately double-bonded–single-bonded carbon atoms. This kind of system is called a conjugated system in which some of the bonding electrons between the carbon atoms are easily delocalised, i.e. they spread themselves across a number of carbon atoms. Such electrons can be easily energetically excited to higher energy levels. The energy to do this is found in the visible region of the electromagnetic spectrum. Chlorophyll a and b both absorb wavelengths in the red and blue regions, which cause plants to look green (Figure 9.4).

There are two light activated systems that take part in photosynthesis **photosystem I (PSI)** and **photosystem II (PSII)**. In PSII light energy is acquired by chlorophyll a molecules in sufficient amounts to cause ionisation,

i.e. chlorophyll a + energy $\longrightarrow$ chlorophyll $a^+ + e^-$

This electron and its associated energy can be passed on via a series of electron acceptors to feed the second PSI. The ionised chlorophyll in PSII obtains a replacement electron from the oxidation of water. Hereafter, the absorption of more light energy causes the chlorophyll a molecule to ionise again and the process of electron transfer to PSI continues.

In PSI chlorophyll a is simultaneously ionised by energy from sunlight and its electron moves down a chain of electron acceptors until it causes reaction by the chemical $NADP^+$ with H^+ ions to form the highly energetic compound NADPH. In addition, it is within the thylakoids that ADP and inorganic PO_4^{3-} undergo photophosphorylation, a condensation reaction, to form ATP. ATP formation and the acquisition of energy are coupled in some way to the flow of electrons from PSII to PSI. This process is known as a **non-cyclic photophosphorylation**. When the availability of $NADP^+$ is low, the electrons moving from PSI are returned to the chain from PSII to PSI, thus forming a cyclic movement of electrons. These electrons may also help to form ATP in **cyclic photophosphorylation**.

The role of ATP in living organisms is to provide energy that will cause various other chemical reactions to occur. It is the main carrier of energy in metabolic processes. NADPH acts as a carrier of hydrogen and electrons. Thus both of these compounds are necessary for the completion of the dark reactions in the stroma, where carbon dioxide takes place in a cycle of reactions (Calvin Cycle) to produce carbohydrates. During these reactions $NADP^+$ and ADP are regenerated and are reused in the phosphorylation light reactions (Figures 9.5 and 9.6).

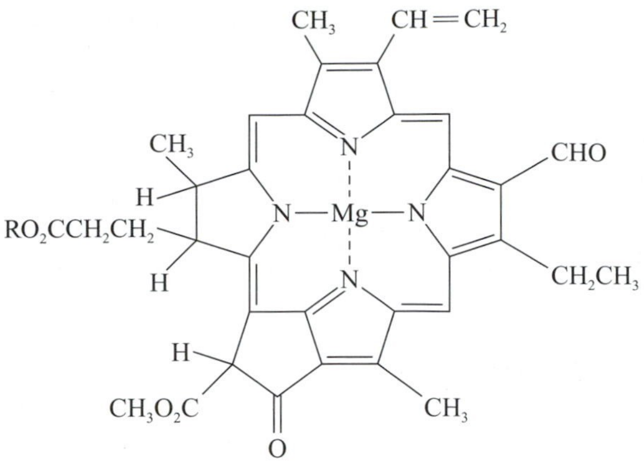

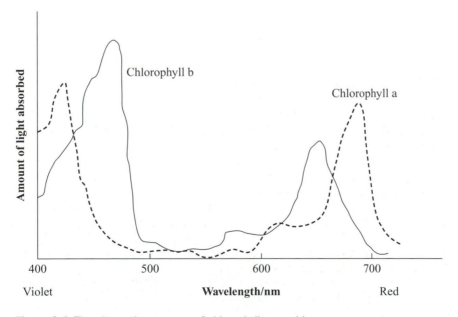

Figure 9.3 Chlorophyll a where R = phytyl, $C_{20}H_{39}$.

$$ADP + PO_4^{3-} \longrightarrow ATP$$

The light reaction is thus the process by which photosynthesising organisms absorb light, store its energy and produce dioxygen. The stored energy is reflected in the form of chemical bonds in NADPH and ATP. Specifically, NADPH is a carrier of hydrogen atoms and electrons and ATP the main source of energy.

In the second process, both ATP and NADPH are used to reduce carbon dioxide to sugars. In doing so, NADPH is oxidised to $NADP^+$ and ATP forms ADP. This reaction does not require light and is therefore referred to as the **dark reaction**,

i.e. $CO_2 + 4H^+ + 4e^- \longrightarrow (CH_2O) + H_2O$

Figure 9.4 The absorption spectra of chlorophylls a and b.

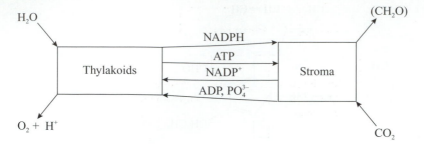

Figure 9.5 The material inputs and outputs of photosynthesis.

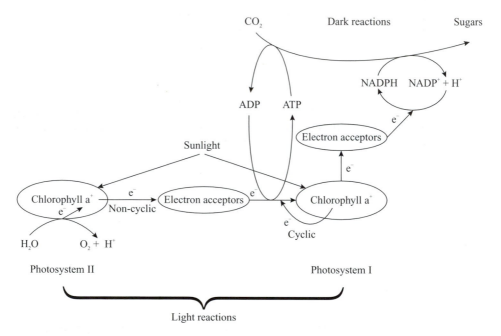

Figure 9.6 Schematic diagram of photosynthesis.

Both the dark reaction and light reaction occur simultaneously, and only as long as $NADP^+$ lasts will the reactions continue in darkness. (For further details on photosynthesis, see Box 9.6 and Figures 9.3–9.6.)

Geochemical cycles

Matter in the environment is undergoing continual chemical and physical changes. The consequence of these is that there is an ever-natural interchange of matter between the atmosphere, lithosphere and hydrosphere. The chemistry of the atmosphere is therefore dependent on what happens in the lithosphere and the hydrosphere. Anthropogenic actions can add or subtract to/from this movement of matter which has, on occasions, had detrimental and advantageous effects on the environment. However, in terms of amounts and geological time human influences have been largely negligible.

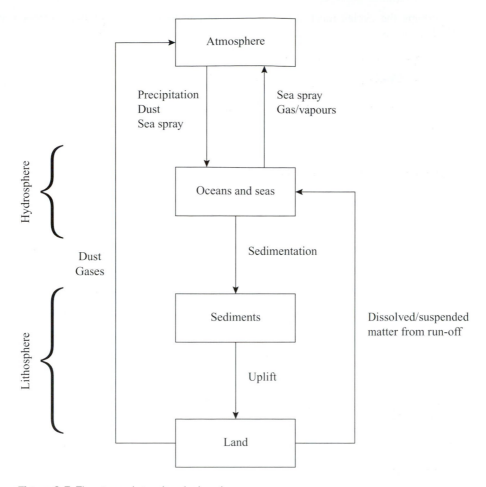

Figure 9.7 The general geochemical cycle.

One way of showing the movement of matter through the global geological environment is via the use of geochemical cycles. Figure 9.7 shows a general geological cycle.

The boxes in these cycles represent **reservoirs** of materials, whilst the lines show the direction of movement (**flux**) of matter between these reservoirs. Providing the system is in dynamic equilibrium, the average time that a particular substance remains in a reservoir is called the **residence time**, such that

$$\text{Residence time} = \frac{\text{amount of a material in a reservoir}}{\text{rate of removal or addition of that material}}$$

Materials that have a long residence time can accumulate to much higher concentrations than those with a short one, and therefore may have a greater influence on the environment. Gases have residence times ranging from days to a few thousand years, e.g. methane gas has a mean residence time of approximately four years in the atmosphere, and substances dissolved in water millions of years, e.g. sodium dissolved in seawater about 210 million years. Residence times in the lithosphere are even longer. This variation reflects the mobility of the materials and their abilities to chemically react with other chemicals.

Perhaps the cycles most relevant to the atmosphere are the carbon, nitrogen and sulphur cycles.

The carbon cycle

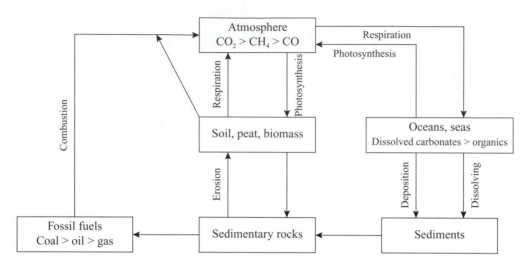

Figure 9.8 The carbon cycle.

The carbon cycle is shown in Figure 9.8. It is clear that the amount of carbon dioxide in the troposphere depends upon the various processes involved in the cycle. Most of the Earth's carbon is found locked away in the form of carbonate rocks and in organic matter. There is relatively very little carbon found in the atmosphere, but what there is has a fundamental effect on living organisms and the environment in general. Carbon dioxide is removed by photosynthesis, whilst respiration returns the gas to the atmosphere.

Carbon dioxide dissolves in water to form a dynamic equilibrium,

$$CO_2(g) \rightleftharpoons CO_2(aq)$$

Cold water is denser than warm water and therefore will sink to great depths taking dissolved carbon dioxide with it. This means that carbon dioxide will be effectively removed from the atmosphere and therefore the oceans act as **atmospheric sinks** for carbon dioxide. Although an overall equilibrium exists, there are some variations within parts of the cycle. For example, there is naturally more carbon dioxide in the atmosphere at night and in winter due to low photosynthesis. Carbon dioxide is one of the results of the combustion of fossil fuels and therefore more can result in colder weather.

The nitrogen cycle

The nitrogen cycle is shown in Figure 9.9. Dinitrogen, N_2, comprises approximately four-fifths of the atmosphere. The enthalpy change of dissociation of dinitrogen is +945 kJ mol^{-1}, and its ionisation energy is +1,505 kJ mol^{-1} for the reaction,

$$N_2(g) = N_2^+(g) + e^-$$

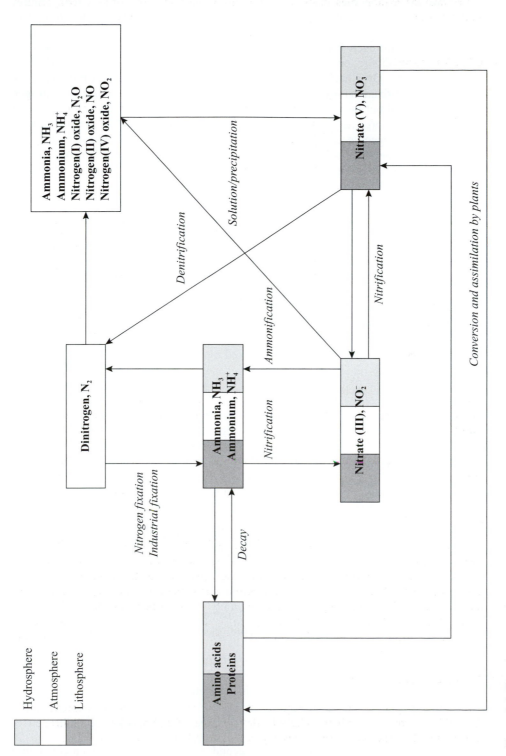

Figure 9.9 The nitrogen cycle.

These two latter properties make dinitrogen chemically inert at ordinary temperatures and pressures. The availability of dinitrogen from the atmosphere is therefore very restricted.

Atmospheric nitrogen becomes part of the biosphere via **nitrogen fixation**. This process converts dinitrogen into a soluble form of nitrogen that plants can take up through their root system. Over half of the yearly dinitrogen fixation occurs via processing by plants, the other fixations occur through industrial and combustion processes. Nitrogen fixation is carried out by anaerobic bacteria and blue-green algae found in the soil and the oceans. These organisms convert dinitrogen to ammonia using the enzyme catalyst nitrogenase. Nitrogenase is also provided by symbiotic organisms found in the root nodules of legumes such as clover and peas. The mechanism involved is very complex and involves the element molybdenum and iron protein molecules of very high relative molecular mass. The ammonia is used in the synthesis of amino acids,

$$\text{e.g. } 2NH_3(aq) + 2H_2O(l) + 4CO_2(g) = 2NH_2CH_2COOH(aq) + 3O_2(g)$$

Nitrification reactions in the soil oxidise ammonia to nitrate(III) (nitrite) ion, NO_2^-, and then to nitrate(V) (nitrate) ion, NO_3^-,

$$2NH_3(aq) + 3O_2(g) = 2H^+(aq) + 2NO_2^-(aq) + 2H_2O(l)$$

$$2NO_2^-(aq) + O_2(g) = 2NO_3^-(aq)$$

Approximately 4 per cent of dinitrogen is fixed by the energy provided by lightning,

$$N_2(g) + O_2(g) = 2NO(g)$$

Because the concentration of the nitrogen(II) oxide (nitrogen monoxide) is so low, its subsequent oxidation to nitrogen(IV) oxide (nitrogen dioxide) is slow.

The amounts of oxides of nitrogen in the atmosphere are decreased by their reaction with water,

$$3NO_2(g) + H_2O(l) = 2HNO_3(aq) + NO(g)$$

Dinitrogen re-enters the atmosphere when nitrate(V) and nitrate(III) are decomposed in the soil. This is done by catalysis using bacterial enzymes,

$$4NO_3^-(aq) + 2H_2O(l) = 2N_2(g) + 5O_2(g) + 4OH^-(aq)$$

Some anaerobic organisms can use nitrate(V) instead of dioxygen as an oxidising agent and provider of energy. This also releases dinitrogen back into the atmosphere,

$$5(CH_2O)(aq) + 4NO_3^-(aq) + 4H^+(aq) = 2N_2(g) + 5CO_2(g) + 7H_2O(l)$$

Sometimes the conversion to dinitrogen is incomplete,

$$6NO_3^-(aq) + 6(CH_2O)(aq) = 6CO_2(g) + 3H_2O(l) + 6OH^-(aq) + 3N_2O(g)$$

Nitrogen(I) oxide (nitrous oxide) is the second most abundant nitrogen-containing species in the atmosphere (see Table 9.1).

The sulphur cycle

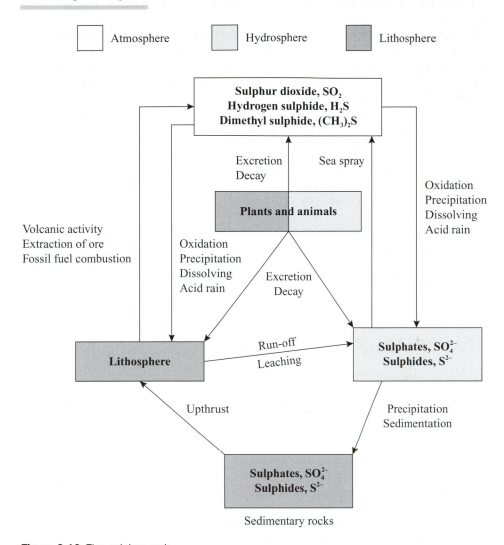

Figure 9.10 The sulphur cycle.

Sulphur and its compounds are not found to any great extent in the atmosphere but it is involved in a complex geochemical cycle. The major reservoir for sulphur is the lithosphere (Figure 9.10).

The decay of dead organic matter by aerobic bacteria gives rise to hydrogen sulphide, H_2S, gas in the atmosphere, which is rapidly oxidised to SO_4^{2-} in the presence of water and oxygen,

$$H_2S(g) + 4H_2O(l) - 8e^- = SO_4^{2-}(aq) + 10\ H^+(aq) \qquad \text{oxidation}$$

$$2O_2(g) + 8H^+(aq) + 8e^- = 4H_2O(l) \qquad \text{reduction}$$

$$H_2S(g) + 2O_2(g) = SO_4^{2-}(aq) + 2\ H^+(aq) \qquad \text{overall}$$

The above reaction takes a few hours to several days to occur in the atmosphere, and probably takes place on aerosol droplets catalysed by metal ions.

Marine phytoplanktons are the major producers of dimethyl sulphide, $(CH_3)_2S$, which is oxidised by dioxygen to sulphur dioxide and subsequently to the sulphate(VI) ion, SO_4^{2-}.

Sulphur dioxide is the product of the oxidation of sulphur-containing compounds. Once it is formed it readily dissolves in water (11 g per 100 g of water at 1 atmosphere pressure and 298 K) to form acidic solutions,

$$SO_2(g) + H_2O(l) \rightleftharpoons SO_2(aq)$$

$$SO_2(aq) + H_2O(l) \rightleftharpoons H_2SO_3(aq)$$

$$H_2SO_3(aq) + H_2O(l) \rightleftharpoons H_3O^+(aq) + HSO_3^-(aq)$$

$$HSO_3^-(aq) + H_2O(l) \rightleftharpoons H_3O^+(aq) + SO_3^{2-}(aq)$$

In the atmosphere, sulphur dioxide is oxidised further to sulphur trioxide by one of three possible ways, catalytic oxidation, photo-oxidation and oxidation by free radicals.

The rate of catalytic oxidation is slow in dry air, but if aerosols are present that contain metal ions such as Cu^{2+} or Fe^{3+} it becomes much faster. Indeed, the higher the humidity, the faster the reaction.

Photochemically, UV radiation energetically excites the sulphur dioxide molecule so that chemical reaction can occur between it and oxygen,

$$SO_2(g) + UV \longrightarrow SO_2^*(g)$$

where SO_2^* represents the excited molecule,

$$SO_2^*(g) + O_2(g) = SO_3(g) + O(g)$$

Oxidation by free radicals depends upon the formation of these species from methane and nitrogen dioxide.

Generally, the reactions of sulphur compounds in the atmosphere result in sulphuric acid or metal sulphates.

An introduction to free radical chemistry

In order to understand many of the chemical reactions that occur in the atmosphere, it is necessary to examine what is meant by a **free radical** and how they undergo chemical reactions.

Ozone (O_3) undergoes decomposition in the gas phase at 293 K, when dibromine (Br_2) is present as the catalyst. The overall reaction follows time independent stoichiometry as follows,

$$2O_3(g) = 3O_2(g)$$

Kinetic investigations show the reaction to be first order with respect to ozone, and one-half with respect to dibromine, i.e.

$$J = \frac{1}{3}\frac{d[O_2]}{dt} = k_r[Br_2(g)]^{1/2}[O_3]$$

This suggests that the reaction is not an elementary one but consists of a number of steps. In addition, the rate of reaction can be accelerated or slowed down (inhibited) by the addition of traces of other chemicals. Such behaviour, i.e. one so sensitive to the addition of trace quantities of materials that do not appear in the stoichiometry, is indicative of a **complex reaction** occurring that involves highly reactive intermediate species. These have been identified as free radicals which are chemical species with an **unpaired electron**. Free radicals normally require large amounts of energy to form (heat, UV, etc.).

$$Br_2 \xrightarrow{k_1} 2Br\bullet \qquad (1)$$

$$Br\bullet + O_3 \xrightarrow{k_2} BrO\bullet + O_2 \qquad (2)$$

$$BrO\bullet + O_3 \xrightarrow{k_3} Br\bullet + 2O_2 \qquad (3)$$

$$2Br\bullet \xrightarrow{k_4} Br_2 \qquad (4)$$

Figure 9.11 A proposed mechanism for the decomposition of ozone using dibromine as the catalyst.

One proposed mechanism by which the above decomposition of ozone may proceed is the chain reaction shown in Figure 9.11. The species $Br^\bullet$ and $BrO^\bullet$ are highly reactive free radicals, the single unpaired electron being indicated. These free radicals are called chain carriers because they start and continue a chain reaction. The mechanism shown is therefore an example of a **radical chain reaction**. All free radical mechanisms require an **initiation step** which starts off the formation of one or more free radicals. This is step 1 in Figure 9.11 where two free radicals are formed. A **propagation step** is one in which a new free radical is formed that can carry on the chain, together with a product molecule. Steps 2 and 3 are propagation steps. Although they are depicted only once in a mechanism, they occur a large number of times (the chain length). Step 4 is the **termination step**, since here the radical chain carrier is removed and the sequence of propagation reactions is stopped. If the propagation steps are summed, then this will result in the stoichiometric equation.

In some free radical reactions, a **branching step** may occur, which increases the number of free radicals causing the rate to dramatically increase. Such a reaction occurs between dihydrogen and dioxygen, which can lead to an uncontrolled reaction called an explosion. This is caused by a number of steps that increase the number of free radicals which leads to a cascading effect and therefore explosion. *Some* of the steps involved in this reaction are shown below:

Initiation	$H_2(g) + O_2(g) = HO_2\bullet(g) + H\bullet(g)$
Propagation	$H_2(g) + HO_2\bullet(g) = HO\bullet(g) + H_2O(g)$
Propagation	$H_2(g) + HO\bullet(g) = H\bullet(g) + H_2O(g)$
Branching	$H\bullet(g) + O_2(g) = HO\bullet(g) + O\bullet(g)$
Branching	$O\bullet(g) + H_2(g) = HO\bullet(g) + H\bullet(g)$

Figure 9.12 shows how an explosion depends upon temperature and pressure.

At low pressures, the dihydrogen/dioxygen mixture can be sparked but there would be no explosion. This is because the chain carriers can reach the walls of the container where they combine and give up their excess energy. As the pressure is raised (p_1), the first explosion limit is reached (move along the dotted line at 800 K) because the chain carriers can react and branching becomes efficient. If the pressure is raised further (p_2),

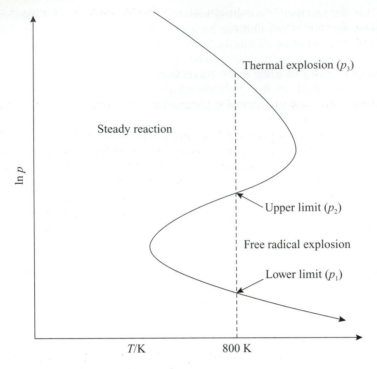

Figure 9.12 The dependency of pressure on temperature.

then the second explosion limit is reached. Here, there are enough molecules to soak up the excess energy and so there is a smooth reaction and no explosion. As the pressure is raised even further (p_3), then the region in which thermal explosion can occur will be reached. In the case of a thermal explosion, there is an exponential dependency of the reaction rate on the temperature. Hence, if the reaction is exothermic as the reaction proceeds the heat liberated 'feeds' the reaction causing it to become uncontrollable.

Ozone and the atmosphere

The amount of ozone in the atmosphere varies and reaches a maximum at an altitude of between 20 and 35 km. How is ozone formed?

Ozone is formed by what is essentially a chain reaction. The first step, initiation, involves the photochemical decomposition of dioxygen by ultraviolet radiation of wavelength equal to or less than 240 nm,

$$O_2(g) + hf = O(g) + O(g)$$

The oxygen atoms may be viewed as chain carriers.
Two propagation steps are,

$$O(g) + O_2(g) = O_3(g) \quad \Delta H^0_m = -106.5 \text{ kJ mol}^{-1}$$

$$O_3(g) + hf = O(g) + O_2(g)$$

In the case of the latter equation, radiation of wavelength equal to or greater than 900 nm causes the dissociation of ozone.

The termination step is,

$$O_3(g) + O(g) = 2O_2(g) \ \Delta H_m^0 = -391.9 \ kJ \ mol^{-1}$$

Two of the reactions are exothermic, and it is this heat that warms up the stratosphere causing the temperature inversion.

The above four reactions comprise the Chapman oxygen-only mechanism in which ozone is being continuously formed and decomposed. This simplistic approach proposes that the steady-state concentration of ozone at any altitude is the result of a dynamic equilibrium being achieved between the production and destruction of the ozone. Unfortunately, the Chapman approach grossly underestimates the rate at which both O and O_3 (referred to as 'odd oxygen') are lost. In addition to the Chapman equations, odd oxygen is lost by other reactions that are catalysed by various free radicals, in particular,

$$X(g) + O_3(g) = XO(g) + O_2(g)$$

$$XO(g) + O = X(g) + O_2(g)$$

where X may be HO˙, NO or Cl˙. All of these species occur naturally.

Groups of molecules form **catalytic families**, i.e. HO_x, NO_x and ClO_x, that provide the species XO. Examples of the nitrogen family are NO_2 and N_2O_5. The former provides NO, an odd electron molecule, which then can react with $HO_2^˙$ or ClO˙ to regenerate NO_2 or to take part in the above catalytic reactions. Note here that a family member is coupling with members of other families, i.e. the hydrogen family and the chlorine family. N_2O_5 will undergo very slow decomposition to NO_2 but it is also an unreactive **reservoir molecule** for NO_x and therefore reduces the rate at which ozone is destroyed. Nitric acid, HNO_3, is also a reservoir molecule in that it holds about 50 per cent of the total stratospheric load of NO_x. The species $ClNO_3$ holds potentially active nitrogen as well as chlorine. Hydrogen chloride acts as the main chlorine reservoir. Both HNO_3 and HCl are readily soluble in water and can therefore be washed downwards. Thus, these act as a natural sink for catalytic species.

The catalytic species described in this chapter are all formed from molecules that have all been acquired from the troposphere. These molecules are decomposed within the stratosphere by photochemical or chemical processes. The main source of chemicals for the nitrogen family is nitrogen(I) oxide and for the hydrogen family, methane. Both gases are formed by naturally occurring processes on the Earth's surface.

The various chemical or photochemical sinks in the atmosphere have varying lifetimes. One way to define this lifetime is the length of time it takes a particular species to reduce its concentration to $1/e$ (the exponential, e = 2.718) of its original value. This time varies from seconds to centuries!

One of today's environmental concerns is the apparent fluctuation in the concentrations of ozone found in the atmosphere. Its removal or decline in concentration will lead to a reduction in the protection it offers from incoming short wave ultraviolet radiation. The addition of pollutants to the atmosphere that have a long lifetime and can help to destroy ozone or influence it adversely in some other ways must be avoided. Pollution of the atmosphere and how it can be managed is the subject of the next chapter.

Summary

- Number density is a way of measuring atmospheric pressure and is given by,

$$\text{Number density} = \frac{\text{total number of molecules/atoms present}}{\text{volume of air}}$$

- The mixing ratio is the fractional abundance of each component of the atmosphere and is given by,

$$\text{Mixing ratio} = \frac{\text{number density of component A}}{\text{total number density of air}}$$

- The atmosphere has four main regions – troposphere, stratosphere, mesosphere and thermosphere.
- At ordinary temperatures and pressures, gases obey Boyle's Law, Charles' Law, Avogadro's Law, Dalton's Law and Graham's Law.
- An ideal gas is one that obeys the ideal gas equation $pV = nRT$.
- The absolute zero of temperature on the Kelvin scale is based on the theoretical assumption that an ideal gas has zero volume at $-278.15\,°C$.
- The solubility of a gas in a liquid is directly proportional to its partial pressure (Henry's Law).
- Photosynthesis is the conversion of carbon dioxide and water to sugars and oxygen, using sunlight as the source of energy and chlorophylls as the catalyst.
- A geological cycle shows the movement of matter between the atmosphere, lithosphere and hydrosphere.
- The cycles most relevant to the atmospheres are the carbon, nitrogen and sulphur cycles.
- All free radical reactions require (i) an initiation step, (ii) a propagation step, (iv) branching steps and (v) termination steps.
- The Chapman oxygen-only mechanism is an attempt to show how ozone is continuously formed and decomposed.
- Groups of molecules form catalytic families. Reactions between families, and to a lesser extent within families, produce relatively unreactive reservoir molecules that act as holding species for reactive radicals. This reduces their ozone-destroying abilities.
- The downward movement of soluble reservoir molecules acts as a natural sink for catalytic species.
- Chemical and photochemical sinks have differing lifetimes ranging from seconds to centuries.

Questions

1 Using the ideal gas equation, calculate the total number of moles, and therefore the number of particles, of gas present in 1 m³ of air at an altitude of 25 km. You may assume that the mean atmospheric temperature at this height is $-54\,°C$ and the prevailing atmospheric pressure 25.0 mbar.

   How many particles are there in 1 cm³ of this air? What is this number called?

2 Estimate the number of cylinders required to store 300 kg of dichlorine at 20 °C.

   The volume of each cylinder is 60.0 dm³ and the pressure the dichlorine is to be stored at is 14.0 MN m⁻².

   Note: M is the SI abbreviation for 'mega' meaning a million.

3   What would be the concentration of 1 mole of pure methane gas at STP in $g\,m^{-3}$? If a sample of air contains 2 ppm of methane, what is its concentration?

4   At 293 K, the vapour pressure of mercury is $0.160\,N\,m^{-2}$. Calculate the concentration of mercury vapour in air at this temperature.

   (*Note*: The maximum permissible concentration of mercury in air is $0.1\,mg\,m^{-3}$.)

5   An unknown gas took 186 s to diffuse through the walls of a porous pot. The same amount of carbon dioxide took 112 s.

   What is immediately obvious about the molar mass of the unknown gas? Calculate its molar mass.

6   Hydrogen sulphide, H_2S, and ammonia, NH_3, both have characteristic, powerful odours. Samples of both, located in the same place in a laboratory, were accidentally allowed to escape. A laboratory worker first noticed the problem via the odour of rotten eggs of hydrogen sulphide. Assuming that the air in the laboratory was essentially stationary, why do you think that relying upon a sense of smell as a warning mechanism is not a sensible idea?

7   In Figure 9.9 no details are given concerning the masses of the nitrogen-containing species involved in the nitrogen cycle. Find out what typical masses of nitrogen are transferred around this cycle.

8   Identify the processes occurring in the following free radical chain reaction mechanism. Show that the addition of certain steps gives the overall stoichiometry of the reaction.

   The reaction is the gas-phase thermal decomposition of a compound called trifluoroethanal, CF_3CHO. The decomposition occurs according to the time-independent stoichiometry,

   $$CF_3CHO(g) = CF_3H(g) + CO(g)$$

   A proposed mechanism for this reaction is as follows:

   $$CF_3CHO(g) = CF_3^{\bullet}(g) + CHO^{\bullet}(g)$$

   $$CF_3^{\bullet}(g) + CF_3CHO(g) = CF_3H(g) + CF_3CO^{\bullet}(g)$$

   $$CF_3CO^{\bullet}(g) = CF_3^{\bullet}(g) + CO(g) \ .$$

   $$CHO^{\bullet}(g) = CO(g) + H^{\bullet}(g)$$

   $$H^{\bullet}(g) + CF_3CHO(g) = H_2(g) + CF_3CO^{\bullet}(g)$$

   $$CF_3^{\bullet}(g) + CF_3^{\bullet}(g) = C_2F_6(g)$$

   Warning! If a product does not appear in the time-independent stoichiometric equation, then it is not part of the main mechanism.

Further reading

Jacobs, D.J. (2000) *Introduction to Atmospheric Chemistry*. Princeton University Press, Princeton, NJ.

   Although this is an introductory text, it also contains up-to-date details on current research and approaches to modelling the atmosphere. Quite a readable text not requiring an advanced knowledge of chemistry.

The Journal of Atmospheric Chemistry

   This journal contains a wide range of aspects covering the chemistry of the atmosphere, many of which are highly specialised. However, if you are interested in the atmosphere, a journal well worth a dip.

The Open University (1996) S342 Physical Chemistry, Principles of Chemical Change, Topic Study 1: *The Threat to the Stratosphere Ozone*. Open University, Milton Keynes.
Although this is a third-level chemistry booklet, it is quite easy to read and understand. It is well illustrated, and is an interesting account of the 'hole in the ozone layer'. An understanding of chemical kinetics and rate equations is necessary to understand it all.

Wayne, R.P. (2000) *Chemistry of Atmospheres*, 3rd edn. Oxford University Press, Oxford.
Again, a good account of the results and kinds of modelling currently going on! The effects of natural phenomena such as volcanic activity as well as anthropogenic effects are included. It provides a basic knowledge of atmospheric chemistry.

10 The pollution of the atmosphere

- The major air pollutants
- Particulate matter – origin, types, primary and secondary particulates
- Sulphur dioxide as a primary pollutant – origin and effects on the environment
- Nitrogen oxides NO, NO_2, N_2O – origin and effects on the environment
- Sulphurous and photochemical smogs – chemistry of the photochemical smog
- Acid rain, its causes and effects
- The ozone layer and CFCs
- The oxides of carbon CO and CO_2
- The greenhouse effect
- The management of air pollution

Introduction

In December 1994, the city of Bhopal, India, experienced the effects of one of the most serious cases of point source atmospheric pollution. The American Union Carbide Company had established a factory in the city, which made pesticides for Third World countries. One of the chemicals involved was liquid methyl isocyanate, CH_3NCO, a highly reactive, inflammable, volatile and toxic organic compound. As well as being toxic and an acute irritant, it causes pulmonary oedema, asthma and nausea. Although the material was stored in underground tanks, the vapours were accidentally leaked into the atmosphere causing the death of at least 3,300 people and affecting over 200,000 others. Many of these people were blinded and suffered kidney failure. The death toll was so great that the Indian Government had to call for volunteers to help the armed forces to clear the corpses of people and animals from the streets. For some time after the accident, the skies over Bhopal were described as glowing red from the hundreds of funeral pyres. It was the worst industrial accident in history.

Unfortunately, this accident occurred at a time when there was a temperature inversion. Temperature inversions commonly occur within the troposphere as a result of some meteorological event and are short term. The effects of air pollution caused either by the intentional or accidental release of materials to the atmosphere are normally greatly reduced by natural atmospheric mixing, i.e. dilution. The degree of mixing depends on a variety of weather conditions, e.g. temperature, wind speed, and the movement of high and low atmospheric pressure systems and their interactions with local topology such as mountains and valleys. Normally, the temperature of air in the troposphere decreases with altitude. Sometimes a cold layer of air can settle under a warm layer and produce a temperature inversion. The result of this is the poor mixing of pollutants with air causing a build-up of the pollutants near ground level. An inversion can last some time

Table 10.1 *Some major air pollutants*

Pollutant	Major sources	Human health effects
Carbon monoxide (CO)	Spark ignition combustion engine (motor vehicle exhausts) Some industrial processes	Displaces oxygen in the blood stream Effects depend upon concentration and exposure time Can include reduction in mental and physical abilities, and eventually death
Carbon dioxide (CO_2)	All combustion/burning processes	Possibly injurious to health only at very high concentrations If in sufficient quantities can cause tiredness and affect judgement Asphyxiation can result Atmospheric levels have risen from about 280 ppm a century ago to a value currently over 350 ppm
Sulphur dioxide (SO_2)	Heat and power generators that use the fossil fuels; coal is the single largest source Smelting of non-ferrous ores Manufacture of sulphuric acid	Short-term exposure to low concentrations affects lung function Higher concentrations cause chemical bronchitis and tracheitis (an inflammation of the trachea or windpipe), and can lead to increased mortality rates
Nitrogen oxides or NO_x (mainly NO and NO_2)	Motor vehicle exhausts Heat and power generators Nitric acid manufacture Use of explosives Welding processes Fertiliser manufacturing plants	Impaired lung function at low concentrations Increase in number of acute respiratory illnesses Lung tissue damage
Lead (Pb)	Motor vehicle exhaust Metal production Thermal power plants and other coal combustion plants	Children are particularly sensitive to lead poisoning Lead has been shown to affect many of the body systems, e.g. renal, reproductive, nervous
Particulate matter	Power plants Industrial processes Motor vehicle exhausts Domestic coal burning Industrial incinerators	Short-term exposure causes respiratory distress, lung impairment and increased mortality
Hydrocarbons (such as ethane, propane, ethene, butanes, pentanes, ethyne, benzene)	Motor vehicle emissions Solvent evaporation Industrial processes Solid waste disposal Combustion of fuels	No generalisations can be made Damage caused is chemical compound specific
Photochemical oxidants (primarily ozone O_3; also peroxyacetyl nitrate (PAN) and aldehydes)	Formed in the atmosphere by reaction of nitrogen oxides and hydrocarbons with sunlight	Eye, nose and throat irritation, chest discomfort, coughs and headache

if it lies under a stationary high pressure system and there is little or no wind. The consequences of the accident in Bhopal were thus far greater than they might otherwise have been.

There have been numerous cases of human activities such as this, which have caused the emission of gases, solid and liquids to the atmosphere. Such emissions have attacked the fabric of buildings and paper, reduced visibility or produced offensive smells.

There are a number of natural sources that cause atmospheric pollution, but only one is currently recognised as a major health threat and that is the radioactive gas radon.

Some major pollutants and their sources are summarised in Table 10.1.

Because of its localised source and restricted nature of the pollution, the accident at Bhopal is described as an example of **point source pollution**. Such cases result in obvious immediate effects and do not usually pose national or international pollution problems. Industrial companies or others can be made accountable, legal action taken and appropriate penalties imposed. The cleaning up of the environment may be problematical but does tend to be confined to a locality. Other emissions, however, cause problems on an international scale and are referred to as **diffuse source pollution**.

The atmosphere contains a range of solid and liquid particulate matter together with gases as a result of both natural and human activities. When materials become concentrated enough to reduce air quality they are pollutants. Many of these pollutants are capable of undergoing chemical reactions with other chemicals present in the atmosphere to produce new harmful products.

Particulate matter

Particulate matter is defined as single particles or aggregates of particles with diameters greater than 2×10^{-10} m. Some particulate matter is natural, i.e. rain, snow, fog, hail and mist, whilst others are often the result of anthropogenic processes, e.g. smoke, soot and fumes. Some natural particulates are affected by human actions, such as fog and wind-blown soils. Smoke and soot are the products of incomplete combustion of coal, petrol and diesel fuels in furnaces, domestic heating systems and vehicle engines.

In the UK smoke can be classified as:

1   the carbonaceous particles produced from the burning of carbon based fuels;
2   the soot, ash, grit and gritty particles in smoke as defined by the 1956 Clean Air Act;
3   all particles collected by filter paper by a British Standard method.

The 1971 Clean Air Regulation defined particles on a size basis. Grit particles have a diameter greater than 75×10^{-6} m but less than 500×10^{-6} m, since particles of greater size than the latter figure are unlikely to become airborne. Dust-sized particles have diameters less than 75×10^{-6} m down to 1×10^{-6} m. Fume particles have diameters less than 1×10^{-6} m. In the chemistry laboratory, fumes are described as any solid/liquid particles that form by condensation from gases.

Aerosols are mixtures of minute solid/liquid particles suspended in air that form a haze or spoil visibility.

The main problem to humans caused by atmospheric particulate matter is how far it is able to penetrate the respiratory system. Particles in the size range 30×10^{-6} to 100×10^{-6} m lodge in the nasal cavity, larynx and trachea. Some examples of particles of this size are pollen, fungal spores, cement dust and coal dust. Particles less than 15×10^{-6} m find their way into the bronchus and bronchioles, e.g. tobacco smoke and fumes. Particles of 4×10^{-6} m and less can enter the alveoli where gaseous exchange takes place between the bloodstream and air, e.g. asbestos dust, glass fibre and viruses.

In the UK and Europe, PM_{10} is the accepted standard measure of particulate matter in the atmosphere. This refers to the measurement of the mass of particles in air of less than 10 μm in diameter. This size was chosen based on epidemiological evidence of the link between these particles and health problems. It is likely that diameter sizes less than 10 μm, i.e. $PM_{2.5}$, will become of more interest as knowledge concerning particulate matter and its health effects increases.

Another way still in use is to measure the blackening effect of smoke, i.e. the 'Black Smoke' measurement. Air is pulled through filter papers and the degree of blackening measured. This is a less accurate technique than the PM_{10} approach since different emissions have different abilities to blacken paper, e.g. diesel emissions on a mass for mass basis have a three times greater blackening ability than those from coal burning.

How long particulate matter stays suspended in air depends mainly upon particle size. Larger particles of the order of 1 mm in diameter can remain in air as long as 10 days.

Particulate matter comes from two major sources. First, those emissions that come directly from sources such as coal combustion, wind-blown dust and quarrying. These are called **primary particulates**. Other particulates can be formed from chemical reactions between pollutant gases such as sulphur dioxide, the oxides of nitrogen and ammonia. Such reactions lead to the formation of solid sulphates and nitrates. Organic aerosols may also be formed by the oxidation of volatile organic compounds. These particulates are termed as **secondary particulates**.

In the UK PM_{10} has fallen steadily from 490 kt in 1970 to 160 kt in 1998 (NAEI 1998, Section 4.1). This is due primarily to a fall in the use of coal for domestic purposes. The fall would be even more dramatic if it wasn't offset by the increase in PM_{10} caused by increased motor vehicle usage, particular those powered by diesel engines. For the same reasons, black smoke emissions have also declined from 1.07 Mt in 1970 to 0.28 Mt in 1998 (NAEI 1998, Section 4.2).

The Air Quality Strategy for England, Wales and Northern Ireland proposals made in the year 2000 list the levels of particulate matter PM_{10} that should not be exceeded:

- 50 μg m^{-3} as a running 24-hour mean, not to be exceeded more than 35 times a year by 31 December 2004;
- 40 μg m^{-3} annual mean by 31 December 2004.

Sulphur dioxide, SO_2

Sulphur dioxide is a colourless, toxic, non-flammable gas at normal temperatures and pressures. It can be tasted at levels between 1,000 and 3,000 μg m^{-3}. At higher concentrations it has an irritating, pungent smell. Sulphur dioxide is very soluble in water (0.166 mole per 100 g of water at 293 K) and is easily adsorbed on to the surfaces of solid matter.

Sulphur dioxide is emitted to the atmosphere from a number of natural sources. Geothermal activity releases vast amounts of sulphur dioxide, together with smaller amounts of sulphur trioxide, elemental sulphur, hydrogen sulphide and particulate sulphates. Such activity accounts for less than 1 per cent of the global atmospheric burden of volatile sulphur compounds. The main source is the oxidation of organic material that contains sulphur or the reduction of sulphur to hydrogen sulphide under anaerobic conditions.

Table 10.2 *Sulphur dioxide emissions in the UK between 1970 and 1998 (kilotonnes)*

Year	Solid fuel	Petroleum	Gas	Non-fuel emissions	Total
1970	3,660	2,564	192	95	6,551
1980	3,130	1,558	92	89	4,871
1990	2,757	814	84	82	3,736
1995	1,628	641	29	59	2,356
1998	1,169	369	30	47	1,615

(*Note*: Data modified from National Atmospheric Emissions Inventory, Section 5.3: SO_2)

Sulphur dioxide also finds its way into the atmosphere by a number of anthropogenic pathways. About 200×10^6 tonnes is emitted each year and is about equal to the amount emitted from natural processes. The main sources of sulphur dioxide emissions are from the combustion of solid fuels and petroleum products. Other contributions originate from the smelting of sulphide ores, in particular those of zinc, copper and lead, refinery processes, sulphuric acid production and transport. Because sulphur dioxide has contributed so much to **acid rain** and other environmental problems in the past, much work has been done on controlling its emission levels. For example, between 1970 and 1998 emissions of sulphur dioxide in the UK fell by about 75 per cent. Energy production using coal-fired and oil-fired power stations contributed about 66 per cent of the total emissions in 1998 (Table 10.2) (NAEI 1998 Section 5.3 SO_2). This is due to the use of the Combined Cycle Gas turbine stations and other gas-fired power stations.

The fitting of desulphination plants at Drax and Ratcliffe power stations has also made significant reductions in sulphur dioxide emissions. This involves the conversion of limestone to gypsum via chemical reaction with SO_2. A limestone/water slurry is sprayed continuously down tall absorber towers against upward-travelling flue gases emitted by the burning of coal. The SO_2 is absorbed by this slurry where it reacts with the calcium carbonate to initially produce calcium sulphate(IV),

$$CaCO_3(s) + SO_2(g) = CaSO_3(s) + CO_2(g) \qquad 10.1$$

At the same time, compressed air is pumped in at the base of the tower and the calcium sulphate(IV) is oxidised by the oxygen in that air to form calcium sulphate(VI),

$$2CaSO_3(s) + O_2(g) = 2CaSO_4(s) \qquad 10.2$$

As crystals of gypsum are formed they are separated from the water in hydrocyclones. Any moisture remaining is removed by centrifuges. About 800,000 tonnes of gypsum are produced per year at the Drax power plant and is sold to make plasterboard, fertilisers, grouting material, etc.

Since 1970, the increased use of nuclear power stations has also reduced the amount of coal and oil that have been used. In the UK, industry has also played its part. Between 1970 and 1998, emissions from industrial processes fell by 90 per cent (NAEI 1998 Section 5.3). Much of this reduction occurred in the period 1970–85 as a result of the decline in the iron, steel and other heavy industries. This has also been accompanied by a decline in the use of coal and oil, and a move towards the use of natural gas as the energy source. Transport contributes about 1 per cent of total emissions of sulphur dioxide. Since 1970 there has been a huge increase in road transport and so emissions from that source have increased. However, the use of low sulphur fuels are now having a serious impact on SO_2 reductions (NAEI 1998 Section 5.3 SO_2).

When sulphur dioxide enters the atmosphere, it can be easily oxidised to sulphur trioxide, SO_3, which dissolves in water to form sulphuric acid. The oxidation of SO_2 can proceed via **catalytic oxidation**. The catalysts are aerosols that contain metal ions (e.g. Mn^{2+}, Fe^{3+} and Cu^{2+}) and metal oxides (Cr, Al, Pb and Ca). This thermodynamically favourable, but kinetically slow, reaction process can be written,

$$2SO_2(g) + 2H_2O(l) + O_2(g) = 2H_2SO_4(aq) \qquad 10.3$$

$$\Delta G_f^0 = -411.4 \, kJ \, mol^{-1}$$

The surfaces of building materials act as catalytic centres, and a high humidity increases the rate of reaction. As the acid concentration increases in water, the oxidation reaction is slowed down. Hence, water having a pH greater than 7.0 will increase the rate of oxidation.

When SO_2 dissolves in water, the following equilibria are formed,

$$SO_2(g) + H_2O(l) \rightleftharpoons H_2SO_3(aq) \qquad 10.4$$

$$H_2SO_3(aq) + H_2O(l) \rightleftharpoons H_3O^+(aq) + HSO_3^-(aq) \qquad 10.5$$

$$HSO_3^-(aq) + H_2O(l) \rightleftharpoons H_3O^+(aq) + SO_3^{2-}(aq) \qquad 10.6$$

$$2HSO_3^-(aq) \rightleftharpoons S_2O_5^{2-}(aq) + H_2O(l) \qquad 10.7$$

In equation 10.4, sulphurous acid is formed, in equation 10.5 the hydrogen sulphate(IV) ion, in equation 10.6 the sulphate(IV) ion and in equation 10.7 the disulphate ion in which the oxidation state of sulphur is (V) and (II).

If the acidity of the water increases, then the positions of the equilibria represented by equations 10.5 and 10.6 will move to the left-hand side, producing more $H_2SO_3(aq)$ and thus affecting equilibrium equation 10.4. The overall result is a decrease in the solubility of SO_2. If ammonia is also present, then this will remove sulphur dioxide to form $(NH_4)_2SO_4$ and NH_4HSO_4 in solution. This causes the rate of reaction to increase.

The oxidation of the sulphate(IV) ion then occurs via the couple,

$$HSO_3^-(aq) + 2H_2O(l) \rightleftharpoons HSO_4^-(aq) + 2H_3O^+(aq) + 2e^- \qquad 10.8$$

$$E^0 = -0.17 \, V$$

$$O_2(g) + 4H_3O^+(aq) + 4e^- \rightleftharpoons 6H_2O(l) \qquad 10.9$$

$$E^0 = +1.23 \, V$$

Sulphur dioxide can also undergo oxidation photochemically. Here, electromagnetic radiation of the appropriate wavelength is absorbed by SO_2 molecules, which become energetically excited. These energetically excited molecules, represented by SO_2^* then react with dioxygen molecules to produce SO_3,

$$SO_2(g) \xrightarrow{hf} SO_2^*(g) \qquad 10.10$$

$$SO_2^*(g) + O_2(g) = SO_3(g) + O^{\bullet}(g) \qquad 10.11$$

$$SO_2^*(g) + SO_2(g) = SO_3(g) + SO^{\bullet}(g) \qquad \text{10.12}$$

$$SO^{\bullet}(g) + SO_2(g) = SO_3(g) + S(g) \qquad \text{10.13}$$

Sulphur dioxide may also react with ozone and nitrogen(IV) oxide,

$$SO_2(g) + O_3(g) = SO_3(g) + O_2(g) \qquad \text{10.14}$$

$$SO_2(g) + NO_2(g) = SO_3(g) + NO(g) \qquad \text{10.15}$$

The former reaction occurs rapidly in water but is slow in the gaseous phase. The nitrogen(IV) oxide in the latter reaction is emitted from car exhausts.

Nitrogen oxides

Three oxides of nitrogen, nitrogen(I) oxide (dinitrogen oxide/nitrous oxide), N_2O, nitrogen(II) oxide (nitric oxide), NO, and nitrogen(IV) oxide (nitrogen dioxide), NO_2, are important in the environment. Two are atmospheric pollutants, i.e. NO and NO_2, and are referred to as NO_x.

Although dinitrogen is not chemically reactive at low temperatures due to the strength of its triple bond, at higher temperatures it will combine with oxygen to form colourless NO gas,

$$1/2\ N_2(g) + 1/2\ O_2(g) \rightleftharpoons NO(g) \qquad \text{10.16}$$

$$\Delta H^0 = +90.3\ \text{kJ mol}^{-1}$$

$$\Delta G^0 = +86.6\ \text{kJ mol}^{-1}$$

As indicated by the sign and magnitude of ΔG^0 at 298 K, the position of the equilibrium lies well over to the left-hand side. Since it is an endothermic reaction, according to le Chatelier's Principle, an increase in temperature will move the equilibrium position to the right and more NO will be produced. For example, in the motor car engine when the spark plug is fired to ignite the petrol–air mixture, high enough temperatures are generated to enable the equilibrium to lie on the right-hand side, and so significant amounts of NO are generated. Upon being passed to the exhaust system and then to the atmosphere where the temperatures are much lower, the NO cannot be decomposed and thus becomes a primary pollutant. There are also natural sources of NO – when lightning occurs both NO and NO_2 are produced,

$$N_2(g) + O_2(g) \rightleftharpoons 2NO(g) \qquad \text{10.17}$$

$$N_2(g) + 2O_2(g) \rightleftharpoons 2NO_2(g) \qquad \text{10.18}$$

Upon exposure to the air, NO combines rapidly with dioxygen to form the secondary pollutant red-brown nitrogen(IV) oxide, NO_2,

$$2NO(g) + O_2(g) = 2NO_2(g) \qquad \text{10.19}$$

$$\Delta H^0 = -246.9\ \text{kJ mol}^{-1}$$

$$\Delta G^0 = -70.5\ \text{kJ mol}^{-1}$$

At lower temperatures, nitrogen(IV) oxide dimerises to form colourless N_2O_4,

$$2NO_2(g) \rightleftharpoons N_2O_4(g) \qquad\qquad 10.20$$

At 25 °C, just above the boiling point of N_2O_4, both the liquid and its gas contain much more of the dimer than the monomer. With increasing temperature, the amount of dimer rapidly decreases rapidly in the gas phase.

Nitrogen(IV) oxide is an odd-electron molecule and is therefore also represented by $NO_2^{\cdot}$. It can thus act as a free radical in the gaseous phase. Because of its reaction with oxygen, the smell of NO is unknown. It is slightly soluble in water (5 cm³ in 100 cm³ of water at 15 °C). Nitrogen(II) oxide is also an odd-electron molecule ($NO^{\cdot}$).

Nitrogen(IV) oxide is a poisonous, choking gas, which causes some of the colour in smog. It dissolves readily in water, where it disproportionates to form nitric(V) acid and nitric(III) acid,

$$2NO_2(g) + H_2O(l) = HNO_3(aq) + HNO_2(g) \qquad\qquad 10.21$$

Hence, NO indirectly contributes to acid rain in that it is a precursor to the formation of two acids.

Nitrogen(I) oxide, N_2O, is a colourless gas which is fairly soluble in water (130 cm³ in 100 g of water at 0 °C). It is tasteless with a slight sweetish smell. N_2O is used as an anaesthetic and, because it is not toxic, is also used as a foaming agent in whipped cream. It is unreactive at ordinary temperatures. Nitrogen(I) oxide is a greenhouse trace gas because it absorbs infrared radiation at 8.6×10^{-6} m and 7.8×10^{-6} m. It is 270 times more effective in this respect than is carbon dioxide. The atmospheric concentration of this gas is currently 310 ppb but appears to be increasing at a rate of about 0.25 per cent per annum. Nitrogen(I) oxide is released to the atmosphere from the oceans, and tropical soils in particular. It is a by-product of both biological denitrification under aerobic conditions and nitrification under anaerobic conditions.

In 1998, the two major sources of N_2O in the UK were from agricultural soils and the manufacture of adipic acid and nitric acid (NAEI 1998, Section 2.4 N_2O). The former was responsible for 50 per cent of emissions, and the latter for 33 per cent of total emissions. Low levels of N_2O were also emitted from the burning of fossil fuels. The use of the catalytic converter causes more dinitrogen(II) oxide to be formed and therefore in future years the petrol engine is likely to contribute more of this pollutant to the atmosphere. In 1990 non-fuel sources contributed some 195.4 kt of N_2O out of a total of 211.9 kt, whilst in 1998 some 158.2 kt out of a total of 186.7 kt³.

Denitrification is the reduction of nitrate(V), NO_3^-, to dinitrogen. A very small amount of N_2O is also produced. In nitrification, ammonia, NH_3, and ammonium, NH_4^+, are both oxidised to nitrate(V) and nitrate(III), NO_2^- and N_2O. The latter process makes more contribution to the atmospheric loading of N_2O. In tropical areas where deforestation occurs via burning and new grasslands are formed, together with the use of fertilisers, the production of N_2O also significantly contributes to the atmospheric load.

N_2O eventually makes its way from the troposphere to the stratosphere where it is rapidly converted by solar UV radiation of wavelengths less than 250 nm into N_2 (95 per cent) and NO (5 per cent),

$$N_2O(g) \xrightarrow{hf} NO^{\cdot}(g) + N(g) \qquad\qquad 10.22$$

$$N_2O(g) \xrightarrow{hf} N_2(g) + O(g) \qquad\qquad 10.23$$

where $NO^{\bullet}(g)$ represents the odd-electron nature of nitrogen(II) oxide. As we shall see, $NO^{\bullet}$ can destroy stratospheric ozone, making the Earth's surface more open to attack from dangerous UV radiation.

In the UK between 1970 and 1998 there was an overall reduction of 30 per cent in NO_x emissions (NAEI 1998, Section 5.3 NO_x). However, in 1984 there was a rise in emission levels, peaking in 1989, due to the increase in road traffic. Since 1989 total emissions have fallen by 38 per cent as a result of a 58 per cent reduction from power stations and a 42 per cent reduction in emissions from road transport (NAEI 1998, Section 5.3 NO_x). One of the contributing factors to a marked reduction in NO_x emissions from petrol-driven engines has been the introduction of the catalytic converter. Stricter emission controls have also been imposed on diesel engines.

Power generation in the same period (1970–98) saw a reduction in NO_x emissions of some 55 per cent. Up to 1989, this was mainly due to the increase in nuclear energy usage. However since 1988 the installation of low NO_x burners in power stations and the move to the use of natural gas has contributed to emission reduction. Industry in general contributed about 10 per cent of the total emissions in 1998. This has also reduced by some 55 per cent since 1970, again due to the shift towards the use of gas and electricity for their energy needs (NAEI 1998, Section 5.3 NO_x) (Table 10.3).

London and air pollution – the smog

Between 4 December and 9 December 1952 occurred the most infamous 'pea-souper' smogs recorded in the UK. This smog caused a range of cardiovascular and respiratory disorders which ultimately resulted in 4,000 deaths above what would normally be expected for that time of year. Ninety per cent of these deaths occurred in people who were 45 years old or over, and the deaths of infants below the age of one doubled. People were killed by inhaling water droplets that were at least as acidic as lemon juice. The smog was again caused by a temperature inversion, which when coupled with no wind held down a vast acidic cloud. By 9 December, the radius of this cloud extended some 30 km from the centre of London. The acid which caused the deaths was probably sulphuric acid since the levels of sulphur dioxide in the smog proved to be very high. The incident prompted legislation to be implemented and resulted in the Clean Air Acts of 1956 and 1968. Even so, London is still not free of smogs.

Between 13 December and 15 December 1991, a severe smog was produced that showed high levels of nitrogen dioxide at ground level. The conditions arose because of a high-pressure system which brought settled weather and created warm dry air at higher altitudes. The air at ground level started to cool and cold, moist air began to rise.

Table 10.3 *NO_x emissions in the UK between 1970 and 1998 (kilotonnes)*

Year	Solid fuel	Petroleum	Gas	Non-fuel emissions	Total
1970	870	1,414	86	128	2,497
1980	858	1,451	1,541	123	2,568
1990	764	1,716	188	121	2,788
1995	461	1,362	213	92	2,054
1998	317	1,093	243	101	1,753

(*Note*: Data modified from National Atmospheric Emissions Inventory)

This was trapped below a layer of warmer air. The cold air stagnated because of the lack of wind. A density inversion was therefore created over London and the pollutants remained trapped near street level to cause a smog. Heavy traffic at the time made matters much worse. Fortunately, no adverse effects on health were reported but it resulted in the worst record of NO_x pollution since measurements began in 1976.

A smog is normally a mixture of smoke and fog. It is formed when the water content of air is high and it is so calm that smoke and fumes accumulate near their emission source. Smog reduces visibility and often irritates the eyes and respiratory tract. **Sulphurous smog** results from high concentrations of sulphur dioxide caused by the burning of sulphur containing fossil fuels. It is aided by damp conditions and the presence of particulate matter. In highly populated urban areas, the death rate may rise considerably during prolonged periods of smog, particularly when, as illustrated above, a temperature inversion creates a smog-trapping ceiling over a city. Smog occurs most often in, and near, coastal cities and is an especially severe air pollution problem in Athens, Los Angeles and Tokyo. Smog prevention requires control of smoke from furnaces and the reduction of fumes from all industrial processes. Smog has also been identified as being a consequence of the noxious emissions from motor vehicles and incinerators. The number of dangerous chemicals found in smog is considerable, and the proportions are highly variable. They include ozone, sulphur dioxide, hydrogen cyanide and hydrocarbons and products formed by their partial oxidation.

Photochemical smog is a whitish yellow to brown haze, which irritates sensitive membranes and damages plants. It contains a variety of potentially harmful organic compounds. This type of smog does not need either smoke or fog to form. It is formed when two of the primary pollutants nitrogen monoxide and hydrocarbons undergo reaction because of the energy supplied by ultraviolet and other types of radiation from the Sun. The conditions necessary for this type of smog to form are the presence of the pollutants themselves, sunlight, a stable temperature inversion and land enclosed by hills. Petrol and diesel engines are regarded as one of the main contributors to this smog problem since they emit large amounts of unburned hydrocarbons and oxides of nitrogen.

The chemistry of photochemical smog

The main pollutants that cause photochemical smog are NO, NO_2, hydrocarbons (RH), PAN (peroxyethanoyl nitrate), ozone and the aldehydes.

As indicated earlier, the high temperatures found in combustion engines can give rise to highly reactive free radicals. Dioxygen and dinitrogen may undergo the following reactions, where M represents some molecule that absorbs the excess energy of the reactions,

$$O_2(g) + M \rightleftharpoons O(g) + O(g) + M \qquad\qquad 10.24$$

$$N_2(g) + M \rightleftharpoons N^{\cdot}(g) + N^{\cdot}(g) + M \qquad\qquad 10.25$$

The resulting nitrogen and oxygen atoms then react to form $NO^{\cdot}$,

$$2N^{\cdot}(g) + 2O(g) \rightleftharpoons 2NO^{\cdot}(g) \qquad\qquad 10.26$$

Other reactions are also possible, such as,

$$H_2O(g) + O(g) \longrightarrow 2\text{·}OH(g) \qquad\qquad 10.27$$

$$N\text{·}(g) + \text{·}OH(g) \longrightarrow NO\text{·}(g) + H\text{·}(g) \qquad\qquad 10.28$$

Atomic oxygen can react with the hydrocarbon molecules that represent unburnt petrol fumes thus,

$$RH(g) + O(g) \longrightarrow R\text{·}(g) + \text{·}OH(g) \qquad\qquad 10.29$$

One of the products of the incomplete combustion of petrol is carbon monoxide, CO. The ·OH radical can react with the CO,

$$CO(g) + \text{·}OH(g) \longrightarrow CO_2(g) + H\text{·}(g) \qquad\qquad 10.30$$

$$\text{then} \quad H\text{·}(g) + O_2(g) \longrightarrow \text{·}OH(g) + O(g) \qquad\qquad 10.31$$

Thus from equation 10.28, the production of the ·OH(g) radical is clearly important in the formation of NO·.

For NO· the slow oxidation to NO_2^- occurs via,

$$2NO\text{·}(g) + O_2(g) \rightleftharpoons 2NO_2^-(g) \qquad\qquad 10.32$$

The free radical HO_2^- is also found in the atmosphere and will react with NO·,

$$HO_2^-(g) + NO\text{·}(g) \longrightarrow 2NO_2^-(g) + \text{·}OH(g) \qquad\qquad 10.33$$

NO_2^- absorbs solar radiation in the visible and near UV regions, i.e. at wavelengths less than 420 nm, and undergoes the reaction,

$$NO_2^-(g) \longrightarrow NO\text{·}(g) + O(g) \qquad\qquad 10.34$$

The O atoms then react with dioxygen in the presence of a third molecule M ($M = O_2$ or N_2),

$$O(g) + O_2(g) + M \longrightarrow O_3(g) + M \qquad\qquad 10.35$$

$$\text{Then,} \quad NO\text{·}(g) + O_3(g) \longrightarrow NO_2^-(g) + O_2(g) \qquad\qquad 10.36$$

Clearly, from the above reactions it can be seen that NO_2^- both helps to make ozone (equation 10.34 shows it producing the necessary O atoms) and to destroy it (equation 10.36 showing the NO produced in equation 32 destroying O_3).

Any hydrocarbons present in the atmosphere can also react with ·OH radicals,

$$RH(g) + \text{·}OH(g) \longrightarrow R\text{·}(g) + H_2O(g) \qquad\qquad 10.37$$

(*Note*: R represents the part of the hydrocarbon chain that remains unreacted!)

The free radical R$^{\bullet}$ then reacts with dioxygen,

$$R^{\bullet}(g) + O_2(g) \longrightarrow ROO^{\bullet}(g) \qquad\qquad 10.38$$

The ROO$^{\bullet}$(g) species is called the peroxy radical. The peroxy radical will react with NO$^{\bullet}$,

$$ROO^{\bullet}(g) + NO^{\bullet}(g) \longrightarrow RO^{\bullet}(g) + NO_2^{\bullet}(g) \qquad\qquad 10.39$$

The species RO$^{\bullet}$ is called the alkoxyl free radical. Thus NO$_2^{\bullet}$ produced in equation 10.39 can react via equation 10.37 to produce more oxygen atoms that will beget more ozone. Aldehydes, RCHO, are formed by the reaction of the alkoxyl radical with oxygen,

$$RCH_2O^{\bullet}(g) + O_2(g) \longrightarrow RCHO(g) + HO^{\bullet}(g) \qquad\qquad 10.40$$

Note in reaction equation 10.40, RO$^{\bullet}$ has been replaced by RCH$_2$O$^{\bullet}$ to emphasise how the aldehyde group arises.

Peroxy compounds, e.g. peroxyethanoyl nitrate (peroxyacetyl nitrate or PAN), are formed by the reaction,

$$ROO^{\bullet}(g) + NO_2^{\bullet}(g) \longrightarrow RCOONO_2(g) \qquad\qquad 10.41$$

Petrol and similar fumes contain the reactive unsaturated alkenes. Ozone readily attacks the double bond to form a range of compounds,

$$RCH{=}CH_2(g) + O_3(g) \longrightarrow \underset{\text{an aldehyde}}{RCHO(g)} + H_2COO^{\bullet}(g) \qquad\qquad 10.42$$

$$\text{or } RCHOO^{\bullet}(g) + \underset{\substack{\text{methanal}\\\text{(formaldehyde)}}}{HCHO(g)}$$

Acid rain – a case of international pollution

Natural rainwater has a pH of about 5.6 due to the reaction of carbon dioxide with water, but atmospheric pollution can lower this value to a pH of 2.0. When the pH of rain, snow, fog or dust falls below about 5.6, then it has been polluted by acid. This 'acid rain' can be a serious threat to the environment. The main culprits are sulphur dioxide and nitrogen(IV) oxide.

Acid rain is a serious problem in industrialised regions and countries where the combustion of fossil fuels releases large quantities of sulphur dioxide into the atmosphere where it reacts with oxygen and water to produce sulphuric acid. The oxides of nitrogen also react with water to form nitric acid. Some regions are polluted because of acidic deposition caused by industrial processes going on in other countries that are up-wind. For example, acid rain damage to Scandinavian lakes and streams appears to be now mainly due to emissions from Central European power plants and motor vehicles. Eastern Canada is polluted by acid deposition carried by prevailing winds from the US.

It is the combustion of coal, oil and petrol which accounts for much of the airborne pollutants involved. For example, some 85 per cent of the sulphur dioxide, 42 per cent of the oxides of nitrogen, and 30 per cent of the particulate matter emitted to the atmosphere in the United States in 1998 were produced by fossil fuel-fired electricity generating power stations, industrial boilers and domestic fires. Seventy-nine per cent of the carbon monoxide, and 53 per cent of the oxides of nitrogen came from the burning of petrol and diesel in on-road and non-road engines and vehicles (USEPA 1998).

What are the effects of acid rain that have given rise to much concern? In the 1980s it was discovered that very large areas of forest in Europe and the United States were showing signs of damage, particularly those forests at higher altitudes.

In Europe, for example, it was estimated that about 5 per cent of the forests were severely damaged and 0.2 per cent had been killed. In what was West Germany, four of the most important tree species (Norway spruce, white fir, scots pine and beech) showed signs of damage. The perceived effect of acid rain was so dramatic that the Germans coined a new term for it, 'Waldsterben' or 'forest death'. The area affected to a greater or lesser extent was estimated to have increased from 8 per cent in 1982 to 54 per cent in 1986. In Eastern Europe, the problem was even more severe, with 16 per cent of Czech forest being badly damaged. This was partly blamed on the burning of brown coal (lignite) which has a high sulphur content.

Similar damage was also being reported in other parts of the world, e.g. in the US, where the Appalachians, Adirondack Mountains, the White Mountains of New Hampshire and the Green Mountains of Vermont were affected. On Whiteface Mountain in New Hampshire, it was estimated that between 20 per cent and 70 per cent of the red spruce trees were showing signs of die-back.

As a consequence of these observations, the United States Government funded an investigative programme called the National Acid Precipitation Assessment Programme which was concluded in 1990. One of the conclusions of this report was that acid rain was not a major cause of plant death. The only trees that were damaged in the US belonged to one species growing at the highest altitudes, i.e. at the limit of their range.

In Europe, it was found that forest decline took more than one form and was caused by a variety of factors. Plant life nearest to the source of pollution will be damaged if the level of pollution is sufficiently high, which is rare in Western Europe. Acid rain was found to contribute to magnesium leaching (an essential plant trace element) in some forest areas but this could easily be redressed by the application of fertiliser. The areas where die-back was worst did not coincide 100 per cent with areas of high acidity. There is thus controversy over the involvement of acidification in forest damage.

Acid rain also does not appear to have any significant effect on crops (National Acid Precipitation Assessment Programme 1990). Other factors have been found to be important, such as agricultural practices, soil type and climate. For example, fast-growing conifers leach out bases from the soil, thus increasing its acidity at a far faster rate than acid rain does. When the trees are cut down, unless lime is added to the soil to increase its pH, then new plant growth will be adversely affected. Even if the rain is not acidic, the run-off from such soil will be and, if it enters an aquatic system, can cause damage.

All types of water resources are very vulnerable to contamination by acids. Unlike soils they cannot as readily adjust to changes in their acidity. For example, in the early 1970s the shores of the lakes of Ontario were found to be strewn with dead fish. Subsequently, it was discovered that, in this province alone, there were over 100 lakes which were so acidified that they no longer contained fish. In 1981, as a result of the high fish mortality due to acidity, the monitoring of 202 lakes in Ontario, Quebec and the Atlantic Provinces was started. Canada introduced the Eastern Canada Acid Rain Programme, which was aimed at reducing the emission figures to less than 50 per cent of those in 1980. In 1994, after years of trying to reduce SO_2 and NO_x emissions, it was found that 33 per cent of these lakes had improved but 11 per cent had worsened in terms of their acidities. The remaining 56 per cent had not changed.

The lakes in Ontario are highly sensitive to acid rain because of the granitic bedrock over which they lie and the poor quality of the soil. Soil has the ability to act as a buffer against acidity, but in the case of the Canadian Shield this proved to be ineffective. It is not the actual level of acidity that kills fish but the aluminium that acid rain is able to leach from soil. Aluminium is lethal – a level of 6.2 mg dm^{-3} being sufficient to kill all fish. The presence of aluminium ions reduces the ability of fish gills to allow ion exchange and therefore reduces a fish's intake of salts necessary for its development and survival. The examination of the blood of dead fish showed, for example, that the concentrations of both sodium and chloride ions were too low. Aluminium is also precipitated out of solution in the gills of fish, effectively blocking the uptake of dioxygen and again interfering with the passage of ions. The fish dies of suffocation! The second way that aluminium acts is to cause the fish to produce more gill mucous, which again clogs them up, preventing the uptake of dioxygen.

The pH range that fish can tolerate is very narrow. At pH 3.5 to 3.0 most fish will die, although some plants and invertebrates will survive. At pH 4.0 to 3.5 trout and salmon will die. Acidity in the range pH 4.0 to 4.5 is harmful to salmon, tench, bream, roach, carp and goldfish. Embryos also fail to develop and therefore the stock of surviving fish gradually dies out. The range pH 5.0 to 4.5 is harmful to the eggs of salmon, fry in general and the carp. The critical pH range where drastic changes start to occur, in that the diversity of both plant and animal life starts to be badly affected and the ecosystem starts to dramatically change, is pH 6.0 to 6.5. The range pH 6.5 to 9.0 is harmless to most fish.

In 1995 it was estimated that sulphur dioxide emissions from the US amounted to some 16.8×10^6 tonnes, and those from Canada, 2.7×10^6 tonnes. More than 50 per cent of the deposition in Eastern Canada came from the US, i.e. about 3.5×10^6 – 4.2×10^6 tonnes. It was realised that Canada alone could not prevent acidification of its lakes and other areas, so in 1991, in conjunction with the US, the Canada–US Control programme was established. Both countries are now working together to reduce emissions (Canada–US Air Quality Agreement 1998). Even so, about 14,000 lakes continue to be acidic at an unacceptable level, and even by 2010 vast areas of Canada will continue to receive harmful levels of acid rain (Canadian Government 2003).

In the US, the Clean Air Act Amendments of 1990 were designed to reduce SO_2 and NO_2 emissions using techniques of energy efficiency and pollution prevention. Phase 1 was started in 1995 and involved 21 eastern and midwestern states. Initially, in 1980, 263 emitting bodies, including 110 mainly coal-burning electricity power stations, were responsible for 9.4×10^6 tonnes of emissions. By 1995 this had been reduced to 4.5×10^6 tonnes and by 1999, 4.3×10^6 tonnes, below the allowed level of 7.4×10^6 tonnes. Phase 2 of the project was started in 2000. This involved a further 2,000 industrial premises where the restrictions on annual emissions were tightened and restrictions placed on smaller, cleaner plants fired by coal, oil or gas (USEPA 2000).

As seen in Chapter 1, building materials such as dolomite and limestone are also attacked by acid rain. Metals are corroded, for example,

$$Fe(s) + SO_2(g) + O_2(g) = FeSO_4(s) \qquad\qquad 10.43$$

$$Fe(s) + H_2SO_4(aq) = FeSO_4(aq) + H_2(g) \qquad\qquad 10.44$$

Metal corrosion is enhanced by the formation of solutions of iron(II) sulphate(VI) because the solution can act as the electrolyte between two metals thus forming an electrochemical cell.

One major problem with atmospheric SO_2 has been its attack on paper. Exposure of paper to polluted atmospheres over long periods of time has led to the paper becoming brittle and discoloured. This is due to the acid hydrolysis of the carbon–oxygen–carbon links in cellulose.

The 'hole' in the ozone layer

The ozone in the stratosphere protects living organisms from dangerous short wavelength ultraviolet radiation. The amount of ozone present is a delicate balance between the making and destruction of ozone that depends upon the existence of naturally occurring trace compounds. If any unnatural compounds are added to this balance that can provide extra catalytic species which can remove monatomic O then the destruction of ozone will be enhanced. This appears to have been the case in what is known as the 'hole in the ozone layer' to be found over the Antarctic ice cap. There is though no literal hole, in that there is a total absence of ozone – there is evidence to suggest that there are fluctuating levels of ozone and a steady depletion in its amount over the last 30 years or so.

The chemicals thought to be mainly responsible for this depletion are the **halocarbons**. These compounds are composed of carbon, fluorine, bromine and chlorine atoms with the occasional hydrogen atom. The halocarbons were used as refrigerants because they were found to be completely non-toxic, non-flammable and non-corrosive. The most widely used compounds were Freon-11 (CCl_3F), Freon-12 (CCl_2F_2), Freon-22 ($CHClF_2$) and BCF (bromochlorodifluoromethane ($CBrClF_2$). In addition to their roles as refrigerants, Freon-11 was also used as a foaming agent in the manufacture of plastics and an aerosol propellant, and both BCF and Freon-12 were used as one of the agents in fire extinguishers. The usefulness of halocarbons lies in the fact that they are so unreactive. It is, however, this lack of reactivity which has caused them to be such a problem in the environment. Freon-11 and Freon-22 in particular are so unreactive that they have been able to pass into the stratosphere. These freons are also referred to as CFCs, or chlorofluorocarbons, because they are composed of chlorine and fluorine combined with carbon. The life-time of Freon-11 in the atmosphere is some 65 years, and that of Freon-12, 130 years. Some CFCs have even longer lifetimes, e.g. Freon-115 ($CClF_2CF_3$) has a lifetime of 400 years!

Although the concentrations of these compounds are very low, e.g. Freon-11 255 pptv and Freon-12 470 pptv, once they are in the stratosphere they are subjected to UV radiation. This causes the breakdown of the molecules producing chlorine free radicals,

$$CCl_3F(g) \xrightarrow{UV} CCl_2F^{\cdot}(g) + Cl^{\cdot}(g) \qquad\qquad 10.45$$

CFCs can thus dramatically increase the chlorine burden of the stratosphere, leading to the destruction of ozone. However, reaction of these free radicals with gases such as $NO_2^{\cdot}$ and CH_4 has locked away much of the chlorine as the relatively unreactive molecules HCl and $ClONO_2$, as depicted in the previous chapter. If CFCs are to blame for enhanced ozone depletion, what then is the cause if they appear to be 'locked up'?

It is currently believed that both chemical and meteorological factors contribute to the way in which chlorine free radicals destroy ozone over the Antarctic. In the Antarctic winter there is a long period of time when the Sun is no longer in the sky and the region is plunged into darkness. This causes rapid cooling of the polar ice cap, which, in turn, causes the development of strong westerly winds in the stratosphere called the **polar vortex**. This vortex is very stable and lasts throughout the winter and spring before it finally breaks down in the summer. Whilst the polar vortex exists, the polar air in the upper atmosphere is essentially sealed off from that at lower altitudes. The temperatures in the lower atmosphere become so low (lower than $-75\,°C$) that **polar stratospheric clouds** form, which are composed of solid acidic ice particles. The presence of these particles enables fast heterogeneous reactions to occur on their surfaces causing the production of dichloride molecules, for example,

$$ClONO_2(g) + HCl(s) \longrightarrow Cl_2(g) + HNO_3(s) \qquad 10.46$$

where HCl is attached to the surface of the solid particles.

As we have seen, dichlorine readily undergoes photolysis to yield $Cl^{\cdot}$, and hence $ClO^{\cdot}$.

Additionally, HNO_3 remains on the solid ice particles and thus removes a major source of $NO_2^{\cdot}$. This prevents $ClO^{\cdot}$ from being locked up again as $ClONO_2$. Since the polar vortex lasts such a long time, the temperature remains low and thus polar stratospheric clouds continue to persist even whilst the pole is in sunlight. Thus the high levels of $ClO^{\cdot}$ continue. As the intensity of the sunlight increases, then the ozone loss will also increase via a catalytic cycle that involves the photolysis of Cl_2O to form $Cl^{\cdot}$,

$$ClO^{\cdot}(g) + ClO^{\cdot}(g) + M(g) \longrightarrow Cl_2O_2(g) + M(g) \qquad 10.47$$

$$Cl_2O_2(g) \xrightarrow{\text{UV}} Cl^{\cdot}(g) + ClO_2^{\cdot}(g) \qquad 10.48$$

$$ClO_2^{\cdot}(g) + M(g) \longrightarrow Cl^{\cdot}(g) + O_2(g) + M(g) \qquad 10.49$$

where M is some other molecule that absorbs any excess energy.

If temperatures were higher, then Cl_2O_2 would simply decompose back to $ClO_2^{\cdot}$ again. It is the lower temperatures and the presence of sunlight that give the right conditions for the generation of the ozone destructive $Cl^{\cdot}$.

The oxides of carbon

Carbon monoxide, together with carbon dioxide, is one of the most common atmospheric pollutants. Carbon monoxide is a colourless, tasteless, odourless gas, flammable and almost insoluble in water. It is mainly of concern because it is both toxic to humans and has a role in the formation of ozone in the troposphere.

The main sources of carbon monoxide are from the incomplete combustion of fossil fuels and wood, the oxidation of methane and the decomposition of organic matter. The combustion of carbon can occur via two steps,

$$C(s) + 1/2\,O_2(g) \;=\; CO(g) \qquad 10.50$$

$$\Delta H^0 \;=\; -110.5 \text{ kJ mol}^{-1}$$

$$\Delta G^0 \;=\; -137.2 \text{ kJ mol}^{-1}$$

$$CO(g) + 1/2\,O_2(g) = CO_2(g) \qquad 10.51$$

$$\Delta H^0 \;=\; -393.5 \text{ kJ mol}^{-1}$$

$$\Delta G^0 \;=\; -394.4 \text{ kJ mol}^{-1}$$

The second reaction is kinetically much slower than the first because of the high bond strength of the carbon–oxygen triple bond. The Lewis structure of carbon monoxide is :C≡O: and thus its lone pairs of electrons make it a strong Lewis base. For example, it readily forms complexes with transition metals such as iron, and is a reducing agent removing oxygen to produce carbon dioxide. More CO is produced naturally by the oxidation of methane, CH_4, in swamps and by the oxidation of organic matter in tropical areas than by anthropogenic actions. The processes are,

$$CH_4(g) + HO^{\bullet}(g) \longrightarrow CH_3^{\bullet}(g) + H_2O(g) \qquad\qquad 10.52$$

$$CH_3^{\bullet}(g) + O_2(g) + M \longrightarrow CH_3OO^{\bullet}(g) + M \qquad\qquad 10.53$$

$$CH_3OO^{\bullet}(g) + NO^{\bullet}(g) \longrightarrow CH_3O^{\bullet}(g) + NO_2^{\bullet}(g) \qquad\qquad 10.54$$

$$CH_3O^{\bullet}(g) + O_2(g) \longrightarrow HCHO(g) + HO_2^{\bullet}(g) \qquad\qquad 10.55$$

$$HCHO(g) \xrightarrow{UV} H_2(g) + CO(g) \qquad\qquad 10.56$$

At higher altitudes, a major source of CO is via the photochemical decomposition of carbon dioxide,

$$CO_2(g) \xrightarrow{UV} CO(g) + O(g) \qquad\qquad 10.57$$

Carbon monoxide has a residence time of between one and four months in the atmosphere. In the tropopause about 50 per cent of the CO is removed by reaction with hydroxyl free radicals,

$$CO(g) + HO^{\bullet}(g) \longrightarrow CO_2(g) + H^{\bullet}(g) \qquad\qquad 10.58$$

The origin of the $HO^{\bullet}(g)$ radicals comes from the decomposition of water molecules,

$$H_2O(g) \xrightarrow{UV} HO^{\bullet}(g) + H^{\bullet}(g) \qquad\qquad 10.59$$

$$O_3(g) \xrightarrow{UV} O_2(g) + O^*(g) \qquad\qquad 10.60$$

$$O^*(g) + H_2O(g) \xrightarrow{UV} 2\,HO^{\bullet}(g) \qquad\qquad 10.61$$

where O* is an energetically excited oxygen atom.

On the surface of soil CO is converted to CO_2 by the rapid uptake and action of bacteria and fungi,

$$2CO(g) + O_2(g) = CO_2(g) \qquad\qquad 10.62$$

$$4CO(g) + 2H_2O(l) = CH_4(g) + 3CO_2(g) \qquad\qquad 10.63$$

The former reaction is caused by *Bacillus oligocarbophilus* and the latter by *Methanosarcina backeri*.

CO is also removed by reaction with dioxygen in the presence of a third molecule M,

$$2CO(g) + O_2(g) + M = 2CO_2(g) + M \qquad\qquad 10.64$$

Plants also convert CO to CO_2 during the hours of darkness, and to amino acids in the presence of sunlight.

In the UK emissions of CO have decreased overall by 43 per cent since 1970 (NAEI 1998, Section 2.1 CO). This has been largely due to a decrease in the use of solid fuels for domestic purposes, and an increase in both the efficiency of internal combustion engines and the use of the catalytic converter for road transport. Road transport is responsible for the vast majority of CO emissions particularly when driving at low speeds. Compared to road transport the emission of CO from other sources is minor. Since 1970, the emissions from the domestic sector have declined by about 80 per cent. The banning of stubble burning in 1993 in England and Wales has caused a decline in CO emissions from agricultural sources, whilst emissions from power stations amount to only about 2 per cent of the total. Table 10.4 illustrates some CO emission trends in the UK since 1970 (NAEI 1998, Section 2.1 CO).

The natural ambient atmospheric concentrations for CO range between 0.01 and 0.23 mg m^{-3} in urban areas (WHO 2000). Mean concentrations over an 8-hour period are usually less than 20 mg m^{-3}, with a one-hour peak level of usually less than 60 mg m^{-3}. The highest concentrations, as expected, are close to busy roads, in road tunnels and in underground car parks. In such regions, levels can rise as high as 115 mg m^{-3} for several hours. The WHO (2000) guideline for CO in the atmosphere is 10 mg m^{-3} for an exposure time of eight hours. Above this figure adverse health effects in humans will start to appear. The UK DoE (1999) guideline is 12 mg m^{-3}, based on an annual eight-hour exposure time. The DoE has set a recommended value of 10 ppm (11.6 mg m^{-3}) for air by 31 December 2003 (DoE, 1999). The USEPA (2000) has its regulations for air set at 9 ppm for an eight-hour exposure time, which should not be exceeded more than once a year. The EU Directive 89/429/EEC (EU 1989) states that the limit of the air content of CO should be 150 mg m^{-3} for 90 per cent of all hourly averages in a 24-hour period. According to the UK National Air Quality Standard the 'standard threshold' for carbon monoxide over an eight-hour running average is less than 10 ppm. If air meets this standard, then it is described as a low level of air pollution. Similarly, the 'information level' is 10–14 ppm for air, which is moderately polluted, and the 'alert level' is 15–19 ppm for a high level of pollution. If levels of CO rise above the latter level, then the level of pollution is very high.

When humans are exposed to CO, it forms carboxy-haemoglobin at the expense of oxyhaemoglobin. Tissues are thus deprived of oxygen and asphyxiation occurs. If the victim continues to receive a high dosage of CO, then permanent brain damage and even death will result. Initial symptoms include dizziness, headache, nausea and faintness. Chronic exposure at 25 mg m^{-3} of CO in air causes cardiovascular problems (Stern et al. 1981), which can be particularly dangerous to a person who already suffers from such problems. The foetus is at considerable risk from the effects of carbon monoxide poisoning since it can cross the placenta (Longo 1976). The inhalation of 35 ppm for eight hours causes a loss in ability to learn and do complicated tasks, reduces awareness, decreases manual dexterity, and disturb sleep activity (Longo 1976).

Carbon dioxide is a non-toxic, colourless gas with a slight odour. It is soluble in water and is about 1.5 times more dense than air. It is used in the production of sodium carbonate, urea and mineral waters, as a refrigerant and in fire extinguishers.

Table 10.4 *CO emission trends 1970–98*

Source	1970	1980	1990	1995	1998	% decline
Road transport	5,427 (64.6)	5,378 (74.1)	5,147 (74.1)	3,969 (74.6)	3,479 (73.1)	36
Domestic	1,251 (8.3)	622 (3.9)	358 (3.9)	260 (4.9)	234 (4.9)	80
Land use	288 (6.0)	449 (3.8)	266 (3.8)	0 (0.0)	0 (0.0)	0
Total from all sources	8,407 (100%)	7,525 (100%)	6,938 (100%)	5,320 (100%)	4,758 (100%)	43

(*Note*: Data modified from National Atmospheric Emissions Inventory)

Carbon dioxide is a major 'greenhouse gas' and is believed to contribute to global warming. Consequently, in 1997 at the Third Conference of the Parties of the United Nations Framework Convention on Climate Control in Kyoto, Japan, it was decided to set targets for the reduction in emissions of CO_2 from anthropogenic sources for all participating countries. For the UK the target is to achieve a reduction of 12.5 per cent on the 1990 figures for the emission of the greenhouse gases CO_2, CH_4, N_2O, SF_6, the perchlorocarbons (PFC) and the hydrofluorocarbons (HFC).

The amount of CO_2 emitted in the UK in 1998 was 148.5 kt, a reduction of some 21 per cent over the 1970 emissions (NAEI 1998, Section 2.2 CO_2). Carbon dioxide emissions arise from the combustion of fossil fuels in power stations, transport and industrial, domestic and commercial sectors.

The greenhouse effect

Without an atmosphere the surface temperature of the Earth would be about $-18\,°C$. The global mean-surface temperature is substantially higher than this at 15 °C *because* of the presence of an atmosphere. The Earth radiates a spectrum of IR electromagnetic radiation mainly between 4 and 100 μm (Figure 10.1). It receives radiation from the Sun in the range 0.1–5.0 μm, i.e. in the UV, visible and near IR regions of the electromagnetic spectrum. About 50 per cent of the Sun's radiation reaches the Earth's surface, whilst 20 per cent is absorbed by gases such as ozone and carbon dioxide, and 30 per cent is reflected back into space by the clouds, snow, desert sands and other reflecting bodies. In order to maintain a constant temperature, the amounts of emitted radiation and the absorbed radiation must be the same.

Mainly because of the presence of water and carbon dioxide molecules (and other trace gases), some of the incoming short-wave radiation and the outgoing longer wavelength radiation are absorbed. Which wavelengths of electromagnetic radiation are absorbed depends upon the molecule, e.g. CO_2 absorbs strongly at about 2.5 μm and 4.0 μm, whilst ozone absorbs at about 9.6 μm. If the absorption caused by CO_2 and H_2O molecules, together with those of naturally occurring O_2, N_2O and CH_4, are all added together, then the atmosphere is found to be largely transparent to incoming visible radiation and outgoing infrared radiation in the region 8–12 μm. Thus in this latter region there is an **atmospheric window**. Ozone absorbs at about 9.6 μm, so even this window is not completely transparent to radiation being emitted from the Earth.

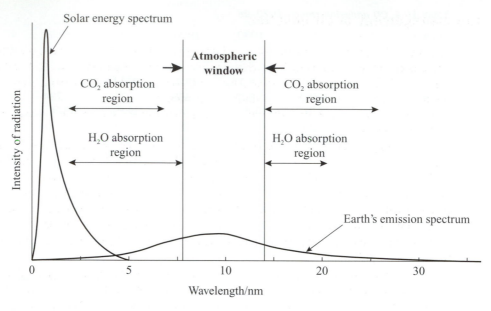

Figure 10.1 The emission spectra of the Earth and the Sun, and the absorption regions for CO_2 and H_2O.

Providing that they are polar and are not involved in symmetric stretching, the chemical bonds in a molecule will absorb infrared radiation of frequencies (and therefore wavelengths!) equal to the frequencies of vibration of those bonds. This absorption of infrared causes the amplitude of the molecular vibrations to increase but their frequencies to remain the same, i.e. the molecules become vibrationally excited. Vibrationally excited molecules can lose their energy by either re-emitting it randomly as infrared radiation, or by collision with other molecules where it is converted to kinetic energy. Many of the longer wavelengths emitted by the Earth are thus absorbed and re-emitted by the components of the atmosphere and are effectively trapped within the atmosphere. Some radiation will be lost to space but some will be radiated downwards. The mean global surface temperature will thus be increased. This effect is referred to as the **natural greenhouse effect**, and those gases that contribute to it are called **greenhouse gases**.

How effective a gas is in contributing to the natural greenhouse effect depends upon how much is present and its particular vibrational frequencies. Water and carbon dioxide are the main contributors and are responsible for about 80 per cent of the total temperature rise. What is to be avoided is an increase in the amounts of natural gases that cause the greenhouse effect, or pollution of the atmosphere by adding 'new' ones that are even better at helping to increase the Earth's temperature. Whilst it is true that any increases in the concentration of greenhouse gases caused by human activities may lead to global warming, the situation is a very complex one. There are many effects occurring in the atmosphere that can have a negative effect on the ability of these gases to produce a greenhouse effect. It is by no means certain what the real result will be. However, if global warming is a reality, then changes may occur in global weather patterns. For example, there could be an increase in the severity of storms and droughts, and an overall rise in sea-level of the order of 50–60 cm by the year 2100, caused by the thermal expansion of oceans and the melting of ice caps.

The management of air pollution

The method currently in general use to minimise the release of pollutants to the atmosphere is called the 'best available techniques not entailing excessive cost' (BATNEEC) approach. Here the word 'best' means the most effective technique(s) which prevents, minimises and renders harmless polluting emissions.

There are many ways of reducing emissions that can be selected according to the problem in hand.

Volatile organic solvents that give rise to much fuming can be replaced with water-based ones. For example, water-based glues/adhesives are much more environmentally friendly in being less toxic, less flammable, easier to store and more easily removed. Halogenated compounds such as CFCs are being replaced by non-halogenated compounds such as propane and isobutane. Pump action valves have also replaced pressurised propellants in hair sprays and roll-on deodorants have become common place.

All of the above are examples of **replacement** and have been made in response to the effects of pollution.

Another way of reducing emissions is to **improve** existing technology so that less waste is produced from a process, or to enable the return of useful 'waste' to the process. For example, nitrogen(II) oxide is always formed in combustion processes, but its rate of formation depends upon how fast the gases pass through the combustion chamber. If the temperature is kept below 1,200 °C, the rate of formation will become so slow that insignificant amounts of nitrogen(II) oxide are produced.

A much more careful approach to the **monitoring** of a process can lead to better control of emissions. This enables the optimisation of the process conditions to ensure its maximum efficiency. In a combustion plant, for example, efficiency ensures minimum fuel usage and a corresponding reduction in pollution.

The control of air pollution is better started at its **source**. There are three main ways that this can be achieved. The first is to modify the processes involved so as to minimise the production of or to avoid altogether the release of wastes to the atmosphere. The second involves the collection of material in the form of particulate matter, and the third the absorption of toxic gases.

The polluting effects of solid particles (grit, dust, fumes) depend greatly upon the size of those particles. Particles that have been produced by mechanical operations such as grinding are usually large in size. These easily settle under the influence of gravity and are more easily controlled and therefore managed. Particulate matter is separated from a gas stream by methods that depend upon the weight of the particles: by using a liquid to wash the gas; by using a fabric filter; or by electrostatic precipitation. If particles are separated by gravity, scraped off fabric filters or removed from electrostatic precipitators, then they are treated as solid waste and dealt with accordingly. Such is the legal demand in the UK on the control of particulate emissions that any inefficiency has to be dealt with immediately and stand-by plants are necessary in case of breakdown. The best available technique for removing particulate matter is normally selected. This decision is based on a review of several factors, including the efficiency of the technique; the size of the particles to be removed; particle sensitivity to temperature; particle resistance to corrosion; electrical sensitivity; and moisture content.

Gaseous pollutants such as sulphur dioxide, the oxides of nitrogen and volatile organic compounds were once allowed to escape from tall chimneys, often as very corrosive and toxic plumes. Now such pollutants are treated according to the their type. Organic materials, for example, are often burned. If they have a commercial value, they may be recovered by absorption in a suitable solvent and adsorption by a suitable surface. Inorganic pollutants are generally absorbed by water, *but* this in turn generates waste water which requires a treatment plant.

The consequences of waste minimisation for air pollution are many. These include savings in energy; costs of cleaning; the avoidance of pollution costs; reduction in the costs of control plants and their operating costs; and the avoidance of disposal costs for dealing with solid wastes or liquid effluents from gas cleaning plants. Clearly, an integrated waste management approach is required so that problems generated by the treatment of wastes are not shifted elsewhere.

Often new research brings to the fore the identity of new and 'unexpected' sources of atmospheric pollution. For example, in 1995 researchers at the University of California reported that fast foods such as hamburgers produced nine times more air pollution in Los Angeles than the city's bus service. It was claimed that the pollution was mainly due to grilling food, which caused the fat from the hamburgers to drip on to the source of heat where it is burned at high temperatures. It has been estimated that fast-food restaurants produce some 14 tonnes of smoke and 19 tonnes of organic chemicals every day. These combustion products contribute to the infamous smogs of Los Angeles, and it is thought that they could cause cancer and respiratory problems. Environmental officers in California are pressing for legislation to force fast-food producers to change their methods of cooking or to install pollution control devices. On the other hand, the producers and restaurant owners are challenging the researchers' findings.

Summary

- A point source of pollution is localised and restricted, and is the subject of easy remedial action.
- Diffuse source pollution has less well-defined origins and is not restricted to a particular location. Such sources are much more difficult to remedy.
- Particulate matter is any airborne particle whose diameter is greater than 20×10^{-10} m.
- Smoke in the UK is classified in three ways: (i) carbonaceous particles from burning fossil fuels; (ii) soot, ash and grit particles in smoke; and (iii) all the particles that can be collected on a filter paper according to a British Standard test.
- Airborne particles may be classified by how fast they are able to penetrate the human respiratory system.
- PM_{10} is the currently accepted UK standard measurement of particulate matter.
- The Black Smoke Measurement depends upon the degree of blackening of filter paper.
- Primary particulates are those that are directly emitted to the atmosphere. Secondary particulates form in the atmosphere as a result of chemical reactions.
- SO_2 is a primary pollutant caused mainly by the burning of fossil fuels.
- SO_2 is easily oxidised to the secondary pollutant SO_3, which forms sulphuric acid with water. It is a major contributor to acid rain.
- The oxides of nitrogen are all primary atmospheric pollutants. NO forms the secondary pollutant NO_2, which is both toxic and acidic. NO_2 forms acids with water, which contributes to acid rain. N_2O is chemically quite unreactive at ordinary temperatures, but it is a 'greenhouse' gas.
- A smog is a mixture of smoke and fog. A photochemical smog is the product of hydrocarbons, NO(g) and sunlight.
- Acid rain may cause the death of plants, the pollution of waterways and corrosion of building materials.

- Halocarbons are mainly responsible for the depletion of ozone above the Antarctic ice cap. Both chemical and meteorological factors contribute to ozone destruction.
- CO and CO_2 are formed from the combustion of fossil fuels, the oxidation of methane and the decomposition of hydrogen sulphide. CO is toxic, whilst CO_2 is a greenhouse gas.
- The Greenhouse Effect is caused by the absorption of radiation by certain molecules in the atmosphere and its re-emission as infrared radiation.
- Air pollution may be managed by (i) the use of the BATNEEC approach; (ii) the replacement of chemicals by less harmful ones; (iii) the improvement of technology; (iv) more rigorous and effective monitoring; and (v) better control of potential pollutants at source.

Questions

1   Draw the Lewis structures for NO, NO_2 and N_2O.

2   A total volume of 5×10^{-6} m³ of spherical particles was collected in a filter system. The mean radius of these particles was found to be 8 μm.

   (a) Calculate the number of particles collected if it is assumed that the volume of the spaces between the spheres can be ignored. (*Note*: Volume of a sphere is $V = 4/3\pi r^3$ where r = the radius.)

   (b) Calculate the total surface area of these particles, given that the surface area of a sphere is $S = 4\pi r^2$.

   (c) Assuming that the same source of the particles reduced the volume to 1/10 of the original, *but* the same total volume was collected, what is the new mean radius of the particles and the total surface area?

   Comment on your answers.

3   The visible region of the electromagnetic spectrum ranges from about 400 nm to 800 nm.

   (a) Calculate the energy of a source of green light of wavelength 560 nm.

      (*Note*: $h = 6.626 \times 10^{-34}$ J s^{-1} and $c = 3.00 \times 10^8$ m s^{-1}.)

   (b) The frequency of a source of UV radiation is 1.25×10^{15} Hz.

      What is the wavelength of this radiation?

4   Calculate the maximum wavelength of electromagnetic radiation that can cause the decomposition of ozone to dioxygen and atomic oxygen from the following data:

$$O_3(g) \xrightarrow{\varepsilon=hf} O_2(g) + O(g)$$
$$\Delta H^0_{f,298K}[O(g)] = +249.2 \text{ kJ mol}^{-1}$$
$$\Delta H^0_{f,298K}[O_3(g)] = +142.3 \text{ kJ mol}^{-1}$$

5   Calculate the wavelengths of the electromagnetic radiation necessary to cause the following reactions:

$$N_2O(g) \xrightarrow{hf} NO^{\bullet}(g) + N(g)$$
$$N_2O(g) \xrightarrow{hf} N_2(g) + O(g)$$
$$\Delta H^0_{f,298K}[NO(g)] = +90.4 \text{ kJ mol}^{-1}$$
$$\Delta H^0_{f,298K}[N_2O(g)] = +82.1 \text{ kJ mol}^{-1}$$
$$\Delta H^0_{f,298K}[N(g)] = +472.7 \text{ kJ mol}^{-1}$$
$$\Delta H^0_{f,298K}[O(g)] = +249.2$$

6   The pH of a lake has been found to be 4.0. In order to raise the pH of the lake to 6.0, it has been decided to add crushed limestone ($CaCO_3$) to neutralise some of the acid. The volume of the lake has been estimated to be $5 \times 10^7 \, m^3$.

   (a)   What is the hydrogen ion concentration in the lake before and after the addition of limestone? What is the total amount of hydrogen ions that will need neutralising?

   (b)   Assuming that the limestone is pure calcium carbonate, how many moles of calcium carbonate will be required to neutralise the hydrogen ions? What mass of limestone will be required?

   (c)   If large amounts of quartz were in the lake, would this have any effect on the amount of calcium carbonate required?

   (d)   What might have been the cause of the lake having a pH of 4.0?

7   According to the UK National Atmospheric Emissions Inventory, benzene is a National Air Quality Strategy pollutant, hydrogen chloride an acidifying gas, whilst the heavy metals nickel and vanadium are hazardous air pollutants.

     What are the origins of each of these chemicals in the atmosphere, and what are the particular hazards associated with each?

References

Canada–US Air Quality Agreement (1998) *4th Progress Report, Environment Canada*, Available at: http://www.ec.gc.ca/special/aircan_e.pdf.

Canadian Government (2003) *Acid Rain and the Facts*. Available at: http//www.ec.gc.ca/acidrain/acidfact.html.

DEFRA (1999) *The Air Quality Strategy for England, Scotland, Wales and N. Ireland*: A consultation document.

European Commission (1989) Directive 89/429/EEC *The Reduction of Air Pollution from New Municipal Waste Incineration Plants*. European Commission, Luxembourg.

Longo, L.D. (1976) Carbon monoxide effects of oxygenation of the foetus in utero. *Science*, **194**, 523–5.

NAEI (National Atmospheric Emissions Inventory) (1998) *Annual Report, UK Emissions of Air Pollutants 1970–98*. Available at: http://www.geat.co.uk/netcen/airqual/naei/annreport/annrep98/naei98. html

National Acid Precipitation Assessment Programme (1990) *Integrated Assessment Report*. Government Printing Office, Washington, DC.

Stern, F.B. (1981) Exposure of motor vehicle examiners to CO: a historical perspective mortality study, *Archives of Environmental Health*, **36**, 59–65.

USEPA (1998) *National Air Pollutant Emission Trends Report 1900–1998*. Available at: http://www.epa.gov/ttn/chief/trends98/chapter2.pdf.

USEPA (2000) *Emissions Data and Compliance Reports*. Available at: http//www.epa.gov/airmarkets/emissions/score00/index.html.

USEPA Office of Air Quality Planning and Standards (2000) *US National Ambient Air Quality Standards* (July). Available at: http://www.epa.gov/airs/criteria.html

WHO (2000) Information Fact Sheet No. 187: *Air Pollution*. World Health Organisation, Geneva.

Further reading

Baird, C. (1998) *Environmental Chemistry*, 2nd edn. W.H. Freeman, New York.
About one-quarter of this book is devoted to the Earth's atmosphere. It covers stratospheric chemistry with particular reference to ozone, the chemistry of the air at ground level and its pollutants, and the Greenhouse Effect and global warming. The chemistry involved is straightforward and quite easy to understand. Its chapters form a more detailed approach than the ones in this book!

Hobbs, P.V. (2000) *Introduction to Atmospheric Chemistry*, 2nd edn. Cambridge University Press, Cambridge.
A companion volume to this author's book, *Basic Physical Chemistry for the Atmospheric Sciences*. It deals very well with air pollution and in particular acid rain, ozone and ozone depletion, and climate change.

National Atmospheric Emissions Inventory (NAEI) *Annual Report, UK Emissions of Air Pollutants 1970–98*. available at: http://www.geat.co.uk/netcen/airqual/naei/annreport/annrep98/naei98.html
Everything you wanted to know about UK atmospheric emissions.

Singh, H.B. (ed.) (1996) *Composition, Chemistry and Climate of Atmospheres*. John Wiley & Sons, Chichester.
This is a collection of writings by world experts on the chemistry of the air and its pollution. It gives an account of the history of air pollution and provides an extensive cover of photochemical smog, acid rain, ozone and the atmosphere, and global warming.

Wayne, R.P. (2000) *Chemistry of Atmospheres*, 3rd edn. Oxford University Press, Oxford.
See Chapter 8.

11 Natural waters and their properties

- The natural water cycle
- Some characteristics of pure water – physical properties
- Temperature variations within a water body – stratification
- The three phases of water – relative humidity and vapour pressure
- Importance of hydrogen bonding
- Characteristics of natural waters
- Water and living organisms
- BOD, COD and permanganate values as a measure of water quality
- Acidity and alkalinity, pH, dissociation constants, buffer solutions
- The $CO_2/H_2O/HCO_3^-/CO_3^{2-}$ system
- Hard and soft water
- Hydrolysis

The Aral Sea – the mismanagement of a water resource

[Levintanus 1993; Voropaev 1993; Precoda 1991] The inland Aral Sea is situated in Central Asia to the east of the Caspian Sea. This immense body of once fresh water was the fourth largest of its type. There are two main tributary rivers, the Syr Darya and the Amu Darya. The Aral Sea is one of the world's most ancient and for centuries its boundaries remained unchanged.

The Aral Sea was an important shipping route and supported a fishing industry. Its environs provided a source of muskrat pelts, whilst its reed beds supplied raw materials for a local papermaking and packaging company. Extensive forests gave shelter to migrating birds, a home for some rare animals and a natural barrier to erosion.

In the 1960s it was decided to grow cotton in this region, which required a heavy withdrawal of irrigation water from both the tributary rivers. By the 1980s so much water was being withdrawn that the Aral Sea was not receiving enough to maintain its size. Thus, it has been estimated that the sea will cease to exist by 2040 (Figure 11.1).

The amount of water taken for irrigation was about two to four times more than what was necessary to grow cotton. This was due to huge water losses caused by seepage from the simple earthen channels before it reached the cotton fields. For example, the 1,200 km-long Karakum Canal, which takes water from the Amu Darya, runs in contact with loose desert sands for hundreds of kilometres. In villages alongside the canal, water is found in cellars, communication lines have been cut and buildings have collapsed because of subsidence. Land has become waterlogged, causing the loss of orchards and vineyards. The Bolshoy Andizhansk Canal in Balykchinek also has no lining, so water seepage has caused more than 2,000 hectares of farmland to become marshes.

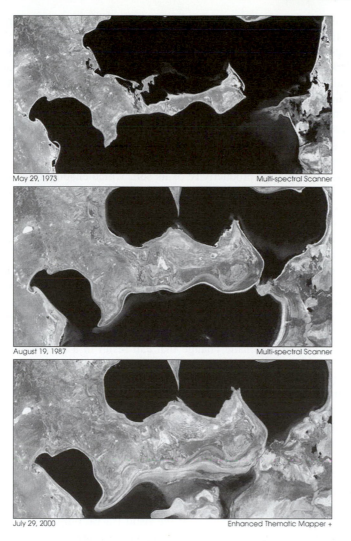

May 29, 1973   Multi-spectral Scanner

August 19, 1987   Multi-spectral Scanner

July 29, 2000   Enhanced Thematic Mapper +

Figure 11.1 The diminishing Aral Sea (images from Earth Observatory)

Source: http://eob.gsfc.nasa.gov

The lost water, together with the careless disposal of collected and drained water from agricultural land, has added to the water table in many areas. This groundwater is saturated with minerals, particularly salt. Because of the added extra water, the groundwater has made its way to the surface in places, bringing with it its dissolved salt. The salt has contaminated soil to such an extent that it has prevented the growing of crops.

In cotton-growing areas, the soil is washed several times with fresh water to remove the salt. This also removes many other soluble materials which are necessary for plant growth, and so more fertiliser has to be added than otherwise would be needed.

By 1985, it was estimated that over 3×10^9 m^3 of water from the Uzbek and Turkmenistan fields, contaminated with chemicals and salt, together with untreated domestic and industrial wastes, were being dumped into the Amu Darya each year. Local soils have thus been contaminated with a variety of harmful substances such as pesticides, nitrate(V) and nitrate(III).

The consequences of this pollution have been dire. In Ashkhabad Oblast it is probable that organic pesticides have caused an increase in numbers of gall bladder, liver and pancreatic diseases. Mineral pesticides have been linked to psychiatric disorders and premature births. Turkmenia has a very high infant mortality rate and very frequent outbreaks of intestinal disorders. These again have been linked to the use of drinking water drawn from the heavily contaminated Amu Darya. The water also regularly contains a high concentration of bacteria, which may be the cause of typhoid cases eight times more numerous than the national average for Russia. In the lower reaches of the Amu Darya, water is so unsuitable for drinking, even after boiling, that it threatens the lives and health of nearly three million people.

Uzbek was renowned for the high productivity of its soil, which enabled three crops per year to be grown. Unfortunately, its sole use for the production of cotton over some 50 years has exhausted the soil. The parallel emergence of plant diseases and pests has resulted in the application to the soil of mineral fertilisers and pesticides, with the consequent death of much of the soil fauna. In this area, yellow jaundice and intestinal diseases have resulted in a soaring death rate, especially amongst children.

In the vicinity of the Aral Sea itself, the levels of kidney and liver diseases are high, and in some areas the infant mortality rate is as high as one in ten. Cancer of the oesophagus, hepatitis and gastric diseases are all common. In some areas, nursing mothers have been advised not to breastfeed their babies because their milk has been found to be toxic.

The Aral Sea once formed a barrier protecting Central Asia from cold north winds. Its reduction in size has caused a climatic change that can result in a maximum daily temperature of 47 °C in summer and −17 °C in winter. Such a temperature variation may lead to the total loss of cotton production in the near future.

The Aral Sea was also a barrier against the drying effects of winds. Now drying winds and dust storms occur in Central Asia where they did not occur before. Wind erosion is common in the Aral region. Salt from the dried seabed is carried as fine particles by the wind and deposited elsewhere to become another problem.

Lower groundwater level in the river deltas plus increased mineralisation of the water have resulted in the death of the forests on the edge of the Aral Sea (about one-fifth of the original forests were left at the end of a 20-year period). The lakes and bogs associated with the sea have nearly all dried up and the reed beds have gone. Of the 178 species of animals previously found in the delta of the Amu Darya, only 38 remain.

Since the early 1960s the Aral Sea's volume has dropped by nearly 70 per cent and the level of the water by 12 metres. This has halved the surface area and increased the salinity of its water to three times its original value. The increase in salinity has adversely affected plant and animal life. The fishing industry is almost non-existent and the waters are no longer transparent or filled with a wide variety of fish.

By early 1990 the loss of water from the sea had caused it to become two bodies of water, separated by a belt of dry land more than 100 metres wide at its narrowest point. The exposed seabed is now covered by a white alkaline soil. The southern Large Aral Sea is fed by the Amu Darya and the northern Small Aral Sea by the Syr Darya. The area covered by the Aral Sea in 1974 was about 24,635 square miles. In 1990 both seas covered a total area of about 14,092 square miles.

The natural water cycle

There are some 1.5×10^9 km^3 of water on the surface of the Earth. Of this 98.3 per cent is in the oceans/seas and 1.6 per cent in the form of ice. The remaining 0.1 per cent

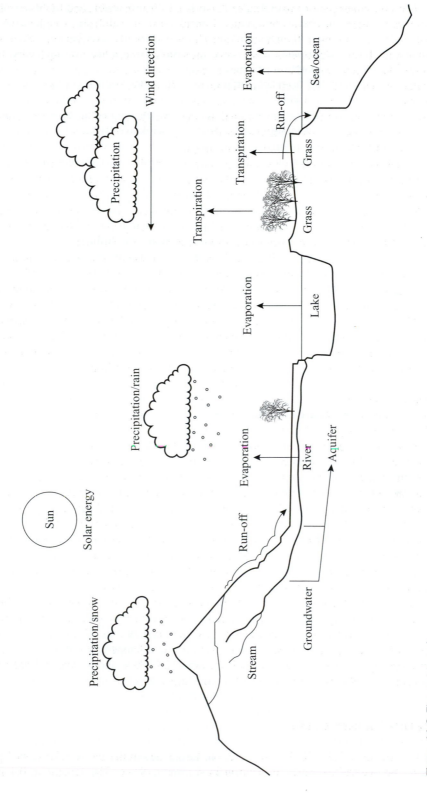

Figure 11.2 The water cycle.

exists as groundwater and in lakes and rivers. All of this water, and the water used, has been a part of the natural water cycle, or global hydrological cycle, at some time (Figure 11.2). This cycle is based on the continuous movement of water between the surface of the Earth and the atmosphere. The water cycle involves a dynamic balance between the two processes of evaporation and precipitation.

Water is evaporated from the surfaces of both water bodies and land surfaces. It is also transpired from living plant cells. The water vapour produced is circulated throughout the atmosphere, where it is eventually precipitated as snow and rain. Snow and rain are thus the ultimate sources of all our drinkable (potable) water.

Depending upon the amount of rain, the water can follow two paths. If it is heavy rain, a very large amount of water will run off the land into streams and rivers, eventually finding its way into lakes and the oceans. How much water becomes run-off will depend upon the porosity, permeability, thickness and previous moisture content of soil. Some of the water will remain in the soil and will be returned to the atmosphere by evaporation and absorbed by the roots of plants to be transpired from leaves. However, some water may manage to move downwards under the influence of gravity (termed infiltration), through porous rock strata, until it reaches an impenetrable layer. Here it collects and becomes the groundwater that is the source of wells and of the springs that feed streams, rivers and lakes. The surface of this groundwater is called the water table. Under natural conditions, a water table will fall or rise according to prevailing weather conditions, and whether or not it is being used as a water supply to a spring or as a reservoir for human use. Since groundwater spends a great deal of time in contact with subterranean rocks, what dissolved materials it contains will be a reflection of the geology of those rocks, as well as the surface material it has passed through.

Some characteristics of pure water

Water is an angular, polar molecule and an excellent solvent for many ionic compounds. When, for example, sodium chloride dissolves in water, it is dissociated into its ions, which on thorough agitation are distributed evenly throughout the water. Thus, since water is a very mobile liquid, it makes salts more readily available and will deliver both unwanted and wanted materials to plants, animals, buildings, etc.

Compared with other materials, water has a high specific heat capacity of $4.18 \text{ J C}^{-1} \text{ g}^{-1}$. This means that it 'protects' living organisms, in that it requires a substantial amount of heat to raise its temperature to levels that might be dangerous. Water also acts as an excellent coolant in industry and elsewhere. Its standard enthalpy change of vaporisation is also high for a liquid at $2,260 \text{ kJ kg}^{-1}$, so therefore its *rate* of vaporisation is not fast. The standard enthalpy change upon freezing, or standard enthalpy change of fusion, is again high at 333 kJ kg^{-1}. This also helps to protect living organisms from the effects of freezing or melting as a consequence of weather conditions.

Water has a maximum density of 1.00 g cm^{-3} at a temperature of 4 °C. When it is in the form of solid ice its density is 0.9 g cm^{-3}, which is about half of what might be expected. This is why ice floats on water. Once a layer of ice forms on the surface of water, because of its poor thermal conductivity, it acts as an insulating layer, which protects the lower layers of water from freezing. The reason why ice has such a low 'unexpected' density is because directional hydrogen bonding between the molecules of water helps to keep these molecules apart and causes the solid to form an open cage-like structure. There are at least nine different structures known for ice, depending upon the conditions under which it has been formed. When liquid water or water vapour

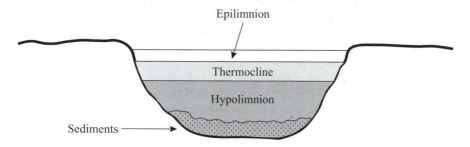

Figure 11.3 The stratification of a water body.

crystallises at 1 atmosphere pressure, the ice with which people are familiar has a hexagonal structure in which each oxygen atom is surrounded by an almost regular tetrahedron of four other oxygen atoms.

When a standing mass of water is heated by the Sun, temperature variations can be caused within that body. This variation is termed **stratification** (Figure 11.3). The surface of the water is warmed first, expands and therefore becomes less dense. This surface layer of warmer water is called the **epilimnion**. The cooler, deeper layers are referred to as the **hypolimnion**. Between these two 'extreme' layers is a narrow region called the **thermocline**, where rapid temperature change occurs. The degree of mixing between these layers of differing temperatures is very poor. The epilimnion is warm and rich in dioxygen because it is next to the atmosphere, whereas the hypolimnion is dioxygen poor. When organisms die and sink to the bottom of a water body they begin to rot. Rotting involves oxidation, so this dead matter will remove dioxygen from the water, depleting it even further. (Thus anaerobic organisms and methanogenic and sulphate-reducing bacteria tend to live at the bottom of a water body.) As decay continues, nutrients build up in bottom waters. In the cooler months of autumn the epilimnium cools and the surface density of the water rises. This denser water sinks and more mixing with lower waters takes place. Stratification disappears – this is called the **overturn**. As layers mix, chemicals are distributed through the water body so that the upper layers become richer in nutrients. This results in more biological activity and bursts of growth of algae called **phytoplankton blooms** can occur. A similar situation occurs during the spring months.

The three phases of water

In the environment there is a continuous exchange of matter at the interface between the three phases of water and the atmosphere. Water evaporates and condenses according to prevailing temperatures and pressures. A measure of the amount of water vapour in the atmosphere is the **relative humidity**. At a particular temperature, the atmosphere can become saturated with water vapour, i.e. the amount of water it can hold is at a maximum. The relative humidity is the ratio of the actual amount of water the atmosphere is holding at the existing temperature to the amount that saturates the atmosphere at the same temperature. As the temperature of the atmosphere increases, so does its relative humidity. Air movements help to prevent equilibrium being achieved between liquid water and vaporised water and therefore high relative humidities are not as common as they might otherwise be.

If pure water is placed in a closed container at a constant temperature, then a dynamic equilibrium is soon established when the number of evaporating water molecules escaping from the surface will equal the number of those that are condensing, i.e. the rate of evaporation is the equal to the rate of condensation. The gaseous water molecules will contribute to the pressure being exerted by the atmosphere in the container. This pressure is called the **vapour pressure** of liquid water. At 25 °C the vapour pressure of water is 23.8 torr. If a non-volatile solid is dissolved in the water, then its vapour pressure will be reduced. This is because the dissolved solid particles will replace water molecules at the surface thus reducing the opportunity for the evaporation of water molecules.

If a solvent A contains a solute B and that solute is not volatile, i.e. it can be assumed it contributes nothing to the vapour pressure exerted above that solution, then Raoult's law will apply. This law states that the vapour pressure of the solution will be proportional to the mole fraction of the A present and is defined by,

$$n_A = n_A / n_A + n_B$$

where n_A = the number of moles of solvent
n_B = the number of moles of solute

Thus, by Raoult's Law, if $p_0(A)$ is the vapour pressure of a pure liquid solvent A, and $p(A)$ the vapour pressure exerted above a solution,

then, $p(A) = \text{a constant} \times \dfrac{n_A}{n_A + n_B}$

To find the constant in the above equation, let $n_B = 0.0$ so that the pure solvent is being considered,

hence, $p(A) = \text{a constant} = p_0(A) = \text{vapour pressure of the pure solvent}$

Therefore, $p(A) = \dfrac{p_0(A) \times n_A}{n_A + n_B}$

Thus for water, the presence of a dissolved substance will lower its vapour pressure. This is because, in a mixture of A (water) and B (e.g. sodium chloride), as the amount of B gets bigger then the mole fraction of A gets smaller. If water contains many non-volatile dissolved substances, then its vapour pressure can be considerably lowered.

If the temperature of a liquid is decreased, its vapour pressure decreases. Eventually, the liquid will start to solidify. A solid also exerts a vapour pressure but it is usually very much lower than that of its liquid and, like a liquid, this vapour pressure also falls as the temperature falls. What happens then at the solidification or melting point? As a liquid solidifies, the vapour pressure of both the liquid and the solid will be the same. Hence, if a non-volatile solid is added to a liquid and its vapour pressure consequently lowered, then the temperature at which the liquid solidifies or the solid melts will be the same and lower. When dealing with pure water, ice and liquid water have the same vapour pressure at 0 °C and 1 atmosphere pressure. However, the presence of dissolved matter in seawater reduces its freezing point to approximately −2 °C.

If they are placed in an open container, ice, liquid water and gaseous water never achieve a dynamic equilibrium. For example, because of the amount of air above an open container, the rate of evaporation will exceed that of condensation. Thus, water will evaporate from a puddle completely!

Water has a low vapour pressure in view of its low molar mass. However, hydrogen bonding causes the liquid to be less volatile. As the temperature of water is raised, its vapour pressure increases until it equals the prevailing atmospheric pressure. When its vapour pressure equals the atmospheric pressure, the water is boiling. When the boiling point of pure water is quoted as 100 °C, this means that its vapour pressure has increased to 1 atmosphere pressure when heated to 100 °C. The higher the atmospheric pressure, the greater is the boiling point temperature, and the lower the atmospheric pressure, the lower the boiling point. The boiling point of any liquid should always be quoted with the prevailing atmospheric pressure at which it was measured.

The boiling point of water is 'unexpectedly' high, again because of hydrogen bonding. The hydrogen sulphide molecule, H_2S ($M_r = 34$), has almost twice the mass of the water molecule, H_2O ($M_r = 18$), but its boiling point is -60 °C (213.2 K) as opposed to 100 °C (273.2 K). In hydrogen sulphide hydrogen bonding is almost completely absent.

Water is an exception to the rule that the freezing point of a liquid decreases with pressure. Water freezes at a *lower* temperature under pressure, again because of the presence of the hydrogen bonding that causes the open cage-like structure to form in ice. As the pressure is increased on water, it tends to remain a liquid because it occupies a smaller volume than ice. Ice tends to melt under pressure because a more compact structure is favoured. Remember, high pressures cause reductions in volume!

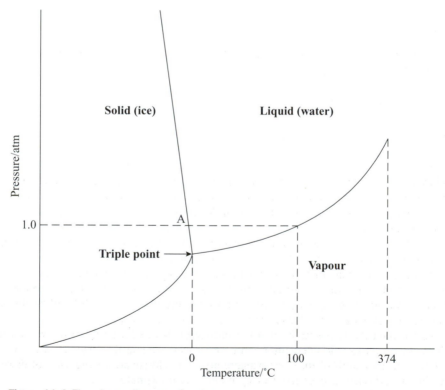

Figure 11.4 The phase diagram of water.

The movement of glaciers is thought to be caused by the weight of ice pressing down on the rocks deep under the surface, causing the ice there to melt. Thus a glacier moves on top of a thin lubricating layer of water.

The relationship between the three phases of water is shown in Figure 11.4. In this **phase diagram**, the lines are phase boundary lines, and any point on these lines shows the conditions under which two phases are present in dynamic equilibrium. For example, point A shows that ice and liquid water are in equilibrium at 1 atmosphere pressure at 0 °C. The triple point is when all three phases coexist at equilibrium, i.e. 4.66 torr at 0.01 °C.

Liquid water can exist only if its vapour pressure can exceed 4.6 torr. On a cold, dry day the partial pressure of water in the atmosphere may be lower than 4.6 torr. In this case, a frost will appear without forming liquid water first.

The characteristics of natural waters

Water has the ability to dissolve a wide range of materials, so that in its natural state (streams, rivers, lakes and underground sources/aquifers) it is never pure. It always contains a variety of soluble inorganic and organic compounds. In addition to these, water can carry large amounts of insoluble materials that are held in suspension. Both the amounts and types of impurities found in natural water vary from place to place and by time of year. These impurities determine the characteristics of a water body.

Suspended solids in a moving body of water such as a river will settle out at various points or be carried longer distances, depending upon their size and the rate of flow of the water. There will be a greater amount of suspended matter in a UK river in winter because of the greater rainfalls normally experienced in the winter months and the melting of snow, which lead to larger volumes of run-off from land. The amount of suspended solid in a water body is measured in $g\,m^{-3}$ and the more there is, the cloudier or more turbid is the water. The colour of the water can also be affected by the presence of suspended matter. Suspended matter can affect the amount of light entering water and therefore restrict the amount of photosynthesis that can occur and therefore the growth of plants. If small particles settle out in large enough quantities on the bottom of a water body, they may prevent some organisms from living there as well as preventing green plants from photosynthesising.

How fast a water body moves affects the degree of mixing of water and how much dioxygen it will carry. Thus, fast-flowing highly agitated streams will not only be saturated with dioxygen but also carry well-mixed nutrients, which will ultimately be carried to a river. Small rapidly flowing streams are nearly always saturated with dissolved dioxygen. Sluggish rivers may well contain 'hot spots' of poorly mixed materials and have oxygen contents well below saturation levels.

The temperature of a natural water body is crucial to the amount of dissolved dioxygen it can contain. The warmer the water, the less dioxygen it contains.

During their condensation and precipitation, rain or snow will dissolve carbon dioxide and any other chemicals that are present in the atmosphere. How much of these chemicals is dissolved will depend on their relative solubilities, the prevailing temperature of the atmosphere and other factors.

In its movement on and through the surface of the Earth, water dissolves and reacts with the minerals that occur in the soil and rocks. The principal dissolved chemicals derived from these minerals in both groundwater and surface water are the chlorides, hydrogencarbonates, and sulphates of sodium and potassium, together with the hydroxides of calcium and magnesium. Almost all natural supplies of water also contain 'controversial' fluorides to a greater or lesser extent.

Table 11.1 *The concentration of some selected ions in various water bodies*

Rainwater		River water		Seawater	
	mg dm^{-3}		mg dm^{-3}		mg dm^{-3}
Na$^+$	2.0	Na$^+$	6.3	Na$^+$	10,540
K$^+$	0.3	K$^+$	2.3	K$^+$	380
Ca^{2+}	0.1	Ca^{2+}	15.0	Ca^{2+}	400
Mg^{2+}	0.3	Mg^{2+}	4.1	Mg^{2+}	1,270
Fe^{2+}	–	Fe^{2+}	0.7	Fe^{2+}	–
Cl$^-$	3.8	Cl$^-$	7.8	Cl^{-1}	18,980
Br$^-$	–	Br$^-$	–	Br$^-$	60
SO$_4^{2-}$	0.6	SO$_4^{2-}$	11.2	SO$_4^{2-}$	2,460
HCO$_3^-$	0.1	HCO$_3^-$	58.4	HCO$_3^-$	140

Seawater contains many soluble compounds as well as a high concentration of sodium chloride. This is because many 'contaminated' streams and rivers are constantly feeding the oceans. At the same time, pure water is being constantly lost from the surface of these vast water bodies by evaporation. Thus the less volatile materials are left behind in ever-increasing amounts which causes the saline nature of the seas and oceans to increase.

Some example concentrations of ions to be found in seawater, river water and rain-water are shown in Table 11.1.

Water and living organisms

Living organisms will exist in natural waters if dissolved or suspended organic matter is present. This organic matter can be utilised by the living organisms if it is biodegradable, i.e. can be decomposed by micro-organisms into simpler inorganic substances.

Once photosynthesising plants and algae are established in a water body, these will initiate food chains and food webs. There should, in any given water body, be an equilibrium established between the amount of living matter produced and the amount of dead, decomposing organic matter produced. If this is not the case, then a water body can either become choked with living organisms or devoid of them. Where the water body is situated and what the geological conditions are like will dictate the range and type of living organisms that can be present. The stability of an equilibrium depends upon the range of living organisms present and the ways in which the food chains and webs are interlinked. A cycle of nutrient movement through the water will be established, which exists in a quite delicate, easily disturbed, ecological balance.

As well as nutrients, a supply of dioxygen is needed to maintain aquatic life. Dioxygen dissolved in natural waters takes part in the biodegradation of organic matter by aerobic bacteria, and is also needed for aerobic respiration by all plants and animals. The dissolved dioxygen comes from two main sources – the atmosphere and photosynthesis. The solubility of dioxygen in water depends upon the temperature, pressure and what else is dissolved in the water. For example, at equilibrium a maximum of 11.28 g of dioxygen will dissolve in 1 m^3 of pure water at 10 °C and a pressure of 1 atmosphere. This maximum amount of dioxygen is termed the saturation concentration. The dissolution of atmospheric dioxygen and that generated by photosynthesisers replaces any oxygen used up in aerobic processes by living organisms. The minimum amount of dioxygen that is required to sustain a variety of desirable living organisms is 5 g m^{-3}. Thus at $10 \,°C$ there is only some 6 g m^{-3} difference in dissolved dioxygen before there is a threat to life. An increase in dissolved salts will also lessen the amount of dioxygen that can be dissolved. Therefore, seawater contains less oxygen than fresh water at the same temperature and pressure.

The time of day and year can also affect the solubility of dioxygen in water bodies. The more hours of daylight there are, the more photosynthesis takes place, resulting in higher concentration of oxygen at midday than midnight, and in the summer months than in the winter months.

Different aquatic organisms have different abilities to withstand temperature variations in the water in which they live. Thus different organisms are found in different parts of the world and in different waters. If the temperature of a water body increases, then the amount of dissolved dioxygen will decrease. In addition, a higher temperature increases the metabolism of an organism. Thus, dioxygen demand goes up at the same time that the availability of dioxygen decreases. Coarse fish such as perch, roach and chub can live in water with temperatures of up to $30 \,°C$ and dissolved dioxygen levels of only 3 g m^{-3}. However, game fish such as salmon or trout die if the oxygen level concentration drops below 5 g m^{-3} or the water temperature moves outside the fairly narrow range of 5–$20 \,°C$.

The determination of a water body's biological characteristics involves not just an examination of macroscopic plants and animals but also of its microscopic species. Many of the species that occur in normal water bodies differ markedly in their sensitivity to different types of pollution. The major groups of organisms that have been used as indicators of the quality or otherwise of the water environment include bacteria, protozoa, algae, macro-invertebrates and fish. The 'bottom'-living macro-invertebrates are particularly suitable as ecological indicators because their habitat preference and relative low mobility cause them to be directly affected by substances that enter their environment.

Using oxygen as a way of assessing water quality

Assessing how much dissolved dioxygen is present in river water or some other water body is one way of evaluating the quality of that water. Removal of oxygen is caused mainly by the biodegradation of organic matter. It is possible to measure the *ultimate* oxygen demand by determining the difference between the amount of oxygen dissolved in a sample of water and the amount of oxygen left after the organic matter has used up as much as it can. The main problem with this is that it is very difficult to know how long to leave the sample to ensure that all of the oxygen has been used up. Hence, two main tests have been devised. The first one is the **biological oxygen demand (BOD)** test, which is standardised on a five-day period of biodegradation. A second test is based on the **permanganate value (PV)**, which has been standardised at periods of three minutes and four hours.

The BOD test measures the oxygen requirement of micro-organisms during the biodegradation of a water sample. The sample is incubated at $20 \,°C$ in a sealed bottle for five days. Both the initial and final oxygen content are determined. This test has been used for well over 70 years and is still the most important indicator of organic pollution. The BOD value forms the basis of discharge standards for effluents and is fundamental to the operation and design of waste treatment plants. The test relies on biological action and is a simulation of actual processes which occur in polluted watercourses or an aerobic treatment plant. It has been internationally adopted as a trustworthy indicator of organic pollution. However, the test is slow and therefore not suitable for rapid process control in a waste treatment plant. In addition, it is not a good indicator of industrial pollutants, since such wastes are toxic and often inhibit the micro-organism activity on which the BOD test relies. It is more sensitive than the PV test for detecting and measuring biodegradable organic wastes.

The PV test uses a known concentration of acidified potassium manganate(VII) (permanganate) to oxidise both organic and inorganic materials. Here, the concentration of the manganate(VII) solution is measured at the start of the test and then again after three minutes and four hours at 27 °C. The difference reflects the uptake of oxygen and is expressed in $g\ m^{-3}$. It has been found that the ratio of the concentration of oxygen uptake after four hours to that used in three minutes gives an indication of the origin of polluting materials. It is an approximate measure of the ratio of organic to inorganic oxidisable materials.

In the **chemical oxygen demand test (COD)**, the water sample is boiled for two hours in a mixture of potassium dichromate(VI) and concentrated sulphuric acid in the presence of silver(I) nitrate as a catalyst. This ensures the complete oxidation of most of the organic and inorganic materials present. It usually gives a higher value for the oxygen uptake than either the PV or BOD tests. In the UK, the costs of receiving and treating industrial wastes by water authorities are established on the basis of the COD values of the wastes after the settlement of suspended solids.

Water, acidity and alkalinity

When acids and bases dissolve in water, they dissociate into ions. The degree to which they dissociate dictates how strong a particular acid or base is. Mineral acids such as hydrochloric (HCl), sulphuric (H_2SO_4) and nitric acid (HNO_3) produce high concentrations of hydronium ions in aqueous solutions and are therefore strong acids. The bases sodium hydroxide (NaOH) and potassium hydroxide (KOH) give large numbers of hydroxyl ions in solution and are therefore strong bases. Such is the degree of dissociation of these strong acids and bases that they are considered to be *fully* dissociated. Many acids and bases found in nature are weak acids and weak bases. Such compounds give low numbers of ions in aqueous solution.

Weak acids and bases often form a dynamic equilibrium in aqueous solution, i.e. both forward and reverse reactions are occurring simultaneously and at the same rate. For example, ethanoic acid (CH_3COOH) donates a proton to water forming the equilibrium,

$$CH_3COOH(aq) + H_2O(l) \rightleftharpoons CH_3COO^-(aq) + H_3O^+(aq) \qquad 11.1$$

All four species are found in a bottle of aqueous ethanoic acid. The position of the equilibrium though lies well over to the left-hand side and therefore ethanoic is a weak acid! Similarly, a solution of the base ammonia contains the dynamic equilibrium,

$$NH_3(aq) + H_2O(l) \rightleftharpoons NH_4^+(aq) + OH^-(aq) \qquad 11.2$$

Again, ammonia is a weak base and therefore the equilibrium lies well to the left. Ammonia is acting as a Lowry–Brønsted base by accepting a proton from water in the forward reaction. In the case of the reverse reaction, the $NH_4^+(aq)$ ion is acting as an acid by donating a proton to the $OH^-(aq)$ ion. Since the ammonium ion is a result of the basic nature of ammonia, it is termed the **conjugate acid** of ammonia. Water in the forward reaction is acting as a Lowry–Brønsted acid in donating a proton to the ammonia molecule. In the reverse direction, the $OH^-(aq)$ ion is acting as a base in accepting a proton. Hence, $OH^-(aq)$ is called the **conjugate base** of water.

In general, we can represent equilibria such as those above by,

$$\text{acid}_1 + \text{base}_2 \rightleftharpoons \text{base}_1 + \text{acid}_2$$

and the equilibrium constant by the equation,

$$K_c = \frac{[\text{base}_1][\text{acid}_2]}{[\text{acid}_1][\text{base}_2]} \qquad 11.3$$

Thus for the ethanoic acid equilibrium we can write,

$$K_c = \frac{[CH_3COO^-(aq)][H_3O^+(aq)]}{[CH_3COOH(aq)][H_2O(l)]} \qquad 11.4$$

where $CH_3COO^-(aq)$ (base$_1$) is the conjugate base of $CH_3COOH(aq)$ (acid$_1$), and $H_3O^+(aq)$ (acid$_2$) the conjugate acid of $H_2O(l)$ (base$_2$). Ions and molecules can be both Lowry–Brønsted acids or bases. For example, the ethanoate ion is a Lowry–Brønsted base and therefore in aqueous solutions of an ethanoate the solution will be basic, providing there is no other species present that is a stronger base or acid.

Pure water acts as an **amphoteric** substance, i.e. one that acts both as an acid and a base,

$$H_2O(l) + H_2O(l) \rightleftharpoons OH^-(aq) + H_3O^+(aq) \qquad 11.5$$

This process is also referred to as the self-ionisation of water or auto-ionisation. At equilibrium, the degree of ionisation is very small,

i.e. $[H_3O^+(aq)] = [OH^-(aq)] = 1 \times 10^{-7} \text{ mol dm}^{-3}$ at 25 °C

Hence, the concentration of the undissociated water has been changed to a negligible extent and can be considered to be constant. Thus,

if, $$K_c = \frac{[OH^-(aq)][H_3O^+(aq)]}{[H_2O(l)]^2}$$

then, $K_c \times [H_2O(l)]^2 = K_w = [OH^-(aq)][H_3O^+(aq)]$ 11.6

The new constant is called the **ionic product** of water or the **auto-ionisation** constant. The constant has the value and units of $1 \times 10^{-14} \text{ mol}^2 \text{ dm}^{-6}$ at 25 °C. The auto-ionisation constant is always equal to this latter value at 25 °C. Hence, if the pure water equilibrium is contaminated by adding acid, i.e. the value of $[H_3O^+(aq)]$ increases, then there will be a corresponding decrease in the value of $[OH^-(aq)]$ to compensate. If the value of $[OH^-(aq)]$ is increased by contamination with a base, then there will be a corresponding drop in the value of $[H_3O^+(aq)]$.

In order to express K_w more succinctly, the negative value of the log to the base 10 of K_w is quoted,

$$K_w = 1 \times 10^{-14} \text{ mol}^2 \text{ dm}^{-6}$$

$$\log \{K_w/\text{mol}^2 \text{ dm}^{-6}\} = 14.0$$

$$- \log \{K_w/\text{mol}^2 \text{ dm}^{-6}\} = -14.0$$

Now, $-\log$ is given the symbol 'p', thus $pK_w = 14$.

If an acid HA is dissolved in water, then we obtain the equilibrium,

$$HA(aq) + H_2O(l) \rightleftharpoons H_3O^+(aq) + A^-(aq) \qquad 11.7$$

for which,

$$K_c = \frac{[H_3O^+(aq)][A^-(aq)]}{[HA(aq)][H_2O(l)]}$$

If the concentration of a weak acid HA is low, then the amount of undissociated water will be much greater than the concentration of either HA and the ions produced. The concentration of water can again be taken to be constant and the new equilibrium constant written,

$$K_a = K_c[H_2O] = \frac{[H_3O^+(aq)][A^-(aq)]}{[HA(aq)]} \qquad 11.8$$

The constant K_a is called the **acid dissociation constant**. The experimental value determined for ethanoic acid at 25 °C is 1.8×10^{-5} mol dm^{-3}.

Again, we can write,

$$\log (K_a/\text{mol dm}^{-3}) = -4.74$$

$$- \log (K_a/\text{mol dm}^{-3}) = 4.74$$

$$pK_a = 4.74$$

The smaller the concentration of the H_3O^+(aq) ions, the bigger will be the pK_a value since larger negative powers of 10 will be involved.

A similar approach can be adopted for a base,

$$B(aq) + H_2O(l) \rightleftharpoons OH^-(aq) + BH^+(aq)$$

for which $pK_b = -\log K_b$. A weak base is a poor acceptor of protons and therefore the equilibrium lies to the left-hand side of the equation. Again, the value of K_b will be small and pK_b large.

The values of pK_a and pK_b enable comparisons to be made concerning the strengths of weak acids and bases. Table 11.2 shows a selection of some of these values.

For the ammonia equilibrium,

$$K_b = \frac{[OH^-(aq)][NH_4^+(aq)]}{[NH_3(aq)]} \qquad 11.9$$

Table 11.2 *Equilibrium constants at 298 K for some selected acids and bases*

Acid	$K_a/$ mol dm^{-3}	pK_a	Base	$K_b/$ mol dm^{-3}	pK_b
Nitrous HNO$_2$	4.7×10^{-4}	3.3	Lead(II) hydroxide Pb(OH)$_2$	9.6×10^{-4}	3.0
Benzoic C$_6$H$_5$COOH	6.3×10^{-5}	4.2	Zinc hydroxide Zn(OH)$_2$	9.6×10^{-4}	3.0
Carbonic H$_2$CO$_3$	4.5×10^{-7}	6.4	Aqueous ammonia NH$_3$(aq)	1.8×10^{-5}	4.8
Hydrogen sulphide H$_2$S	8.9×10^{-8}	7.1	Hydrazine N$_2$H$_4$	1.2×10^{-7}	6.1
Hydrocyanic HCN	4.9×10^{-10}	9.3	Beryllium hydroxide Be(OH)$_2$	5.0×10^{-11}	10.3

For the conjugate acid of ammonia, i.e. NH_4^+, we can write,

$$NH_4^+(aq) + H_2O(l) \rightleftharpoons H_3O^+(aq) + NH_3(aq) \qquad 11.10$$

$$K_a = \frac{[H_3O^+(aq)][NH_3(aq)]}{[NH_4^+(aq)]}$$

The product of K_a and K_b is thus,

$$
\begin{aligned}
K_a \times K_b &= \frac{[H_3O^+(aq)][NH_3(aq)]}{[NH_4^+(aq)]} \times \frac{[OH^-(aq)][NH_4^+(aq)]}{[NH_3(aq)]} \\
&= [H_3O^+(aq)] \times [OH^-(aq)] \\
&= K_w
\end{aligned}
$$

A similar analysis of all conjugate acids and bases will show that,

$$K_w = K_a \times K_b$$

Hence,

$$\log K_w = \log K_a + \log K_b$$

$$pK_w = pK_a + pK_b$$

To illustrate the usefulness of this latter equation we can use it to calculate the pK_a of the ammonium ion. The value of pK_b for ammonia is 4.75. Thus,

$$pK_a = pK_w - pK_b = 14.00 - 4.75 = 9.25$$

Hence, the ammonium ion is a weaker acid than ethanoic acid.

Another look at pH

In Chapter 1, pH was defined as the negative of the log to the base 10 of the magnitude of the hydrogen ion concentration,

i.e. $pH = -\log$ (concentration of $H_3O^+/mol\ dm^{-3}$)

Thus, for an acidic solution which contains $3.5 \times 10^{-4}\ mol\ dm^{-3}$ of $H_3O^+(aq)$ ions the pH will be given by,

$$pH = -\log (3.5 \times 10^{-4}) = 3.46$$

By taking the inverse log (or anti-log) of a measured pH it is possible to calculate the hydronium ion concentration. For example, a solution with a pH of 5.6 has a hydronium ion concentration of $2.51 \times 10^{-6}\ mol\ dm^{-3}$.

It is possible to determine a very approximate value of the pH of a solution using universal indicator paper. This strip of paper will change colour according to the acidity of the solution it is dipped in, and reference can then be made to a colour chart for comparative purposes. For more accurate work, a pH meter is used. Note that the pH scale is a logarithmic scale and therefore a small error in pH will cause a larger error in calculating the hydronium ion concentration. If a pH meter reading showed a value of pH = 5.5 for the solution that recorded a previous value of 5.6, then the former figure would yield a hydronium ion concentration of $3.16 \times 10^{-6}\ mol\ dm^{-3}$, and the latter $2.51 \times 10^{-6}\ mol\ dm^{-3}$.

In the case of a strong base it is equally easy to calculate its pH. For example, a $0.025\ mol\ dm^{-3}$ solution of sodium hydroxide will give a solution containing the same concentration of OH^- ions if fully dissociated. From the ionic product of water the concentration of the hydronium ions can be calculated and hence the pH,

$$K_w = [OH^-(aq)][H_3O^+(aq)]$$

$$\log K_w = \log [OH^-(aq)] + \log [H_3O^+(aq)]$$

Hence, $pK_w = pOH + pH$

Therefore, $pH = pK_w - pOH$

Hence, for the $0.025\ mol\ dm^{-3}$ sodium hydroxide solution,

$$pH = 14.00 - 1.60 = 12.40$$

Calculating the pH of a weak acid is a little more problematical! To calculate the pH of a 0.100 mol dm^{-3} solution of ethanoic acid the equilibrium equation 11.11 is required,

$$CH_3COOH(aq) + H_2O(l) \rightleftharpoons CH_3COO^-(aq) + H_3O^+(aq) \qquad 11.11$$

Let the initial concentration of $CH_3COOH(aq)$ be 0.100 mol dm^{-3} and the initial concentrations of both $CH_3COO^-(aq)$ and $H_3O^+(aq)$ be zero. When equilibrium is established, it can be assumed that x mol dm^{-3} of $CH_3COO^-(aq)$ and x mol dm^{-3} of $H_3O^+(aq)$ will be formed. Since it takes one molecule of ethanoic acid to produce one of each ion type, then x moles of ethanoic acid will give x moles of each of the ionic species. Thus at equilibrium the concentration of undissociated ethanoic acid will be $(0.100 - x)$ mol dm^{-3}. The equilibrium concentrations can now be substituted into the expression for K_a for ethanoic acid,

$$K_a = \frac{[CH_3COO^-(aq)][H_3O^+(aq)]}{[CH_3COOH(aq)]}$$

$$K_a = \frac{x^2}{(0.100 - x)}$$

or $\quad x = (K_a\, 0.100)^{1/2}$

Now, $\quad x = [H_3O^+(aq)]$ and $K_a = 1.8 \times 10^{-5}$ mol dm^{-3}

Therefore, $\quad [H_3O^+(aq)] = (1.8 \times 10^{-5} \times 0.100)^{1/2} = 1.34 \times 10^{-3}$ mol dm^{-3}

Hence, the pH is 2.87.

A similar type of calculation can be carried out for a weak base. For the equilibrium (equation 11.2) the expression for K_b would become,

$$K_b = \frac{x^2}{0.100 - x}$$

If x is very much smaller than 0.100, we can ignore it with respect to this latter value.

Therefore, $\quad K_b = \dfrac{x^2}{0.100}$

Hence, $\quad x = (K_b\, 0.100)^{1/2}$ where $K_b = 1.8 \times 10^{-5}$ mol dm^{-3}

Since $\quad x = [OH^-(aq)]$ then $[OH^-(aq)] = (1.8 \times 10^{-5} \times 0.100)^{1/2}$

$$= 1.34 \times 10^{-3} \text{ mol dm}^{-3}$$

pOH $= 2.87$

Using pH $= pK_w - pOH$, then pH $= 14.00 - 2.87 = 11.13$

Buffer solutions

When a small quantity of a strong acid such as hydrochloric acid is added to pure water, the pH of that water changes dramatically from a pH of 7.0 to a lower pH. For example, if 0.1 cm^3 of 1.00 mol dm^{-3} were added to 1 dm^3 of pure water, the pH would change from 7.0 to 4.0, a change of 10^3 in terms of the hydrogen ion concentration. Thus the pH of water is very sensitive to 'contamination' by acids. This is also true if a small quantity of sodium hydroxide solution is added. Living organisms are very susceptible to changes in pH of their environment and of their own biological systems. However, both animals and plants are able to protect themselves against sharp changes in pH caused by *small* additions of acids and alkalis because of the presence of **buffers**. For example, human blood is a buffer. Its pH is normally 7.4 and deviations from this of only 0.5 will result in death. A buffer then is a solution which can resist changes in pH when small amounts of acids or alkalis are added to it. If too much acid or alkali is added to the environment that an organism lives in, or it ingests in some way excess acid or alkali, then, well before any acid/alkali burning occurs, the organism will die.

Some industrial processes that are pH sensitive rely upon the use of buffers to control the pH within set limits. The food and cosmetic industries in particular have to ensure that the pH of their products is buffered to protect against a too severe change in pH during their utilisation.

A buffer normally consists of either a weak acid in the presence of one of its salts, e.g. ethanoic acid and sodium ethanoate, or a solution of a weak base in the presence of one of its salts, e.g. ammonia and ammonium chloride. How does a buffer work?

If HA represents the formula of a weak acid, then as on p. 289, it will be weakly dissociated according to the equation,

$$HA(aq) + H_2O(l) \rightleftharpoons H_3O^+(aq) + A^-(aq)$$

If MA represents the salt of this acid, then it will be fully dissociated into its ions,

$$MA(aq) \longrightarrow M^+(aq) + A^-(aq)$$

Thus, when the two substances are mixed, there is a relatively high concentration of anions in solution and a relatively high concentration of undissociated acid molecules.

If an acid is added to the solution, then the hydrogen ions of the acid will combine with some of the anions present to form unionised acid. If there is a large enough number of anions, then nearly all of the hydrogen ions that have contaminated the solution will be removed. This means that the hydrogen ion concentration will only be increased by a minute amount, resulting in a very small change in pH.

If an alkali is added to the original mixture, then the hydroxide ions OH$^-$(aq) will react with the hydrogen ions of the dissociated acid to form water. This removes hydrogen ions from the acid equilibrium, which will then cause the acid to dissociate to re-establish the equilibrium (le Chatelier Rules OK!). If there are enough undissociated acid molecules available to undergo dissociation, then there is very little change in the concentration of the hydrogen ions and therefore again the pH does not significantly change.

Thus a solution that is a buffer is so called because it is able to provide a buffering effect against contamination by acids or alkalis.

Most acids found in the environment are weak acids and therefore there are many situations where buffers operate. However, pollution by the addition of strong acids or alkalis may overcome this buffering effect and cause serious damage to the **biota** and even death to its organisms.

Terrestrial buffering and acid rain

Terrestrial ecosystems have been observed to be better buffered against acid rain than aquatic ecosystems. When acid rain falls on land, it first meets surface features such as buildings, roads, forests, grassland, and so on. This enables the level of acidity to be modified by chemical reactions with these features. In addition, acid rain will also be modified by soil and leaf litter, e.g. it may be neutralised by limestone or its acidity increased by acidic litter. Conifer litter is known to increase the cation and anion content of rainwater as it percolates through that litter, but to have a minimal effect on hydrogen ion concentration. Some hardwoods are known to increase the anion and cation concentrations in rainwater less than conifers, but to increase the hydrogen ion concentration. Once the rainwater has passed through the canopy and litter, it is the nature of the soil and bedrock that affect the acidity of that water. Igneous rocks are overlaid by thin poorly weathered regolith and soil, which are the least buffered against acid attack. This type of situation exists in the north-eastern US, south-eastern Canada and central Europe. Sedimentary rocks which are rich in cations such as Ca^{2+} and Mg^{2+} are better buffered against the effects of acidification.

If H^+ ions are added to a soil, they will undergo ion exchange with the important nutrient ions such as K^+, Ca^{2+} and Mg^{2+}. This will cause these latter cations to become subject to leaching. How susceptible a soil is to acidity will depend upon the type of buffering that is present. Carbonate buffering is an important buffering system in the range pH 6.5 to 8.3, e.g.

$$CaCO_3(s) + H_2O(l) + CO_2(g) = Ca^{2+}(aq) + 2HCO_3^{2-}(aq)$$

Acid rain will be neutralised by silicates causing buffering in the range pH 5.0 to 6.2,

$$Ca\ Al_2Si_2O_8(s) + 3H_2O(l) + 2CO_2(g) = Ca^{2+}(aq) + 2HCO_3^{2-}(aq) +$$
$$Al_2Si_2O_5(OH)_4(s)$$

In soil cation buffering takes place in the range pH 4.2 to 5.0,

$$soil—Ca^{2+} + 2H^+(aq) \longrightarrow soil—2H^+ + Ca^{2+}(aq)$$

Aluminium(III) buffering in the range pH 2.8 to 4.5 occurs via the reactions,

$$Al(OH)_3(s) + 3H_2O(l) \rightleftharpoons [Al(H_2O)_3(OH)_3](aq)$$

$$[Al(H_2O)_3(OH)_3](aq) + H_3O^+(aq) \rightleftharpoons [Al(H_2O)_4(OH)_2]^+(aq) + H_2O(l)$$

$$[Al(H_2O)_4(OH)_2]^+(aq) + H_3O^+(aq) \rightleftharpoons [Al(H_2O)_5(OH)]^{2+}(aq) + H_2O(l)$$

$$[Al(H_2O)_5(OH)]^{2+}(aq) + H_3O^+(aq) \rightleftharpoons [Al(H_2O)_6]^{3+}(aq) + H_2O(l)$$

(Compare the above reactions with those on p. 300.)

Iron(III) ion in a soil will also undergo ion exchange buffering.

When a particular buffering system is exhausted, the soil pH falls and cation leaching starts to occur. At low pH the solubility of metals such as aluminium and manganese also increases, which may cause their toxic levels to be reached.

The carbon dioxide, water, hydrogencarbonate and carbonate system

Carbon dioxide is fairly soluble in water and so all natural waters contain dissolved carbon dioxide, either as a result of its presence in the atmosphere or from decaying organic matter found in the water. Seawater and other surface waters will also contain the results of being in contact with carbonate rocks such as limestone.

When it dissolves in pure water, carbon dioxide forms the weak acid carbonic acid, H_2CO_3. Carbonic acid is an example of a **polyprotic acid**, i.e. one that has more than one acid dissociation constant. An equilibrium is established between gaseous carbon dioxide, water and carbonic acid,

$$CO_2(g) \rightleftharpoons CO_2(aq) \qquad\qquad 11.12$$

$$CO_2(g) + H_2O(l) \rightleftharpoons H_2CO_3(aq) \qquad\qquad 11.13$$

Applying Henry's Law to equilibrium equation 11.12,

$$[CO_2(g)] = K_H p$$

where p = partial pressure of $CO_2(g)$ in the atmosphere.

The equilibrium constant is thus,

$$K_H = \frac{[CO_2(g)]}{p} \qquad\qquad 11.14$$

The partial pressure of CO_2 in the atmosphere is about 3.5×10^{-4} atmosphere at 25 °C, $K_H = 3.38 \times 10^{-2}$ mol dm^{-3} atm^{-1}. The concentration of carbon dioxide in pure water would thus be 1.18×10^{-5} mol dm^{-3}.

Carbonic acid forms two successive equilibria,

$$H_2CO_3(aq) + H_2O(l) \rightleftharpoons H_3O^+(aq) + HCO_3^-(aq) \qquad\qquad 11.15$$

for which the first acid dissociation constant is given by,

$$K_1 = \frac{[H_3O^+(aq)][HCO_3^-(aq)]}{[H_2CO_3(aq)]} = 3.98 \times 10^{-7} \text{ mol dm}^{-3} \qquad\qquad 11.16$$

and $\quad HCO_3^-(aq) + H_2O(l) \rightleftharpoons H_3O^+(aq) + CO_3^{2-}(aq) \qquad\qquad 11.17$

for which the second acid dissociation constant is,

$$K_2 = \frac{[H_3O^+(aq)][CO_3^{2-}(aq)]}{[HCO_3^-(aq)]} = 5.0 \times 10^{-11} \text{ mol dm}^{-3} \qquad\qquad 11.18$$

From equation 11.16,

$$\log K_1 \quad = \quad \log \frac{[HCO_3^-(aq)]}{[H_2CO_3(aq)]} + \log [H_3O^+(aq)]$$

$$\text{Hence, } pK_1 \quad = \quad -\log \frac{[HCO_3^-(aq)]}{[H_2CO_3(aq)]} + pH$$

$$\text{or} \quad pK_1 - pH \quad = \quad -\log \frac{[HCO_3^-(aq)]}{[H_2CO_3(aq)]} \qquad \qquad 11.19$$

What is the significance of equation 11.19? If pH is less than the value of $pK_1 = 6.4$, then $(pK_1 - pH)$ will be positive and the term on the right-hand side will also be positive. This means that $[HCO_3^{2-}(aq)]$ must be less than $[H_2CO_3(aq)]$ so that the log of their ratio is a negative number. Below a pH = 6.0 the concentration of $H_2CO_3(aq)$ will be almost 100 per cent. When $pK_1 = pH = 6.4$, then the concentrations $[H CO_3^{2-}(aq)]$ and $[H_2CO_3(aq)]$ will be the same. When the pH is greater than the pK_1 value, then the concentration of $H_2CO_3(aq)$ will decrease towards zero. Below a pH of 6.4, $HCO_3^{2-}(aq)$ will be very small but above pH = 6.4 it will be the predominant species. A similar analysis using equation 11.18 will show that the predominant species above a pK_2 value of 10.3 will be $CO_3^{2-}(aq)$ and the concentration of HCO_3^- will be negligible. Figure 11.5 illustrates the relationship between all of the species present in water due to the dissolving of carbon dioxide. The term 'α_x' in this figure is the fraction present of a particular species in a water/carbon dioxide/carbonate/hydrogencarbonate system.

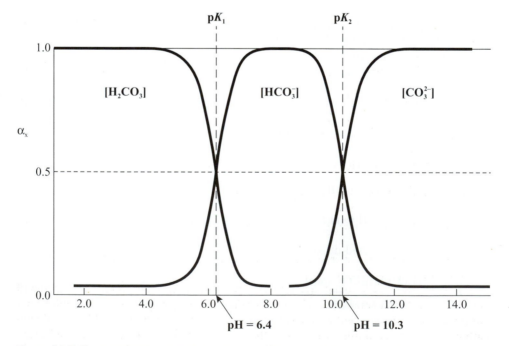

Figure 11.5 The species present at different pHs as a result of the dissolution of carbon dioxide in water.

Rainwater is acidic because it dissolves atmospheric carbon dioxide – its pH is of the order of 5.6. If distilled water in a laboratory is tested, then it will also be found to be acidic due to the presence of carbon dioxide.

Calcium carbonate has a very low solubility in water and forms the following equilibrium,

$$CaCO_3(s) + H_2O(l) \rightleftharpoons Ca^{2+}(aq) + CO_3^{2-}(aq) \qquad\qquad 11.20$$

For this reaction the equilibrium constant is,

$$K_c = \frac{[Ca^{2+}(aq)][CO_3^{2-}(aq)]}{[CaCO_3(s)]}$$

Since the concentrations of both solids and solvents in such equilibria are taken to be unity,

$$K_c[CaCO_3(s)][H_2O(l)]^2 = K_{sp} = [Ca^{2+}(aq)][CO_3^{2-}(aq)] \qquad\qquad 11.21$$

The new constant K_{sp} is called the **solubility product** of calcium carbonate.

At 25 °C, $K_{sp}\{CaCO_3(s)\} = 5.0 \times 10^{-9}$ mol^2 dm^{-6} at 25 °C.

The carbonate ion can act as a base with water, producing hydrogencarbonate ions and hydroxyl ions,

$$CO_3^{2-}(aq) + H_2O(l) \rightleftharpoons HCO_3^{-}(aq) + OH^{-}(aq) \qquad\qquad 11.22$$

When both carbon dioxide and calcium carbonate dissolve in water, a complicated set of interrelated equilibria is formed. The key equilibria are,

$$CO_2(g) \rightleftharpoons CO_2(aq) \qquad\qquad 11.23$$

$$CO_2(g) + H_2O(l) \rightleftharpoons H_2CO_3(aq) \qquad\qquad 11.24$$

$$H_2CO_3(aq) + H_2O(l) \rightleftharpoons H_3O^{+}(aq) + HCO_3^{-}(aq) \qquad\qquad 11.25$$

$$HCO_3^{-}(aq) + H_2O(l) \rightleftharpoons H_3O^{+}(aq) + CO_3^{2-}(aq) \qquad\qquad 11.26$$

$$CaCO_3(s) + H_2O(l) \rightleftharpoons Ca^{2+}(aq) + CO_3^{2-}(aq) \qquad\qquad 11.27$$

$$CO_3^{2-}(aq) + H_2O(l) \rightleftharpoons HCO_3^{-}(aq) + OH^{-}(aq) \qquad\qquad 11.22$$

together with that of water itself, $H_2O(l) + H_2O(l) \rightleftharpoons OH^{-}(aq) + H_3O^{+}(aq)$.

These equilibria are illustrated in Figure 11.6.

It is clear from these equilibria that the species that will be present in water will depend upon the predominance of an equilibrium, i.e. the magnitude of its equilibrium constant, and to what extent each equilibrium contributes to that water. If only calcium carbonate was in equilibrium with water, then from equilibria equations 11.20 and 11.22 the water would be expected to be alkaline. If only carbon dioxide and water were in equilibrium, then from equilibria equations 11.15 and 11.17 an acidic solution would

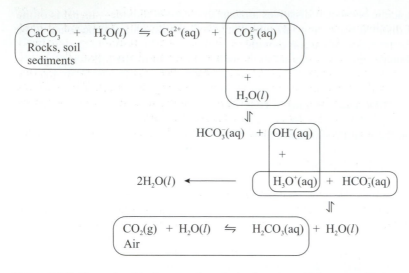

Figure 11.6 The natural hydrogencarbonate/carbonate acid/base equilibria.

be expected. If calcium carbonate, carbon dioxide and water are all in equilibrium, then some of the hydronium ions will neutralise some of the hydroxyl ions and the pH of the solution will be affected. The solubility of calcium carbonate is increased in the presence of dissolved carbon dioxide because hydroxyl ions will be removed from the right-hand side of equilibrium equation 11.22, thus causing it to move to the right. This will remove carbonate ions from equilibrium equation 11.20 and, according to le Chatelier's Principle, more calcium carbonate will dissolve. The position of each equilibrium will also be affected by an increase or decrease in pH caused by sources of additional hydroxyl or hydronium ions from any sources. Appropriate mathematical analysis of the contributing equilibria can be done and therefore quite detailed explanations of the condition of unpolluted waters attempted (Stumm and Morgan 1996).

Most waters that are free from pollution have a pH = 8.2.

Hard and soft water

When Ca^{2+} and Mg^{2+} ions dissolve in water, 'hard water' is produced. The source of these ions is the erosion of carbonate rocks by water containing carbon dioxide. Limestone is more readily attacked than dolomite,

$$CaCO_3(s) + H_2O(l) + CO_2(aq) \rightleftharpoons Ca^{2+}(aq) + 2HCO_3^-(aq) \qquad 11.23$$
limestone

$$CaCO_3.MgCO_3(s) + 2H_2O(l) + 2CO_2(aq)$$
dolomite
$$\rightleftharpoons Ca^{2+}(aq) + Mg^{2+}(aq) + 4HCO_3^-(aq) \qquad 11.24$$

The larger the concentration of these two ions, the harder is the water.

If it is necessary, it is relatively easy to reduce or remove hardness from water. The first way is to precipitate the metal ions from solution by causing the reforming of insoluble carbonates. Boiling the water will reverse the two equilibria equations 11.23 and 11.24 and precipitate the calcium and magnesium carbonates. This kind of hardness associated with the hydrogen carbonate ion is termed **temporary hardness**. However, this is also the cause of furring in kettles, pipes and boilers with the associated problems. If the accompanying cation is sulphate, then this gives rise to **permanent hardness**, which cannot be removed by simply boiling the water. The solubility of calcium sulphate in water declines with rising temperature so some very slow deposition does occur. Fortunately, this latter type of hardness is rare.

If washing soda, $Na_2CO_3.10H_2O$, is added to hard water the following reaction occurs,

$$CO_3^{2-}(aq) + Ca^{2+}(aq) = CaCO_3(s)$$

Here the soluble sodium carbonate has provided the carbonate ions necessary for insoluble calcium and magnesium carbonates to precipitate out of solution. The addition of slaked lime, $Ca(OH)_2$, will also precipitate calcium carbonate out of solution,

$$Ca^{2+}(aq) + HCO_3^-(aq) + OH^-(aq) = CaCO_3(s) + H_2O(l)$$

A second way of softening water is to add detergents containing polyphosphates, $P_3O_{10}^{5-}$, which form soluble complexes with calcium and magnesium ions,

$$Ca^{2+}(aq) + P_3O_{10}^{5-}(aq) = Ca\,P_3O_{10}^{3-}(aq)$$

Surfactants in detergents can be represented by the formula $ROSO_3^-\,Na^+$, where R stands for a long hydrocarbon chain. Surfactants also help to soften water by forming water-soluble complexes,

$$Ca^{2+}(aq) + 2ROSO_3^-(aq) = Ca(ROSO_3^-)_2(aq)$$

The final way of softening water is using ion exchange resins in which the calcium and magnesium ions are exchanged for hydronium ions or sodium ions,

$$Resin–H_2(s) + Ca^{2+}(aq) = Resin–Ca(s) + 2H_3O^+(aq)$$

One of the major problems with hard water is that it is difficult to produce a good lather with soap. A grey-white scum is often seen to collect on the surface of water, which is deposited around the inside surfaces of baths and sinks, and is left on the fibres of washed fabrics. The reason for this is that the magnesium and calcium ions form an insoluble solid with the hydrocarbon chain of the soap molecules. Hardness is an important factor in the taste of water, i.e. above $500\ mg\ dm^{-3}$ $CaCO_3$ equivalent water starts to taste unpleasant. The EU sets no limits on the hardness of water but sets a minimum concentration of $150\ mg\ dm^{-3}$ $CaCO_3$ equivalent for softened water. Limits have, however, been set for calcium and magnesium levels in water.

However, hard water does help to develop strong bones and teeth, and the incidence of heart disease is lower in hard water regions. It also protects against lead entering solution in lead pipes by coating the inside of the pipes with insoluble lead(II) carbonate or sulphate. In order to prevent lead from the lead pipes entering drinking water, it has been necessary to increase the hardness of the water by, for example, adding lime.

Hydrolysis

When water reacts with another substance, resulting in the breaking of one or both of its O—H bonds, then this is called **hydrolysis**.

Metal ions in water usually start by being solvated either by the formation of co-ordinate bonds, e.g. $[Cu(H_2O)_4]^{2+}$, or by ion–dipole interaction, e.g. $[Na(H_2O)_6]^+$. Solvation of the metals in Group 1 and Group 2 of the Periodic Table does not proceed any further than ion–dipole interactions. In the case of other metal ions, M^{n+}(aq), hydrolysis can produce acidic solutions,

$$M^{n+}(aq) + xH_2O(l) \rightleftharpoons [M(H_2O)_x]^{n+}(aq)$$

$$[M(H_2O)_x]^{n+}(aq) + H_2O(l) \rightleftharpoons [M(H_2O)_{x-1}OH]^{(n-1)+}(aq) + H_3O^+(aq)$$

$$[M(H_2O)_{x-1}OH]^{(n-1)+}(aq) + H_2O(l) \rightleftharpoons [M(H_2O)_{x-2}(OH)_2]^{(n-2)+}(aq) + H_3O^+(aq)$$

This can proceed until there is no further charge on the cation, i.e. when $n = x$, and usually results in the formation of an insoluble hydroxide. For example,

$$Al^{3+}(aq) + 6H_2O(l) \rightleftharpoons [Al(H_2O)_6]^{3+}(aq)$$

$$[Al(H_2O)_6]^{3+}(aq) + H_2O(l) \rightleftharpoons [Al(H_2O)_5(OH)]^{2+}(aq) + H_3O^+(aq)$$

$$[Al(H_2O)_5(OH)]^{2+}(aq) + H_2O(l) \rightleftharpoons [Al(H_2O)_4(OH)_2]^+(aq) + H_3O^+(aq)$$

$$[Al(H_2O)_4(OH)_2]^+(aq) + H_2O(l) \rightleftharpoons [Al(H_2O)_3(OH)_3](aq) + H_3O^+(aq)$$

The extent of hydrolysis depends upon the charge on the cation, its ionic radius and the pH of the solution. An ion with a high charge and a small radius will have a high surface charge density. When a water molecule approaches such an ion it will become easily polarised,

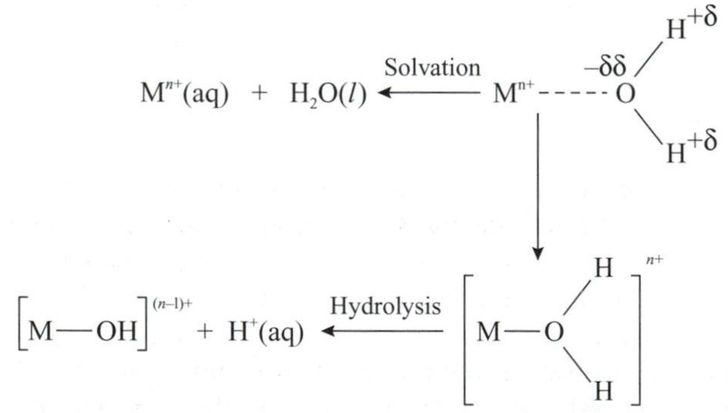

How easily the above process proceeds depends upon the ability of the M^{n+} ion to polarise the water molecule. Small, highly charged ions such as the beryllium(II) ion, Be^{2+}, and the Al^{3+} ion would therefore be expected to, and do, undergo hydrolysis very readily. If the equilibria for aluminium shown above are examined, then it can be seen that adding acid to the water (le Chatelier's Principle!) would suppress the hydrolysis of the metal cation. When metal hydroxides are formed by hydrolysis, then the more polarising the metal cation, the lower the pH will be when the insoluble hydroxide precipitates out of solution. For example, Mg^{2+} is a larger ion with less charge than Fe^{3+} and hence the former will precipitate out of solution as the hydroxide at a pH of 10.0 whilst Fe^{3+} precipitation occurs at a pH of 2.1. This means that some metal ions will not exist as free ions in river water or in seawater, since both have a pH of approximately 8.0.

Covalent compounds can also undergo hydrolysis if the atom that the water can attack is not well shielded by its attachment to large atoms/groups. For example, tetrachloromethane, CCl_4, is not hydrolysed by water because the carbon atom is small and is shielded by larger chlorine atoms. In addition, carbon has no available atomic orbitals, e.g. 3d, that water can use to form co-ordinate bonds via the lone pairs of electrons on the oxygen atom. On the other hand, the silicon atom in silicon tetrachloride, $SiCl_4$, is larger than the carbon atom and also has available 3d atomic orbitals for bonding, thus hydrolysis can occur,

$$SiCl_4(l) + 4H_2O(l) \rightleftharpoons Si(OH)_4(s) + 4HCl(aq)$$

Metal oxides (basic oxides) on the left-hand side of the Periodic Table undergo hydrolysis to give alkaline solutions, for example,

$$CaO(s) + H_2O(l) \rightleftharpoons Ca(OH)_2(s) \rightleftharpoons Ca^{2+}(aq) + 2OH^-(aq)$$

Non-metallic oxides form acidic solutions via hydrolysis, for example,

$$SO_2(aq) + H_2O(l) \rightleftharpoons H_2SO_3(aq) \rightleftharpoons H^+(aq) + HSO_3^-(aq)$$
$$\rightleftharpoons 2H^+(aq) + SO_3^{2-}(aq)$$

Some oxides are amphoteric in that they can either react as an acid or a base, according to the pH of the water,

Aluminium oxide acting as a base: $2Al_2O_3(s) + 6H^+(aq) \longrightarrow Al^{3+}(aq) + 3H_2O(l)$

Aluminium oxide acting as an acid: $2Al_2O_3(s) + 2OH^-(aq) \longrightarrow AlO_2^-(aq) + H_2O(l)$

Conclusions

A good indication of a 'safe' natural water is its ability to support complex, fragile ecosystems. This ability of natural watercourses to sustain aquatic life depends on a variety of physical, chemical and biological conditions. Biodegradable nutrients and dissolved oxygen must be available for the metabolic activities of the algae, fungi, protozoa and bacteria which are at the bottom of the food chain. In addition, plant and animal growth cannot occur outside narrow ranges of temperature and pH. As we have seen, stratification, both thermal and saline, can hinder the transport of necessary nutrients. Dissolved carbon dioxide, hydrogencarbonates, carbonates, nitrates, phosphates and salts that cause hardness must all be present in the amounts necessary for the successful functioning of the life forms in a natural water body. Any variation of the kinds and amounts of materials found in natural waters can have a serious adverse effect on its biota, and on humans and other living organisms that may depend on drinking water derived from that water.

The pollution of waterways can lead to very serious consequences and therefore the monitoring, identification of pollutants and the subsequent treatment of water must be ensured. There is no doubt, for example, that major sources of river pollution have their beginnings in the daily life of a community and in industrial operations within the water catchment area.

Summary

- Water's thermal properties help to protect living organisms from extreme weather conditions.
- Hydrogen bonding causes a reduction in volatility of water, the boiling point to be 'unexpectedly' high, and ice to float on water.
- Stratification is the variety of temperature found in a water body and affects the degree of mixing of that body.
- Relative humidity is the ratio of the amount of water in the atmosphere at a particular temperature to the maximum amount it can hold at the same temperature.
- Dissolved substances depress the vapour pressure of water and the melting point of ice.
- The types and amounts of dissolved solids and suspended solids determine the characteristics of water. They also help to determine the range of organisms present.
- Dissolved dioxygen is necessary for living organisms and the amount of dioxygen present determines, for example, the fish species present.
- BOD, PV and COD are all measures of water quality.
- The strength of a weak acid or a weak base depends upon the degree of dissociation.
- pK_a and pK_b are ways of expressing the strengths of a weak acid and a weak base, respectively.
- The $H_2O/CO_2/HCO_3^-/CO_3^{2-}$ system is composed of a complex set of interrelated equilibria. The species present in a natural water depends upon the magnitude of the equilibrium constants involved and to what extent each equilibrium contributes to that water. The system is very important in the control of the pH of natural waters.
- Hard water is due to the presence of Ca^{2+} and Mg^{2+} ions. Hardness can be removed by boiling or adding sodium carbonate to precipitate insoluble carbonates. Temporary hardness is associated with the HCO_3^- ion and permanent hardness with the SO_4^{2-} ion.
- Water hardness can also be removed by the addition of polyphosphate and surfactants, or by the use of ion exchange resins.
- Hydrolysis is the breaking of an O—H bond in water. The extent of hydrolysis depends upon the charge on any cations present, their ionic radii and the pH of the solution. Hydrolysis can be the cause of acidity in water.

Questions

1   Concerns have been expressed about the decline in the state and size of the Aral Sea for some years. Find out what the current situation is and what can be done to improve its situation.

2   (a) Infiltration is an important part of the natural water cycle. How would you expect it to be affected by (i) roads and buildings, (ii) agricultural land, (iii) dense vegetation, and (iv) topology?

    (b) Which components of the natural water cycle can affect the availability of ground water resources?

3   (a) What amount of energy would be required to raise 20,000 kg of pure water from 5 °C to 10 °C?

(b) Pure water boils, providing the prevailing atmospheric pressure is 1.01×10^5 N m^{-2}, at 100 °C. How much energy would be required to boil 10 m^3 of water?

(c) At 0 °C water freezes. How much energy would be required to melt 100 g of ice and raise its temperature to 15 °C?

4   (a) An aqueous solution contains 8.00 g of carbonyl diamide (urea), NH_2CONH_2, dissolved in 100 g of water at 80 °C. Calculate the vapour pressure of this solution given that the vapour pressure of pure water at 80 °C is 4.73×10^4 Pa.

(b) The vapour pressure of pure water at 20 °C is 23.4×10^2 Pa. What would be the vapour pressure exerted by a 4 per cent by mass solution of fructose, $C_6H_{12}O_6$, at 20 °C?

(c) An aqueous solution of iron(III) hydroxide contains 0.01 mol of $Fe(OH)_3$ per kilogram of water. What is the vapour pressure of this solution at 65 °C? The vapour pressure of pure water at 65 °C is 250×10^2 Pa.

5   Calculate the pH of (a) a 0.125 mol dm^{-3} aqueous solution of ammonium chloride, and (b) a 0.125 mol dm^{-3} sodium ethanoate aqueous solution.

6   A buffer solution can be made by mixing 50 cm^3 of 0.100 mol dm^{-3} potassium dihydrogen phosphate with 5.6 cm^3 of 0.100 mol dm^{-3} sodium hydroxide, at 25 °C. What is the pH of this buffer solution?

7   (a) pK_{sp} for $Fe(OH)_3$ is 39.1 at 25 °C. Calculate the solubility of this compound in mol dm^{-3}.

(b) Write down the expression for the solubility product of copper(I) sulphide, and give its units.

The solubility of the compound in water at 25 °C is 1.29×10^{-15} mol dm^{-3}. Calculate the value of K_{sp} and thus pK_{sp}.

8   What is meant by an amphoteric compound?

Aluminium hydroxide is amphoteric. Explain what would happen if acid or alkali were added to a beaker containing saturated solution of the hydroxide, which contains undissolved solid at the bottom.

References

Levintanus, A. (1993) On the fate of the Aral Sea. *Environments*, **22**, 1.
Precoda, N. (1991) Requiem for the Aral Sea, *Ambio*, **20**, 3–4.
Stumm, W. and Morgan, J. (1996) *Aquatic Chemistry*. J. Wiley Interscience, New York.
Voropaev, G.V. (1993) Can the Aral Sea be recovered today? *Water Resources*, **19**, 2.

Further reading

Virtually any of the standard A-level and undergraduate physical and inorganic text books will give a sound account of the general properties of water. Such texts are already listed in earlier chapters.

Andrews, J.E., Brimblecombe, P., Jickells, T.D. and Liss, P.S. (1996) *An Introduction to Environmental Chemistry*. Blackwell Science, Oxford.

The chemistry of water and its physical properties have to be searched for! There is an excellent, well-illustrated and explained section though on the oceans, which forms about one-fifth of this book.

Chapman, D. (ed.) (1996) *Water Quality Assessments: A Guide to the use of Biota, Sediments and Water in Environmental Monitoring*. Spon Press, London.
An essential guide to the design and practicalities of monitoring the quality of fresh water.

Howard, A.G. (1998) *Aquatic Environmental Chemistry*. Oxford Science Publications, Oxford University Press, Oxford.
One of the Oxford Chemistry Primers and therefore a very cheap and short read! A straightforward account of topics such as the acidity of water, metal complexes in solution, oxidation and reduction, etc.

12 Natural waters and their pollution

- Sources of water pollution
- Indirect effects of water pollution – dioxygen sag, antagonistic and synergic toxicity
- Micro-organisms in water
- Self-purification of water
- The biotic index as a measure of water quality
- Drains, waste water and sewage treatment
- Waste water disposal
- Sludge treatment and disposal
- Water supplies and the purification of water
- Nitrate(V) in drinking water

Water pollution

Water pollution is the contamination of water by foreign matter such as micro-organisms, chemicals, industrial or other wastes, or sewage in amounts likely to cause harm to living organisms. Polluted water can also chemically and physically damage industrial plant and equipment. Pollution causes deterioration in the quality of the water, making it unfit for its chosen use.

Natural waterways normally contain micro-organisms, which enable them to undergo self-purification. However, some pollutants can overload this self-purification process. For example, the discharge of domestic sewage effluent in small quantities to rivers can be beneficial in acting as a source of organic materials. If too much is added, it can lead to the dissolved oxygen being used up to such an extent that no animal or plant life can live in the water and a foul, extremely offensive smell is produced. In the UK, industrial wastes that find their way into watercourses tend not to be toxic, but if they are organic in nature and in large enough quantities they can lead to oxygen depletion.

Because rivers are moving bodies of water, any pollution can be flushed out into the sea. Lakes in particular are very vulnerable to pollution because of their enclosed nature. A lake will stay polluted for a long time because there is no flushing-out effect and the volume of water is too small to cause effective dilution. Lakes rely mostly on their living organisms for their self-purification. Lakes are prone to eutrophication, i.e. nutrients contained in the discharges and run-offs to lakes build up and encourage the excessive growth of algae. When the algae die, their subsequent decomposition uses up dissolved dioxygen to the detriment of other living things. In the case of estuaries, a wide range of factors affects the probability and the extent of pollution. These include dissolved dioxygen, nutrients, indigenous plant and animal life, salinity, waves, tides, currents, sediments and mud. The pollution here is caused mainly by the direct input of both domestic sewage and industrial wastes, and other materials brought to the estuary by the rivers and by the tides of the sea itself.

Sources of water pollution

Although large in number, sources of water pollution can be placed in one of ten groups (Table 12.1).

Table 12.2 lists some types of pollutants and their main sources, what their general effects are on the natural characteristics of a water body, and how they affect the biota and water supplies.

In water the presence of heavy metal cations will directly poison living organisms. In order of toxicity, the common heavy metal pollutants are: mercury > copper > cadmium = zinc > tin > aluminium = nickel > iron > manganese. The main inorganic anions that can also cause toxic effects if present in large enough amounts are ammonium, cyanide, fluoride, nitrate(V), sulphide and sulphate(VI). The ammonium ion often enters water via landfill leachate, fluoride and nitrate(V) from fertilisers containing phosphate, and sulphide and sulphate(VI) from mine effluents.

Organic compounds in water can be either natural or synthetic. The former may result from the exploration, transportation and refining of oil. Synthetic organic compounds

Table 12.1 *Sources of water pollution*

Source of water pollution	Comment
Water and sewage company works	Organic wastes and sometimes industrial wastes Aluminium residues from water treatment
Washing of equipment and plant in the food and drink industries	Large, dilute volumes of effluent containing carbohydrates, proteins and fats May cause depletion in dissolved dioxygen in water
Industrial wastes from paper, wool, leather industries	Organic effluents containing proteins, fats, oils and putrescible solids Also lime, potash and chromium salts Sulphides from leather industry
Electroplating and other metal industries	Effluents containing metals and cyanides
Petrochemical, oil refining and pharmaceutical industries	A diverse chemical content and therefore difficult to treat
Seepage from landfill sites	Industrial and domestic waste containing wide variety of chemicals May be difficult to treat
Run-off from land, agricultural wastes and fertilisers	Intensive farming causes concentration of waste in small areas – it causes effects and is treated similarly to domestic sewage Excessive use of fertilisers can pollute rivers with nitrates via run-off
Petroleum industry	Oil spills from ships, oil supertanker disasters and offshore drilling operations
Acid rain	Formed by combination of SO_2 and NO_x with water in the atmosphere
Radioactive materials	Present in wastes and (i) uranium and thorium mining and refining, (ii) nuclear power plants and (iii) industrial, medical and scientific use

include detergents and pesticides. In the late 1950s and early 1960s the use of detergents to replace soaps for the washing of clothes, textiles, etc. caused severe foaming in rivers. This resulted in manufacturers replacing branched chain hydrocarbon parts of their surfactants with longer straight chain hydrocarbon stems. The latter are much more prone to degradation in water than the former, thus reducing their long-term foaming abilities.

Pesticides do not just kill their target organism. A major problem can be their persistence in fresh water. Herbicides and fungicides tend not to be a problem in the UK because of the amounts and types that are now used.

It is clear from Table 12.2 that domestic, industrial and agricultural effluents have a variety of effects on the natural characteristics of water resources into which they are discharged. Organic pollutants can cause eutrophication and remove dissolved dioxygen necessary to sustain life from the water. Toxic chemicals have an insidious effect at all levels of the food chain. Water-borne pathogens spread disease. Effluents can also change the physical properties of water. The measurement and control of water quality are therefore of crucial importance in the interests of public health and in maintaining the quality of the environment.

The indirect effects of water pollution

The vast majority of pollutants poison aquatic organisms directly but many others cause indirect effects according to the prevailing circumstances. Organisms are adapted to waters of a particular salinity, i.e. levels of dissolved salts. If these levels are exceeded, adverse effects on the osmotic processes taking place in a living organism can occur, which can result in its death. This occurs irrespective of the ions contributing to the salinity.

In general, freshwater organisms are unaffected by pH changes in the range pH 5.0 to pH 9.0. The toxicity of many pollutants is pH dependent, e.g. nickel(II) cyanide is 500 times more toxic to fish at pH 7.0 than at pH 8.0, and ammonium ion is ten times more toxic at pH 8.0 than pH 7.0.

The hardness of water is determined by the concentration of Ca^{2+} and Mg^{2+} ions in solution. These concentrations are related to the pH of the water. In general, the softer the water the more toxic any heavy metal ions present are likely to be.

Ion interactions are also important in determining the toxicity of waters. Organisms are adapted to the natural ion concentrations in water, and to the natural ratios that exist between these ions. If these ratios are affected by pollution, enhanced toxicity may result. If the effects of ions in a system cancel each other out, then the system is described as an **antagonistic** one. When two toxic ions are present, e.g. Zn^{2+} and Cu^{2+}, the effects of the two heavy metals together is greater than had they been present in isolation. This type of system is described as being **synergic**.

Effluents containing reducing agents or biodegradable organic matter will deoxygenate the receiving water. Sewage, food-processing effluents, agricultural effluents and paper production effluents are all rich in organic matter. Raw sewage is sent to sewage treatment works (see pp. 315–18), which reduces its polluting ability. Treated effluent can then be released into rivers and the sea. The main effects of the release of effluents containing organic matter are on the levels of dissolved dioxygen, BOD and suspended solids.

Table 12.2 Types of water pollutants and some of their effects

Pollutant	Anthropogenic sources	Natural sources	General effect	Effect on biota	Effect on water supplies
Heat	Steel plants Cooling towers Power generation	Unlikely	Decrease in concentration of dissolved dioxygen Increase in metabolism of living organisms	Possible reduction in the ability to breed and growth rate	None
Suspended solids	Quarrying Paper mills Run-off from roads	Soil erosion Storms Floods	An increase in turbidity and therefore a reduction in light penetration Discolours water Covering/blanketing bottom after settling	Reduction in ability of plants to photosynthesise Clogging gills of fish Blanketing of bottom-living plants and animals	Blocks filtering systems Need to treat water to remove solids
Surfactants	Detergents Oils	Unlikely	Formation of foams which could prevent oxygen and carbon dioxide exchange Changes surface tension of water	Reduction in dissolved dioxygen Life cycle of some insects affected	Interfere with treatment process May need extra treatment if particularly stable foams formed

Types of pollutant	Sources		Effects		Problems for water treatment
Biodegradable wastes	Domestic sewage Animal wastes from farms Food processing companies	Run-off and seepage through the soil	Oxygen demand increases Provide food for organisms lower down in the food chain	Can be useful to organisms as source of food If too much is present can reduce dioxygen to dangerous/fatal levels	Will need extra treatment
Nitrates, phosphates and other possible plant nutrients	Detergents Fertilisers Tanneries Intensive animal husbandry Ammonia-containing industrial wastes	Nitrogen cycle	Excessive plant growth	Heavy demand on dissolved dioxygen	Will need extra treatment
Inorganic chemicals (e.g. acids, alkalis, salts)	Steel, chemicals and textiles industries Coal and salt mining	Naturally acid or alkaline rocks	Raise or lower the pH	Plants and animals can only tolerate a narrow range of pH	Corrosion of equipment and pipes Silting
Toxic chemicals (e.g. heavy metals like mercury and lead, phenols, PCBs, etc.)	Detergents Pesticides Tanneries Pharmaceuticals Oil refineries	Rare	Poison living organisms	Can cause death in animals and humans	Water cannot be used until levels of toxic materials are at acceptable levels May require extensive extra treatment
Pathogenic bacteria and viruses	Raw sewage	Rare in UK	Bacteria can cause diseases Action of viruses uncertain	Can prove fatal	Will need extra treatment

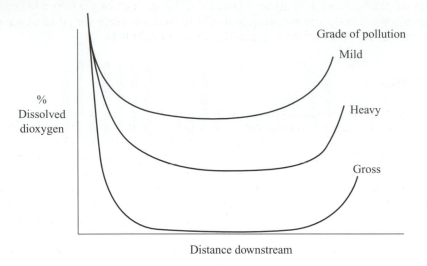

Figure 12.1 The de-oxygenation of a river receiving organic effluent.

Dioxygen **sag** can occur as the effluent is degraded by the action of micro-organisms. This can result in the death of sensitive animals and their exclusion from a water body. Figure 12.1 shows how the concentration of dioxygen can vary along a water body such as a river.

Both BOD and the amount of suspended solids increase where the discharge of organic-rich effluents occurs. From then on, the BOD declines as organisms break down the organic matter, and the amount of suspended solids decreases fairly rapidly by sedimentation. This deposition of solid matter can cause a change in the composition of the riverbed and the smothering of benthic species, especially plants.

There are very large numbers of micro-organisms in sewage effluents. These, together with the naturally occurring bacteria that already exist in water, cause what is known as **sewage fungus**. This is a grey to brown encrustation or streaming formed below sewage outlets, extending downstream from the source of pollution. It is not a fungus but an assembly of several different types of organisms, predominantly bacteria.

Although heavy pollution will eliminate plants from a river, as the river recovers nutrients are released by the decomposing organic matter which leads to an increase in algal diversity. For example, *Cladophora* is a filament-like algae which is very abundant where recovery from organic pollution is taking place. Once algal diversity builds up, an associated faunal community may develop which includes species tolerant to pollution such as freshwater lice, leeches, snails and alderfly larvae.

The removal of dioxygen will cause the death of invertebrates that cannot tolerate pollution, e.g. stoneflies, mayflies and caddis fly larvae. These may be replaced by more tolerant species such as bloodworms and tubifuid worms.

Although they can be 'caught out', fish are mobile animals and can usually avoid water deficient in dioxygen or of poor quality.

The recovery of a water body from organic pollutants occurs through the eventual natural dilution and breakdown of the organic matter. This is termed **self-purification**. It is the changes that occur during this self-purification process that form the basis of the biological monitoring of water quality.

When using the presence of biological species to indicate water quality, it must be remembered that, for any given river, the type and number of organisms are not the same at all points. There is a natural variation. A living organism, which inhabits the bottom of a water body and remains essentially in the same place for all of its natural life, is particularly useful in determining the effects of pollution.

The biotic index

A measure of pollution of a particular site is the biotic index. A visual examination is made of a sample of river sediment and the species present identified. One way of determining the biotic index is by using the Biological Monitoring Working Panel (BMWP) method which is in common use in the UK. An arbitrary score is allocated to each species according to their resistance to pollution, i.e. the more sensitive the species, the higher the score and the more tolerant the species the lower the score. Only the family is identified and no account is taken of the number of family members. When all scores have been added up, the overall score gives an indication of the pollution at the site (Box 12.1).

Biotic indices can be used as an instantaneous measure of pollution, as a means of observing changes at a site over a long period of time, and as a way of comparing different sites. One of the main advantages of using a biotic index compared with chemical methods of analysis is that living organisms continue to show the effects of intermittent pollution over a long period of time. The sample(s) is a test only at that time and at that place. Subsequent checks will normally show a variation. It is possible to sample a stream for chemical testing when the pollution is temporarily not present or when a pollution source has passed by. Organisms in a stretch of water also respond to all pollutants both known and unknown. Their response may tell an investigator that something is present in the water that they had not initially thought about. However, care must be taken when interpreting a biotic index because pollution may not be the only effect acting on the organisms, e.g. how fast a river is flowing will also affect the species present.

Heptonstall Slack, West Yorkshire, UK

(Howard, 1844) In the nineteenth century, a surgeon called Robert Howard described in great detail the sanitary condition of Heptonstall Slack, a village just north of Hebden Bridge in West Yorkshire, UK. In the winter of 1843–44 there had been a typhus epidemic (See Box 12.2), and Howard's description clearly makes the link between this disease and both the drinking water supply and toilet facilities.

Concerning the drinking water supply he describes one water spring thus, 'At its origin this is capital water, but along its passage it becomes, to a certain extent, loaded with vegetable matter, and during the summer and autumn, is converted into a nursery of loathsome animal life, which, aided by solar heat, is highly injurious to its quality. From the commencement to the termination of its track . . . it is open and exposed, and runs over a bed of mud and slime; and the crystal stream is further polluted by the offal of a slaughter house; thus, that which is essential to the health and comfort is converted into an agent of disease'. He then turns his attention to another supply of water '. . . in the "pump house", opposite the Smithy, this water appears to hold in solution carbonate and sulphate of lime, with traces of iron; and besides containing these saline ingredients, its abominable contiguity with a privy – the partition between the well and it being merely the wall which conjoins the two buildings – ensures that . . . scarcely a doubt can be entrained that, in some modified form, a portion of human excrement will filter

Box 12.1

An example of the use of biotic indices – the BMWP method

Organism type	Score
Mayfly nymphs (e.g. *Ephemeridae*) Stonefly nymphs (all families)	10
Damselfly and dragonfly (all families) Freshwater crayfish (*Astacidae*)	8
Cased caddis larvae (all families) Caseless caddis larvae (e.g. *Rhyacophilidae*)	7
Freshwater shrimp (*Gammaridae*) Freshwater limpet (*Ancylidae*)	6
Water bugs (all families) Water beetles (all families) Flatworm (all families)	5
Alderfly larvae (*Sialidae*) Mayfly nymphs (*Baetidae* only)	4
Snails (e.g. *Lymnaediae*) Leeches (all families) Water mites (all families)	3
Fly larvae (*Chironomidae* only)	2
True worms (all families) Fly larvae (e.g. *Culicidae*)	1

In the above table a particular family of a species is identified in some cases, e.g. the two types of fly larvae. This is indicative of how sensitive different families are to the presence of pollution.

Assume that a sample of water and sediment is taken from the bottom of a lake. The species found in this sample together with their respective scores are shown below:

Mayfly nymphs from the *Ephemeridae* and *Baetidae* families	10 and 4
4 types of stonefly nymphs	$10 \times 4 = 40$
Freshwater crayfish	8
1 type of cased caddis larvae	7
Freshwater shrimps	6
3 types of snails	$3 \times 3 = 9$
Fly larvae from the *Culicidae* family	1

The total score is therefore $10 + 4 + 40 + 8 + 7 + 6 + 9 + 1 = 85$

Such a high score would indicate that the water was clean.

The above shows how biotic indices may be used to calculate an overall score. There are other types of biotic indices apart from the one listed here.

Box 12.2

Typhus

Typhus and typhoid are two different diseases though the latter was long confounded with the former. It was probably typhoid fever that Howard was referring to in the year of 1843. Typhoid fever has been one of human beings' greatest scourges. Before 1875 the disease was widespread in Britain, but the Public Health Act of that year ensured improvements in sanitation and water supplies, which led to a dramatic fall in the frequency of the disease. The disease is still a major problem in tropical and subtropical countries, and in many parts of Southern and Eastern Europe. Typhoid is caused by bacteria called *Salmonella typhii*. This bacterium is found in the faeces of a person who has the disease, or in someone who is a carrier of the bacteria. Typhoid is usually spread by the contamination of drinking water with sewage, or by flies carrying the bacteria from infected faeces to food. It can also be transmitted by typhoid carriers who handle food. For example, in 1937 in Croydon, UK, there was an outbreak of typhoid caused by a carrier who excreted the bacteria in his urine. He worked in the town's wells and the urine-contaminated water supply caused an explosive outbreak causing 310 cases and 43 deaths.

Once a person has contracted the disease, the bacteria pass through the wall of the bowel and spread to the lymphatic glands where they multiply. During this period the victim shows no symptoms – this is the incubation period of the disease. After about ten days the bacteria begin to enter the bloodstream from the lymphatic glands. The victim now starts to feel ill, with headaches and various muscular aches and pains. A fever develops, which rises in a regular fashion until it reaches its peak after about a week. The abdomen is uncomfortable and tender, and there is usually constipation. A faint rash may appear. In the second week of illness, the victim's condition rapidly deteriorates. Constipation gives way to diarrhoea. Mental confusion appears and the victim becomes apathetic with a pinched-looking face, flushed cheeks and dilated pupils. The illness reaches its peak in the third week. The victim may now progressively deteriorate and die. This disease is now rarely seen in the developed countries because appropriate treatment with drugs cuts short the illness.

into the well, and some deleterious gas or gases float in the cavity of the well unoccupied by the water . . . Therefore until the privy is razed to the ground, the water can scarcely be deemed acceptable for human uses.'

Howard's description of the sewers and privies leaves the modern reader very thankful for the flush toilet and the sewage works! In one instance he states that a stream of water '. . . discharges itself into a sewer; it is an open one; and previous to reaching the cottages . . . runs through the cesspool of a privy, driving before it the agglomeration of human excrement. On its arrival at the cottages, it meets another open sewer; they now unite in front of these habitations, and the commingled filth and detritus then pass through a sewer under one of the dwellings – the flags of the floor being its only covering – and the effluvia which permeates the seams is occasionally suffocative to the inmates. The refuse now makes its exit behind the house, and re-accumulates in a hole prepared for its reception.' In the cottages alluded to there were seven cases of fever, which resulted in one death. In a second example, Howard points out the existence of a covered sewer whose opening was in the porch of a farmhouse. 'It had no proper outlet at its termination in the field behind the house, the field being considerably above the level of the sewer, and not having been opened for twenty years, the stench emitted from its large aperture in the porch was extremely noisome. . . . The sewer was opened, and the exhalations from it well-nigh overwhelmed the bystanders: it acted, physically, as a powerful depressant, producing nausea, vertigo, and sickness.' Six people who lived in this house contracted typhus, three subsequently dying of the disease. In the third example, Howard draws his reader's attention to 'the well formerly supplied with a pump, situated in a front unoccupied room, in the centre of two hovels . . . The successive inhabitants of these abodes, for the space of thirty years, have made the well the reservoir of all manner of refuse. The augmenting mass of material, in its dark and loathsome location, by chemical decomposition, became converted into offensive and putrid matter, and having, in the course of years, attained a more concentrated power, the constitution of the atmosphere in its vicinity was deeply contaminated, which, assisted by the summer and autumnal heat, rendered the inmates the victims of typhus.' Although the well was covered with a stone trough, it had a hole in it through which the matter had found its way. Howard reports a number of fever victims from these hovels, together with a few fatalities.

The privies used by the villagers were crude in the extreme. The walls were made of rough, unmortared stone and most were without a roof. None were fitted with a door and the seat was a simple pole set into the stones on either side about two feet above the ground in front of the cesspool.

London and the Year of the Big Stink

It might be expected that there would be problems with water supplies and sewage in an obscure village somewhere in the north of England in the middle of the nineteenth century. Unfortunately, the nation's capital city was also badly affected by the same problems.

In 1750, London's population was around 75,000 and the River Thames was teeming with fish. The river had been used as a sewer for many years but there had not been enough of it to cause serious pollution. By 1840, the population had risen to over two million. Thus sewage, together with industrial wastes, exceeded the river's natural ability to clean itself. Only eels were able to survive in its waters. In 1858, referred to as the Year of the Big Stink, government matters had to be adjourned in the Houses of Parliament next to the River Thames. This was because of the extremely unpleasant smell coming from the river. The smell was caused by the combination of a long, hot summer and the reduced flow of river water, together with high rates of sewage discharge into the Thames. There had also been three major epidemics of cholera in London (1831–2, 1848–9 and 1853–4). Eventually, after more cholera outbreaks (see Box 12.3) and the continuing smell, a Royal Commission was set up in 1881 to investigate these problems. A major result of this Commission's work was the opening of London's first sewage treatment plant in 1889. By 1900, the population of London was over six million and six of the less sensitive fish species had returned to the river. Between the First and Second World Wars, London's population rose to about eight million and industry in the city increased. The sewage works could no longer cope with the demands made on it. By 1945, there were again no fish to be found in the River Thames in the vicinity of London. In the 1950s, new efforts were made to clean up the river and new sewage works were built. By 1970, fish of many kinds had returned to the River Thames.

Box 12.3

Cholera

Cholera has been known in north-east India for centuries, where it still breaks out regularly today. In the nineteenth century, because of increased international travel and exploration, cholera spread throughout the world, causing millions of deaths. During the first half of the twentieth century, the disease was confined to Asia. A new pandemic started in 1961 in Indonesia and spread to much of the rest of Asia, Africa, the Mediterranean and the Gulf Coast of North America. A few cases occur each year in Britain but mainly as a result of travellers returning from Asia or Africa. Cholera is a serious and often fatal disease to humans and is caused by a comma-shaped bacteria called *Vibrio cholerae*. The main source of infection is water contaminated by human faeces containing the bacteria. This disease was once endemic in England but has now disappeared because of effective sanitation and treatment of water. Poor water hygiene is a particular contributor to the spread of the disease. Carriers of the disease, who show no symptoms, are of even greater importance than the actual cases of cholera themselves in the spreading of this disease.

The main feature of the disease is severe diarrhoea due to the irritation of the bowel by a toxin produced by the bacteria. The diarrhoea is so profuse and liquid that it is given the name of 'rice water'. These stools have to be collected so that the amount of fluid lost by the victim can be measured before the stools are disposed of in a sanitary fashion. Loss of water with contained mineral salts is the main cause of death. Protection against cholera can, to a certain extent, be done by vaccination, but the main method of prevention is in the use of proper sanitation to dispose of human excreta, supported by public health measures and health education.

The drains, waste water and sewerage treatment

It is clear that the original objective in treating sewage was to prevent water bodies becoming so polluted that disease, the death of aquatic organisms (and indeed human beings!) or unpleasant smells were caused. One major source of pollution of our natural water resources today can still be the discharge of effluents from water authority/ company sewage works. Where then do the flushings from toilets, and contents of sinks, baths and washing machines go? How are these waste waters treated and made safe for feeding back into the natural water cycle? Not only are a variety of processes used in the collection, treatment and sanitary disposal of water-borne wastes from households but special treatments may be necessary for effluents from industrial plants.

The drains

Drains carry away excessive amounts of water from buildings and other areas. If the water is drained from fields, roofs and paved surfaces, it is normally relatively clean waste water. Most people will relate the term 'drain' to the pipes which carry foul drainage water which comes from toilets, hand-basins, baths, etc. to the public sewer.

In most buildings, when the toilet is flushed, the contents of the bowl are washed down the soil pipe and into the main underground drain. Households are usually connected to the sewer mains by clay, cast-iron or PVC pipes 8–10 cm in diameter. Larger diameter pipes are located along the centre line of a road about 1.8 m or more below the surface. In the UK most above-ground pipes are installed inside buildings to prevent frost damage. It may be noticed that there is an extension to the soil pipe which protrudes through the roof of a building and vents to the atmosphere. This helps to reduce air pressure fluctuations in the pipes which stops the build up of disgusting smells inside the drains. It also helps to dry them out, thus encouraging the flaking off of any deposits on the drain walls.

Gutters, down-pipes and drains carry rainwater away from all roofs and paved areas and, in the UK, constitute a separate system from that used for sewage. If the system carries both domestic and storm-water sewage, then it is called a combined system. Such systems are found in older sections of urban areas. Often, the system of keeping sanitary sewage separate from storm sewage is continued to the final treatment stages. This arrangement is more efficient because it excludes the very large volumes of storm sewage from the treatment plant. It permits flexibility in the operation of the plant and prevents pollution caused by combined sewer overflow, which occurs when the size of the sewer is not sufficient to transport both household and storm water. The two systems are, however, likely to be combined when draining factories and abattoir yards, markets, etc.

Sewers

After the sewage has left the building, it is carried away by the drains into the main sewers. These are large pipes or masonry-lined tunnels, which carry the combined flow to the treatment works, or in an isolated or coastal area to some other point of disposal.

Raw domestic sewage when fresh is a light grey to brown colour and has a sweetish smell. However, it must be remembered that domestic sewage is organic in nature and therefore subject to biodegradation. Provided there is enough dioxygen dissolved in the waste water, this process will remain aerobic and the sewage will remain fresh. Should the rate of dioxygen demand exceed that at which it is replenished, then anaerobic conditions will result. The waste stream will then rapidly turn a black colour and hydrogen sulphide will be formed. Hydrogen sulphide is a very dangerous gas to sewage workers since it is very toxic and a major source of corrosion. It is also the source of offensive smells in the primary phase of treatment at the sewage works. This can be a particular problem in hot climates because of the reduced solubility of dioxygen in water at the higher temperatures and the more rapid biological action. Aerobic conditions can be maintained in a sewer by providing proper ventilation, ensuring that sewers have the right gradients and by reducing the time that sewage remains in the collection system.

The sewage flow in drains and sewers closely follows the consumption patterns of water in domestic and industrial premises. This flow varies both with time of day and with distance along the sewer network. The combined total of average daily flows due to sewage and trade wastes to be expected in a sewer is called the dry weather flow and other flows may be expressed in these terms. Nearly all drainage systems rely on gravity flow and partially filled pipes. The only exceptions are surface water drains, which are assumed to be full when water flow is at its peak. Foul drains and sewers slope down-wards towards the treatment works or pumping stations, and the sewage flows in the bottom of the pipe rather than filling it completely. It is necessary to design a drainage and sewage system that allows a rate of flow that enables water to sweep the solids along the pipes and thus be self-cleansing. Sometimes it is not possible to rely on gravity alone to cause the flow of sewage because the treatment works is at a higher point than the source of sewage. Here, it is necessary to pump the effluent to the works. Such a pumped system is known as a rising main.

The nature of sewage

When waste matter enters water, the resulting mixture is called sewage or waste water. The origin, quantity and composition of waste are related to a society's existing life patterns. Waste water originates mainly from domestic, industrial, groundwater and meteorological sources. These forms of waste are commonly referred to as domestic sewage, industrial waste, infiltration and storm-water drainage, respectively.

Domestic sewage results from people's day-to-day activities, such as bathing, going to the toilet, food preparation and recreation. On the other hand, the quantity and character of industrial waste water are highly varied, and depend upon the type of industry, the management of its water usage, and the degree of treatment before the waste water is discharged. Infiltration occurs when sewers are placed below the water table or when rainfall leads to percolation down to the depth of the pipe. It is undesirable and should not occur because it imposes a greater load on the sewers and treatment plant. The amount of storm-water drainage to be carried away depends on the amount of rainfall as well as on the run-off or yield of the watershed. The typical composition of raw sewage in terms of percentage by mass is shown in Figure 12.2.

The composition of waste water is characterised by several physical, chemical and biological measurements. The most common analyses include the measurements of solids, biochemical oxygen demand (BOD), chemical oxygen demand and pH. The functions of sewage treatment are (i) to reduce the total biodegradable material, including suspended solids, to acceptable levels as measured by BOD; (ii) to remove toxic materials; and (iii) to eliminate pathogenic bacteria.

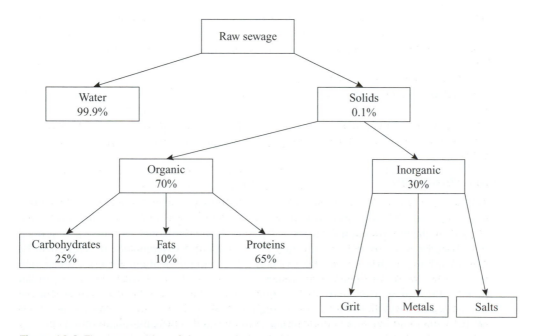

Figure 12.2 The composition of raw sewage.

The solid wastes include dissolved and suspended solids. Dissolved solids are classified as those materials which will pass through a filter paper, whilst suspended solids are those that will not. Suspended solids are divided into those that will or will not settle. This is measured in terms of how many mg of the solids will settle out of one litre of waste water in one hour. All of these solids can be sub-divided into volatile or non-volatile solids, the volatile solids generally being organic material and the other solids being inorganic or mineral matter.

One of the most important measures of overall water quality is the amount of dissolved dioxygen that it contains. As indicated earlier, removal of oxygen is caused mainly by the biodegradation of organic matter.

The composition of industrial waste cannot be readily characterised by a typical range of values because its makeup depends on the type of manufacturing process involved.

The composition of infiltration depends on the nature of the groundwater that seeps into the sewers. Storm-water sewage can contain significant concentrations of bacteria, trace elements, oil and organic chemicals. Because of the unpredictability of sewage flow rates, one of the problems with the sewage system is estimating the load it is capable of taking. One estimate is to use the Dry Weather Flow (DWF). The DWF is the base flow rate through the sewers at times of low rainfall, i.e. it is the flow rate after a period of seven days in which less than 0.25 mm of rain has fallen on any one day. It should be estimated twice a year, once in the summer months and once in the winter months. The DWF is given by,

$$DWF = PQ + I + E$$

where P = population served
Q = mean daily domestic waste water generated ($m^3 \, day^{-1}$)
I = mean rate of infiltration
E = mean flow of industrial effluent discharge ($m^3 \, day^{-1}$)

Sewage treatment works cope with DWFs from 0–3. If prolonged rain occurs, then DWFs for raw sewage in excess of 6 are achieved, and the sewage may be released into receiving tanks without further treatment because they are highly diluted.

Waste water treatment

The processes involved in municipal waste water treatment plants are usually classified as being part of primary, secondary or tertiary treatment.

Primary treatment

This is the removal of coarse and fine solids. The waste water that enters a treatment plant may contain fairly large solid items such as rags, paper and wood, that might clog or damage the pumps and machinery. Such materials are removed by screens of vertical steel bars from which they are manually or mechanically scrapped off. These can be burned or buried, but more usually are chopped up and returned upstream for reinsertion into the waste water. The waste water then passes through some kind of grinder (macerator), where leaves and other organic materials are reduced in size so that efficient treatment and removal can occur in subsequent processes. In some works, the functions of the screen and macerator are combined in one unit called the comminutor and the subsequent screenings are not removed from the sewage flow.

The waste stream then goes to a chamber called the detritor, which separates inorganic materials like sand, silt, gravel and grit. In the past, long and narrow channel-shaped settling tanks, known as grit chambers, were used to remove this kind of inorganic matter. These chambers were designed to permit inorganic particles 0.2 mm or larger to settle at the bottom while the smaller particles and most of the organic solids that remain in suspension passed through. Today, sewage is made to flow along V-shaped or parabolic-shaped channels at a constant low speed to allow the settling out of sand and grit. The grit is removed mechanically and disposed of as sanitary landfill. Grit accumulation can range from 0.02 to 0.20 m^3 per 1,000 m^3 of sewage.

After grit removal, primary sedimentation takes place. The waste water passes into a sedimentation-holding tank where organic materials and finer particles are allowed to settle out and form a liquid sludge. This is drawn off for disposal. Any scum that forms is removed by a device that skims the surface of the water. The process of sedimentation removes about 60 per cent of the suspended solids, and because this contains oxidisable material lowers the BOD by about 30 per cent. In small treatment plants this takes place in septic tanks.

The rate of sedimentation is increased in some industrial waste-treatment stations by incorporating processes called chemical coagulation and flocculation in the sedimentation tank. Coagulation is the process of adding certain chemicals to the waste water, which causes the surface characteristics of the suspended solids to be altered and thus the solids to attach to one another and precipitate. Flocculation is the operation that causes the suspended solids to coalesce by colliding with each other. Coagulation and flocculation can lead to the removal of suspended solids in excess of 70 per cent.

An alternative to sedimentation in the treatment of waste water is flotation. In this process air is forced under pressure into the waste water. The waste water, which is now supersaturated with air, is then discharged into an open tank. Here rising air bubbles lift the suspended solids to the surface from where they are removed by skimming. Flotation can remove more than 75 per cent of the suspended solids.

By the end of the primary treatment, the effluent contains something like 100 g m^{-3} of suspended solids and 150–200 g m^{-3} of BOD. The secondary treatment uses aerobic biological processes to break down the remaining organic material to a BOD and suspended solid levels, which allows discharge of the effluent to rivers and lakes.

Secondary treatment

This involves the biological oxidation of organic matter and the removal of solids resulting from biological treatment. Usually, the microbial processes employed are aerobic. Secondary treatment involves harnessing and accelerating nature's process of waste disposal. Aerobic bacteria in the presence of oxygen convert organic matter to substances like carbon dioxide, water, nitrates and phosphates, as well as new organic materials. This new organic matter must itself be removed before the waste water is discharged into the receiving stream.

There are two main ways of bringing micro-organisms into contact with the effluent. The first involves the use of 'biological filters' where the micro-organism grows on a solid support in the form of a bed of loose materials. The second method suspends the micro-organisms in the effluent and is referred to as the 'activated sludge process'.

The 'biological filters' are known as trickling filters, bacterial beds or percolating filters. These work by oxidation; filtration is not literally involved. In this process, a waste stream is distributed intermittently over a bed of some kind of porous material, e.g. blast furnace slag, gravel, crushed rock. A gelatinous film of micro-organisms (bacteria, fungi, algae, protozoa, micro and macro invertebrates) is coated on to the support material. The effluent from the primary process is sprayed on to the surface of the bed and allowed to trickle through. The organic matter in the waste stream is absorbed by the microbial film and converted to carbon dioxide and water. This trickling-filter process, if carried out slowly, can remove about 70 per cent of the BOD of the effluent and 99 per cent of the pathogenic bacteria.

The use of the activated sludge technique involves an aerobic process. Air is introduced into the effluent from the primary process using compressed air or mechanical agitation. The oxygen contained in this air 'feeds' the aerobic bacteria already present in the effluent enabling them to utilise the organic material present to form new cells. The resulting suspension is thus called the activated sludge. The activated sludge particles, known as floc, are composed of millions of actively growing bacteria bound together by a gelatinous slime. Organic matter is absorbed by the floc and converted to aerobic products. The reduction of BOD fluctuates between 60 and 85 per cent.

An important companion unit in any plant employing activated sludge or a trickling filter is the secondary clarifier that separates bacteria from the liquid stream before discharge.

By the end of the secondary processing, the effluent contains a maximum of 30 g m^{-3} of suspended solid and a BOD of 20 g m^{-3}. If however the river to which the effluent is being discharged is being used as a water supply, then the water must contain a maximum of 10 g m^{-3} of suspended solid, a BOD of 10 g m^{-3} and ammonia at 10 g m^{-3}. When such a high standard is required, then a tertiary stage of treatment will be included.

Tertiary treatment

Microstraining can be used to clean up the sewage effluent further. Such microstrainers are composed of rotary screens with a very fine mesh that retain solids with diameters larger than about 0.045 mm.

One form of tertiary treatment is the use of the stabilisation pond or lagoon, which requires a large land area and thus is usually located in rural areas. Lagoons that function in mixed conditions are the most common, being 0.6 to 1.5 m in depth, with a surface area of several acres. Anaerobic conditions prevail in the bottom region, where the solids are decomposed; the region near the surface is aerobic, allowing the oxidation of dissolved and colloidal organic matter. A reduction in BOD of 75–85 per cent can be obtained. Effluents can also be passed through rapid gravity sand filters. Solid matter is trapped in the bed and is subsequently removed by backwashing.

Sometimes, more advanced waste water treatment is necessary. Processes are available to remove more than 99 per cent of the suspended solids and BOD. Processes such as reverse osmosis and electrodialysis reduce dissolved solids. Ammonia stripping, denitrification and phosphate precipitation can remove nutrients. Activated charcoal is used to remove non-biodegradable organic materials such as pesticides and detergents, or to remove odours, tastes and colour.

Sludge treatment and disposal

In the UK, the first step in sludge treatment is digestion. The sludge from the primary and secondary treatments is a foul-smelling, highly putrescible, thick liquid containing large quantities of water. This is pumped into 'dewatering tanks', where it partially separates into an upper aqueous layer and an even thicker lower layer on standing. The water at the top has a BOD of 1,200–1,600 g m^{-3} from chemically conditioned sludge

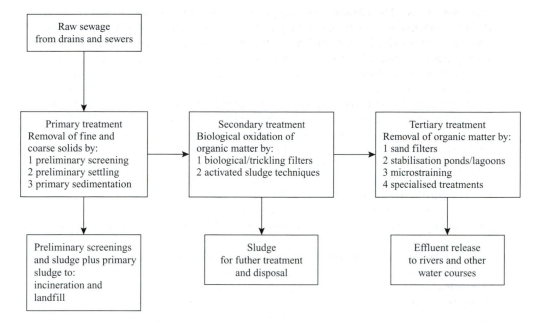

Figure 12.3 The components of raw sewage treatment.

and 4,000–7,000 g m^{-3} from heat-treated sludge. This has to be reduced, so it is pumped back to the start of the treatment process and mixed with incoming sewage. The thickened part of the sludge is heated in a closed tank or digester and stored at 35 °C for about three weeks. Inside the digester, microbiological processes occur, which convert the complex organic chemicals present in the sludge to methane, carbon dioxide and an inoffensive black, creamy material that smells of tar. Digestion reduces the organic material by between 45 and 60 per cent and also destroys pathogens. The methane gas can be collected and used to heat the digester. The sewage sludge contains too much water, so in the second step it is dewatered. This will reduce its mass and volume, and prevent further decomposition and foul smells. The sludge is air dried in open beds until it is dry enough to be removed (sludge cake). This is weather dependent and does give rise to smells. Other methods used to dewater sludge use the expensive methods of vacuum filtration and pressure filtration. Figure 12.3 shows the major components of raw sewage treatment.

The ultimate disposal of the treated liquid stream is accomplished by direct discharge into a river or lake. In some parts of the world, waste water is reused for a variety of purposes, i.e. groundwater recharge, irrigation of non-edible crops, industrial processes, recreation, etc. The main problem is the disposal of the solid sludge, which is the most expensive part of the entire waste water treatment.

The final solid material is disposed of in a number of ways. The first is by land dumping, where it is spread on agricultural land to condition it, and ploughed in after further drying has occurred. It is also dumped in ground depressions called sludge lagoons, which can cause unpleasant smells, and in trenches where it is covered with soil. It can also be mixed with household refuse and composted to produce organic manure. When sludge is mixed with soil the main problem is the presence of pathogens, such as bacteria (e.g. *E. coli*, *Salmonella*), fungi, protozoa (e.g. *Cryptosporidium*) and parasites (e.g. tapeworm and roundworm eggs). Hence, there are strict rules about when and where sludges can be applied to agricultural land. For example, 90 days must elapse before livestock can use the land, and salad crops must not be grown on treated land for one year. Higher concentrations of the heavy metals are also present in sewage sludges, e.g. 7.2 mg of copper per kilogram of dried sludge.

Sludge is also dumped in selected refuse tips, where problems involving methano-genesis can be met. The second method involves dumping at sea – now banned in the UK. The solid waste is carried well out to sea by purpose-built vessels. The problem here is that the fate and effects of the dumped waste is not yet fully understood. The third method involves incineration. This process reduces the volume of the solid waste, destroys toxic organic compounds but leaves toxic inorganic materials in the ash.

Water supplies and the purification of water

Drinking or potable water in the UK and in many other developed countries is normally collected from either underground sources or by exploiting surface water. Such water must be free of pathogens and any toxic organic and inorganic chemicals. In addition, it must be colourless, contain no suspended solids, i.e. must not be turbid, and have an acceptable taste and smell. The corrosive nature of natural water must also be reduced in order to protect the delivery systems. Natural water must therefore be treated before it is potable. Where the water is extracted from dictates the degree of treatment it requires.

Underground water is readily and cheaply accessed by drilling down to the water table and pumping. Alternatively, deep-lying pressurised water can be tapped by the borehole of an artesian well. About 25 per cent of the drinking water in the UK is obtained from groundwater. Normally such water is chemically and bacteriologically of good quality but because it exists under anaerobic conditions may contain Mn^{2+} and Fe^{2+} salts. These latter salts must be removed by aeration to oxidise them to their less soluble higher oxidation states. The anaerobic bacteria must also be destroyed by disinfection.

There are three ways of obtaining surface water: pumping from rivers and lakes; building a barrage across a river and diverting its flow through a canal system; or building a dam across a valley at the lower end of a natural catchment area. The long-term storage of this collected water is via large, open reservoirs or in artificial lakes. Ideally, to prevent water shortages, the rate of extraction of the water must equal its rate of replenishment. The longer water remains in a lake or reservoir, the cleaner it becomes due to the bacterial breakdown of organic matter, sedimentation and flocculation, and the disinfecting ability of UV-radiation. However, because of stratification the epi-limnion may become green coloured because of the development of algal blooms. Special treatment may then be necessary. In the hypolimnion, anaerobic bacterial action on the organic material falling from above may cause the formation of Fe^{2+}, Mn^{2+}, NH_3, S^{2-}, PO_4^{3-} and SiO_2. Denitrification may occur in which nitrate(V) is reduced to dinitrogen. The presence of sulphide ions is particularly unwanted because it chemically depletes the water of dissolved dioxygen, interferes later with any chlorination process, and causes foul smells and taste. Stratified waters can therefore cause problems for water treatment works. To overcome this, the depth of abstraction may have to be varied or the water mixed by pumping it from the hypolimnion to the epilimnion. Normally, water from reservoirs is very clean and usually aerobic. It does not require further treatment except microfiltering and disinfecting. If the water is eutrophic, then further more complex treatment will be required.

Rivers obtain their water via surface run-off, direct precipitation, interflow (i.e. excess water in soil drains into the river) and water table discharges. The composition of a river thus reflects recent rainfall patterns. The quality of river water is affected by the quality of rainfall, e.g. if it contains suspended dust particles and/or dissolved acid gases such as SO_2 and NO_2. Surrounding land use will contribute to water quality as will the geology of an area. Hard water is the consequence of water passing over and through calcareous catchment areas, whilst soft water with high turbidity is the result of impervious rock such as granite. River water is therefore much more variable in quality than either groundwater or water taken from a reservoir and thus requires more extensive treatment.

Before river water in particular is allowed to enter a water treatment plant, it must be screened to remove any large floating and suspended objects. This is done by using rubber booms, which float on the surface of the water near its extraction point thus acting as a physical barrier. These booms also absorb any oil that may be floating on the surface.

Further screening is necessary to remove floating and suspended debris that has passed under the booms. There are three grades of screens used: metal bars that remove coarse material; stainless steel drums/discs/belts with 25,000 holes per centimetre, which are self-cleaning to remove fine material; and microfilters which contain holes of 20–50 μm in diameter.

Screened water is then stored to help reduce the number of bacteria present in the water before 'real' purification begins. Oxidation helps to degrade organic matter that may be present, and hydrolysis precipitates any iron and manganese. After storage of up to seven days, the water is cascaded down a series of steps, which enables more oxygen to be dissolved whilst allowing the escape of dissolved gases such as carbon dioxide and hydrogen sulphide as well as volatile organic compounds to the atmosphere. The removal of carbon dioxide from the water disturbs the carbon dioxide–carbonic acid equilibria, and therefore more carbonic acid decomposes to form carbon dioxide and water. This reduces the acidity of the water and hence its corrosive effects. A final preliminary screening may occur before the water undergoes the main purification process.

The water is now purified by the four principal processes: sedimentation, filtration, aeration and sterilisation. Figure 12.4 shows the general lay-out of a typical water purification plant.

Before sedimentation can occur, all particles in the water must first be of the right size to undergo settling. Gravity does not cause the sedimentation of all suspended solids – some particles are so small that they form a **colloidal** suspension. The presence of colloidal particles is the main reason why the water will appear turbid even on prolonged standing. These particles originate from clays, proteins, metal oxides and organic matter. They all carry a negative charge and therefore electrical repulsion prevents their coagulation. Chemicals called flocculants, e.g. aluminium sulphate, iron(II) sulphate, are added which contain positively charged ions, e.g. Al^{3+}, Fe^{2+}, that neutralise the negative charges on the colloidal particles, enabling them to clump together. If the water contains high concentrations of Al^{3+} and/or Fe^{2+} ions and is sufficiently alkaline, then precipitates of hydroxides and oxides will form. Aluminium sulphate is gradually being replaced by iron(II) sulphate because of concerns over the toxicity of the Al^{3+} ion and its possible involvement in Alzheimer's disease (Garvey 2001; CSIRO 1999). Colloidal particles will be trapped by these precipitates and removed by settling. Calcium hydroxide or sodium carbonate may be added to the water to increase its alkalinity to encourage precipitation. Occasionally additional coagulants are added to water. Such coagulants include clay, silica and **polyelectrolytes**. Polyelectrolytes are long chain organic compounds, which have groups along their length that become charged on placing in water. Once coagulation has occurred, the water is gently agitated to enable

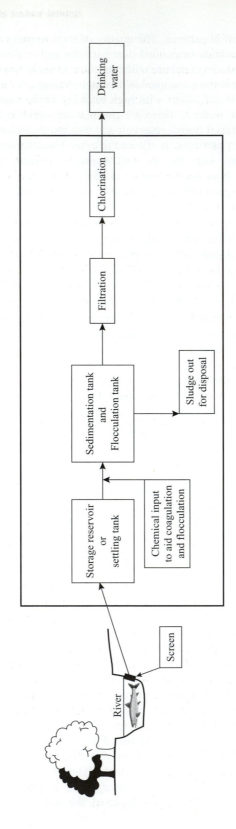

Figure 12.4 The water purification plant.

the particle size to grow bigger. This process is called flocculation and, when it is over, the water is allowed to undergo sedimentation.

Sedimentation occurs in sedimentation or flotation tanks. In some sedimentation tanks, water is forced in from the bottom and up through the tank, which overflows at a weir before continuing on its way for further treatment. The suspended material settles at a particular height in the tank to form a sludge blanket, which is regularly removed. In flotation tanks compressed air is forced in at the bottom and air bubbles rise to the surface carrying with them the flocculated material. The sludge is then removed from the surface by scrapers. If the water is considered to be too hard, a water softener (slaked lime, $Ca(OH)_2$, and washing soda, sodium carbonate) is added. The calcium and magnesium ions that cause the hardness are precipitated as insoluble carbonates. If any Fe^{3+} or Mn^{2+} ions are present, these will also be precipitated as $Fe(OH)_3$ and MnO_2, respectively. The precipitates can be removed by sedimentation. Activated charcoal can also be added to remove unpleasant smells, tastes or colour.

Although filtration removes about 90 per cent of all bacteria, the last process involves disinfecting the water. This is done by adding dichlorine (or ozone) to the water in order to sterilise it and then adding sulphur dioxide to remove any excess dichlorine once it has done its job of killing harmful pathogens. Dichlorine is very soluble in water and reacts with it to form chloric(I) acid (hypochlorous acid), HOCl, and hydrochloric acid,

$$Cl_2(aq) + 2H_2O(l) \rightleftharpoons HOCl(aq) + H_3O^+(aq) + Cl^-(aq)$$

Further reaction yields,

$$HOCl(aq) + H_2O(l) \rightleftharpoons H_3O^+(aq) + ClO^-(aq)$$

Sedimentation removes about 90 per cent of the turbidity of water. The remaining suspended matter must be removed by filtration. The water is filtered by passing it through sand beds that contain harmless bacteria, which decompose any organic matter to form unobjectionable inorganic compounds. Some filter beds are made of a layer of anthracite or activated charcoal on top of a deep layer of coarse sand. These latter types of beds are used for the fast filtering of water that has undergone flocculation and enables a faster rate of filtration because the floc adheres to anthracite particles rather than to sand particles. Fine sand is also removed at this stage. After filtration, the water is again aerated by passing it over a cascade. This increases the amount of dissolved oxygen in the water and reduces the amount of dissolved carbon dioxide, thus aiding natural purification of inorganic material by aerobic bacteria.

Both the chloric(I) acid and the chlorate(I) anions supply the necessary oxygen to cause the oxidation of bacteria and hence their death. Thus dichlorine is a powerful oxidising agent. It reacts with many organic compounds and therefore a higher dosage must be used to take this into account. Some of these chlorinated organic compounds are toxic. Dichlorine is ineffective though against viruses. Sometimes sodium chlorate(I) or calcium chlorate(I) are used to supply chlorite(I) anions. Although the sterilisation of water at the waterworks is paramount, it is still necessary to ensure that the potable water continues to be sterile as it enters the distribution system for human use. This is done by the addition of ammonia to the chlorinated water which produces chemicals called chloroamines,

$$NH_3(aq) + HClO(aq) \rightleftharpoons NH_2Cl(aq) + H_2O(aq)$$

$$NH_2Cl(aq) + HClO(aq) \rightleftharpoons NHCl_2 + H_2O(aq)$$

$$NHCl_2 + HClO(aq) \rightleftharpoons NCl_3(aq) + H_2O(aq)$$

Box 12.4

The nature of colloids

Sodium chloride solution is an intimate and uniform mixture of water and sodium chloride. This solution is clear and colourless but other solutions may be coloured. A true solution will pass through filter paper. If colourless lead(II) nitrate solution is added to colourless sodium chloride solution, then a precipitate of white lead(II) chloride will be produced. This suspension can be separated by filtration.

Substances like mayonnaise or paint are not like a pure liquid or a true solution and cannot be filtered. Mayonnaise is a liquid (vinegar) suspended in another liquid (olive oil), i.e. it is a fine suspension of one state in another **continuous** state. This is an example of a material called a colloid. As well as not being able to be filtered, colloids also have peculiar optical and flow properties. The viscosity of a colloid is higher than that of the continuous phase used.

If egg white (liquid) is beaten (mixed with air), then a thicker liquid is produced. Initially, the egg white is transparent but on the addition of air it becomes translucent, i.e. the optical properties have changed because the refractive index of the dispersed phase is different from that of the original continuous phase and more light is scattered. There are eight possible combinations of two phases that will produce colloids (Table 12.3). The size of particles found in colloids ranges from 10^{-9} m to 10^{-6} m. These do not settle easily under the influence of gravity, even with the aid of a centrifuge. Colloidal particles also carry an electrical charge on their surfaces. The surface charge is balanced by an equal number of counter ions. This leads to the formation of an electrical double layer. Since like charges repel, electrostatic interaction causes the particle to remain dispersed and unable to grow to a size that would enable precipitation.

Table 12.3 *The classification of colloids*

Colloid type	Example	Name
Solid in gas	Smoke	Solid aerosol
Liquid in gas	Mist	Liquid aerosol
Solid in liquid	Paint	Sol
Liquid in liquid	Mayonnaise	Liquid emulsion
Gas in liquid	Foam on beer	Foam
Gas in solid	Foam rubber	Solid foam
Solid in solid	Stained glass	Solid sol
Liquid in solid	Butter	Solid emulsion

Chloroamines react with water gradually releasing chlorate(I) ions, which continue to act as a bactericide.

Ozone, O_3, is a very powerful oxidising agent and will destroy viruses, bacteria, spores and the protozoan *Cryptosporidium*. The main problem with ozone is that it rapidly breaks down into dioxygen and therefore has no long-term effects as a disinfectant. Hence, after the addition of ozone it is necessary to chlorinate the water. Ozone is also made by subjecting dry air or oxygen to electrical discharge and is expensive to make. Small volumes of water can be sterilised using ultraviolet radiation but the water must be free of potentially protective particulate matter.

The water is usually by this stage fit for drinking and is stored in water towers or reservoirs.

Sometimes it is necessary to subject the water to further treatment. The presence of nitrates in water is a serious problem. Solid anion exchange resins, RCl, are used to remove nitrates to safe levels, where R is a complex resin. Thus, if contaminated water is passed over the resin, then far less dangerous Cl^- ions will exchange for the NO_3^- ions,

$$RCl(s) + NO_3^-(aq) \longrightarrow RNO_3(s) + Cl^-(aq)$$

Once the ion exchangers have become saturated with nitrate they are taken out of surface and regenerated by passing strong solutions of brine through them to reverse the above process.

If the water contains an abnormal load of organic materials, these can give rise to bad odours and tastes. To remove such compounds, the water is passed over granulated activated carbon after the filtration process. This form of carbon is obtained by heating vegetable or animal matter to 800–990 °C in the absence of air. The resultant carbon is very porous and has a high surface area. Activated carbon is an excellent adsorber of organic molecules. Once the carbon has become saturated, then it can be regenerated by heating.

Is our drinking water as safe as we think? In many Third World countries, where the sterilisation (chlorination) of water is not carried out, there is a high child mortality rate as a consequence of disease. In the developed countries safe drinking water is taken for granted. However, the chlorination of water has caused the formation of traces of chlorinated organic compounds and there appears to be a link between these chemicals and cancer of the bladder. However, there is far greater danger from drinking untreated water than that from the slight risk of bladder cancer.

In parts of Great Britain, fluoride (optimal concentration 1 µg cm^{-3}) has been added to the drinking water to improve dental health, especially to reduce the number of fillings. The addition of fluoride is the last process to be carried out with water and is added in the form of sodium fluorosilicate, Na_2SiF_6, sodium fluoride, NaF, or fluorosilicic acid, H_2SiF_6. This has met with opposition from a number of people who see it as a dangerous practice and one tantamount to pollution. It should be remembered though that many potable waters have a natural load of fluorine in them. Tea drinkers might like to know that tea leaves are a relatively rich source of fluorine! People who live in areas where local water supplies have a naturally high concentration of fluorine do tend to have mottled teeth (concentrations greater than 1.5 mg dm^{-3}). Excessive intakes of fluorine greater than 3–6 mg dm^{-3} can cause deformed limbs (skeletal fluorosis) (McDonagh *et al.* 2000), which can become crippling at concentrations greater than 10 mg dm^{-3}.

Nitrate(V) and drinking water

Extensive farming causes soil to become deficient in nitrate(V), NO_3^{-1}, the main source of nitrogen for plants, and therefore farmers have to apply nitrate(V) fertilisers. Unfortunately, nitrates are amongst the most water-soluble inorganic compounds and can be leached into groundwater and be present in run-off. If excess fertiliser is used, then wells, rivers and other water bodies can become polluted. Nitrate(V) itself is not toxic to humans and other animals. However, it is readily converted to the toxic nitrate(III) ion, NO_2^{-1}, in the digestive system of human infants and some animals. Indeed, nitrate(V) is one of the few chemicals that can cross the human placenta and be absorbed by the foetus. Foetal haemoglobin has a high affinity for nitrate(III). In the first few months of their lives, human babies are very susceptible to acute nitrate(V) poisoning. This is because the acidity of their stomachs is not sufficiently high enough to destroy the bacteria found there, which causes the formation of nitrate(III). This latter ion reacts with haemoglobin to produce methaemoglobin which does not carry dioxygen. This condition is called methaemoglobinaemia. The skin consequently turns a bluish colour, particularly around the eyes and mouth; hence, it is also called the 'blue-baby syndrome'. The level of oxygen carried by the blood may be so reduced that the baby may suffocate. Poisoning by nitrate(V) is now extremely rare in the UK, the last recorded case being in 1972. It is most common in other areas of the world where water is drunk from wells. In older children and adults nitrate(V) is absorbed and excreted and therefore methaemoglobinaemia is not a problem.

Adult cows, horses and sheep together with piglets, foals and chickens are also affected by nitrate(V) poisoning in the same way as human infants. In the case of these animals, there is a bluish or brownish colouration around the eyes and mouth or of the mucous membranes. This is accompanied by a slow, staggering gait, rapid heartbeat, frequent urination and laboured breathing. In severe cases, collapse, convulsions and coma may occur, leading to death.

Research (Parslow *et al*. 1997) has also indicated a link between child diabetes and nitrates. Where there are increased levels of nitrate(V) in water in Yorkshire, UK, the incidence of diabetes in children has increased. In particular, it has been found that East Yorkshire and rural areas of North Yorkshire are the worst hit. The average nitrate(V) level in these areas is 30 mg dm^{-3} of nitrate(V) compared with just 7 mg dm^{-3} in West Yorkshire.

Nitrate(V) in drinking water (and in foodstuffs) is known to cause diabetes in animals and has also been linked to stomach cancer, thyroid problems and birth defects.

For drinking water, the EU (EC 1980) guideline level for nitrate(V) is 25 mg dm^{-3} of NO_3^{-1}, with a maximum admissible level of 50 mg dm^{-3} of NO_3^{-1}. In the US (USEPA 1996), the maximum contaminated level of nitrate(V) expressed in terms of nitrogen is 10 mg dm^{-3} or 45 mg dm^{-3} of NO_3^{-1}.

How are excessive nitrate(V) levels reduced in water? Although it is relatively easy to remove or reduce nitrate levels in water, because they are so soluble it is not cheap. There are three methods that can be used: (i) demineralisation via simple distillation or reverse osmosis, (ii) ion exchange, and (iii) blending.

Demineralisation removes all soluble minerals from water. Simple distillation involves the boiling of water, the condensation of the gas on a cold surface and the collection of the distilled water. The involatile soluble matter remains in the boiler. In reverse osmosis pressurised water is forced through a membrane that filters out the minerals, including nitrate(V). Most of the water though remains on the 'wrong' side of the membrane and therefore becomes rejected water. Both techniques used in demineralisation are energetically demanding and produce low yields of potable water. A household might be sufficiently served by demineralisation, but for livestock it would be prohibitively expensive.

Ion exchange resins can be used which are charged with Cl^{-1} ions that can be exchanged for nitrate(V) ions. To regenerate the ion exchanger, the resin is simply backwashed with salt solution. Large volumes of water can be treated by this method but the disadvantage is that not just NO_3^{-1} ions are exchanged. For example, if SO_4^{2-} ions are present in the water, then more resin will be required to take this into account.

In both demineralisation processes and ion exchange methods there will be the need to dispose of contaminated waters/residues.

Blending is the mixing of water which contains a high level of nitrate(V) with water that does not until an acceptable level is reached. This is useful in supplying livestock with water but not human beings.

Summary

- Water can be polluted by micro-organisms, chemicals, industrial or other wastes, or sewage.
- Micro-organisms help natural waters to undergo self-purification.
- Pollution can lead to deoxygenation of waters.
- Enclosed water bodies are particularly prone to eutrophication.
- Any effluent rich in biodegradable organic matter will reduce dioxygen levels and increase BOD and suspended solids.
- The biotic index at a particular site can be used as a measure of water quality.
- Waste matter plus water constitutes sewage.
- Waste water is domestic sewage, industrial wastes, infiltration and storm-water drainage.
- The composition of waste water is reflected in the amount of solids present, BOD, COD and pH.
- Solid waste is the sum of the dissolved and suspended solid waste.
- Waste water undergoes primary, secondary and tertiary treatments.
- Sludges are treated by digestion to remove biological matter and destroy pathogens. They are then dewatered and dried.
- Solid sludge is disposed of by land dumping, as organic manure, in refuse tips, by dumping at sea and by incineration.
- The long-term storage of water increases its cleanliness by the bacterial breakdown of organic materials, sedimentation and flocculation, and the sterilising effects of UV-radiation.
- Algal blooms and stratification may necessitate the special treatment of waters stored over a long period of time.
- The quality of river waters is affected by rainfall patterns – the quality often depends upon the run-off from land.
- The four principal processes used in the purification of water are (i) sedimentation, (ii) filtration, (iii) aeration and (iv) sterilisation.
- Sedimentation involves the coagulation of colloidal particles using flocculants.
- Disinfecting water involves the addition of dichlorine or ozone.
- Because of their usage and solubility, nitrates have been a particular problem in drinking water in the past.

Questions

1 Find out the origin of your drinking water and how it is prepared for human consumption.

2 Hydrogen sulphide can be a threat to sewer workers. What is the source of this gas? Describe the effects of this gas on humans.

Complete a risk assessment for hydrogen sulphide if you were going to use it in the laboratory.

3 A small town has a population of 20,000. The total per capita water consumption (domestic and industrial) is 250 dm^3 per person per day. Infiltration accounts for 20 per cent of DWF.

Calculate the maximum sewage flow rate to the treatment plant that serves this population.

4 What are the sources and types of organic compounds that are potential pollutants of fresh water in the UK? Compare this with a selected country in the EU.

5 A river is subject to seasonal pollution by fruit canning factories which are active between June and September every year. A survey carried out on two different days to investigate the flow of water in the river and the nitrate(V) concentrations at a point downstream of the factories gave the following results:

	Discharge/m^3 s^{-1}	[NO_3^-]/mg dm^{-3}
Day 1	5.3	0.29
Day 2	2.2	0.83

Calculate how much N is flowing down the river on each occasion (kg N s^{-1}). Which is the higher value and why?

6 The area of Europe is about 4.9×10^{12} m^2 and has a population of 480 million.

(a) If the average precipitation rate over Europe is 600 mm per year and the mean rate of evaporation is 360 mm per year, what is the mean run-off of water from Europe in millions of cubic metres per day?

(b) If all this run-off were collected, what would be the equivalent per capita water available in dm^3 per day?

(c) Why do calculations such as those above not give reliable indication of water availability?

7 What are the main sources of pollution in the sea?

Why do you think the dumping of domestic wastes on the sandy shores of Tunisia may not be seen by some as a threat to the Mediterranean Sea?

References

CSIRO (1999) *Aluminium Study puts Drinking Water in the Clear*. Available at: http://www.det. csiro.au/Apr99-e.html.

EC (1980) *Relating to the Quality of Water Intended for Human Consumption*. Council Directive 80/778/EEC. Official Journal of the European Communities, Luxembourg.

Garvey, L. (2001) *Is Aluminium a Risk Factor in Alzheimer's Disease?* Communications Office, Institute of Food Research, Norwich. Available at: http://www.ifr.bbsrc.ac.uk.

Howard, R. (1845) A History of the Typhus of Heptonsall Slack which prevailed during the winter of 1843–1844. A document in the possession of the author.

McDonagh, M.S., Whiting, P.F. and Wilson, P.M. (2000) *A Systematic Review of Public Water Fluoridation*. NHS Centre for Reviews and Dissemination, University of York, York.

Parslow, R.C., McKinney, P.A., Law, G.R., Staines, A., Williams, R. and Bodansky, H.J. (1997) Incidence of childhood diabetes mellitus in Yorkshire, UK, is associated with nitrate in drinking water: an ecological analysis. *Diabetologia*, **40**, 550–6.

USEPA (1996) *Water Environmental Indicators*. Office of Water. Washington, DC.

Further reading

Harrison, R.M. (ed.) (2001) *Pollution, Causes, Effects and Control*, 4th edn. Royal Society of Chemistry, Cambridge.
Recently revised and therefore an up-to-date account of pollution in general. In particular the chapters on the chemical pollution of freshwater and marine environments, drinking water standards, the biology of water pollution and the treatment of sewage will supplement and extend this chapter!

Hester, R.E. and Harrison, R.M. (eds) (2000) *Chemistry in the Marine Environment*. Royal Society of Chemistry, Cambridge.
This book covers a wide range of topics including the pollution of the oceans and its useful-ness, e.g. as a source of pharmaceuticals. It has an international approach, and is an interesting account of the influences of chemical reactions on the seas and oceans.

Laws, E.A. (2000) *Aquatic Pollution*. John Wiley, New York.
An expensive but easy-to-read book! It covers pollution in lakes, rivers, streams, underground aquifers and the oceans. Amongst the topics covered are the polluting effects of urban run-off, acid rain and sewage disposal. Numerous case studies are also included.

Mason, C.F. (1996) *Biology of Freshwater Pollution*, 3rd edn. Longmans, London.
This is a very well-written and comprehensive account of this subject. It includes some very interesting case studies. The fourth edition is about to be published.

The Open University (1993) T237, Water Quality, Analysis and Management, Units 5–6; *Water Supply and Sewage Treatment*, Unit 7: *Environmental Control and Public Health*. Open University, Milton Keynes.
Both these booklets may be a little long in the tooth but they still provide excellent reading in the areas of water quality, water supply and sewage treatment, with particular reference to the UK. A minimal amount of chemical knowledge is required to understand the content.

13 Organic chemicals and the environment

- Dioxins and furans. Persistent organic pollutants
- Toxic equivalence factor
- Pollution by benzene and other aromatic compounds
- PAHs and PCBs
- Pesticides – their nature and problems. Toxic effects on humans
- Herbicide types and their toxic effects
- Insecticide types and mode of action – organochlorines and organophosphate insecticides
- Toxic organic waste disposal

A south-east Asian war

At the start of the US involvement in the Vietnam War, a programme initiated in January 1962 was to raise some of the most heated controversy concerning the use of chemicals on the environment that continues to this day. In 1965, under the programme named 'Ranch Hand', US forces used aircraft to spray selected areas of jungle with large amounts of defoliating agents. The job of these herbicides was to strip the leaves off vegetation, thus denying cover for the enemy. This programme of defoliation proved not to be very successful.

The major material used was **Agent Orange** (Westing and Pfeiffer 1995; Buckingham, 2000), named after the orange stripe on the drums in which it was stored. Agent Orange was thought to contain two herbicides. Depending upon the source and the degree of quality control during the manufacture of the herbicides, Agent Orange was found to be contaminated with chemicals known as dioxins (levels ranging from 2–50 μg/g). Although there are no precise figures available, it is widely believed that these dioxins caused innumerable animal deaths. In addition, because of the both deliberate and accidental delivery of Agent Orange to agricultural land, dioxins have been blamed for the increase in prenatal deaths and in the birth of deformed Vietnamese children.

After the Vietnam War, US veterans who had been exposed to Agent Orange claimed that the illnesses they have experienced, including cancer and genetic disorders in their children, were caused by the presence of dioxins. So far, little relationship has been determined between these claims and Agent Orange. The problem has been the lack of details concerning the length of time veterans were exposed to the material and the dosage levels involved. In 1998 the US Institute of Medicine (1998) reported on the continuing investigations carried out to establish if Agent Orange has caused health problems such as Hodgkin's disease, lymphoma, prostate cancer, etc. The findings are still not clear.

Dioxins and furans

By the late 1960s scientists were aware of the presence of dangerous by-products called dioxins in certain herbicides, particularly those containing or derived from the chemical 2,4,5-trichlorophenol (TCP). Several accidental exposures of large groups of people to substances containing dioxins led to their investigation as hazardous chemicals.

2,4,5-trichlorophenoxyacetic acid (2,4,5-T)

One of the earliest accidents occurred in 1949 when a container used for the production of TCP blew up at the Monsanto plant in West Virginia, US. Exposed workers were affected by chloracne, a severe and persistent form of acne, which can lead to permanent disfigurement. The culprit was later identified as a dioxin, 2,3,7,8-tetrachlorodibenzo-*p*-dioxin (TCDD), present in the reaction mixture.

In November 1953 a runaway chemical reaction at BASF's Ludwigshafen plant in former West Germany caused a vat containing TCP contaminated with TCDD to boil over. Hundreds of workers were contaminated, some developing chloracne. Since so many workers were involved, there was the ideal opportunity to study a large sample. Unfortunately, the then West German legislation worked against the acquisition of epidemiological information and thus the possibility of a useful risk assessment was lost. The medical histories of the victims, which were held by West Germany's state-run health insurance companies, were not available due to the enforcement of a strict data protection law. In 1990 the law changed and access to some data was allowed, providing that no individual was identified. Examination of this data has shown an 18 per cent increase in general illness for the exposed workers when compared with a control group. In particular, there has been a dramatic increase in thyroid diseases, intestinal and respiratory infections and disorders of the peripheral nervous system. There has been no increase in illness from cancer. This German study would indicate that dioxins disturb the human immune system and hormonal control mechanisms.

Since 1979, in the US, there has been a federal ban on the use of some herbicides containing TCP. North American studies have shown dioxins to be present in many industrial wastes, paper-mill effluents being a notorious example. Areas around some chemical plants are heavily contaminated and dioxins have been found in parts of the Great Lakes. Dioxins have also entered the environment through the use of contaminated oil as a roadway dust suppressant, so much so that the town of Times Beach, Montana, was abandoned in 1983 because tests indicated that highway spraying had left high levels of dioxins in the soil.

Dioxins are a group of related aromatic organic chemicals, specifically the polychlorinated dibenzo-*p*-dioxins (PCDDs), and the furans are the polychlorinated dibenzo-furans (PCDFs). There are 75 known compounds in the former category and 135 in the latter. Both families of compounds are **persistent organic pollutants (POPs)**. Of these

Table 13.1 *The toxic equivalence factors for some dioxins and furans*

	International TEF	WHO TEF
Some dioxins		
2,3,7,8-TCDD	1.0	1.0
1,2,3,7,8-PCDD	0.5	1.0
1,2,3,4,7,8-HCDD	0.1	0.1
Some furans		
2,3,7,8-TCDF	0.1	0.1
1,2,3,7,8-PCDE	0.05	0.05
2,3,4,7,8-PCDE	0.5	0.5

210 compounds, 17 are defined by the NATO/CCMS (1998) international toxic equivalent (I-TEQ) scheme as being involved in emissions – seven PCDDs and ten PCDTs. The chemical of most concern was 2,3,7,8-tetrachlorodibenzo-*p*-dioxin (TCDD) since this was identified as being the most toxic. This compound is taken as the 'standard' to refer other dioxins and furans to, i.e. it is given a **toxic equivalence factor (TEF)** of 1.0. All the other 16 compounds' toxicities are expressed as a fraction of that of TCDD. The WHO (Van der berg *et al.*1998) has suggested a modification to these values to calculate the TEF for some PCDDs and PCDFs (Table 13.1).

dibenzofuran
(the carbon atoms are numbered)

dibenzo-p-dioxin

2,3,7,8-tetrachlorodibenzo-*p*-dioxin (TCDD)

Although TEF values are estimates only, they can be used to give a useful idea as to the 'potency' of emissions. If the amounts of dioxins and furans are determined from any emission sources, then their relative dangers can be estimated. For example, if a source is emitting 0.5 g of 2,3,7,8-TCDD per day and 2.0 g of 1,2,3,7,8-PCDE, then the total mass of emissions is 2.5 g. The TEQ would be (0.5 × 1.0) plus (2.0 × 0.05) or 0.6 g per day. However, if a second source were emitting 0.75 g of 2,3,7,8-TCDD and 1.0 g of 2,3,4,7,8-PCDE, then its total mass of emissions is 1.75 g per day and its TEQ (0.75 × 1) plus (1 × 0.5) or 1.25 g per day. The latter is thus emitting less mass than the former, but its TEQ is greater and thus a greater hazard.

These chemicals are not easily broken down physically or biologically. Usually, the more chlorine atoms that are present, the more stable is the compound and therefore the greater the persistence. They are only very slightly soluble in water, but are readily soluble in organic solvents, fats and oils. The chemicals are adsorbed on the surface of soil particles and are not readily desorbed. They thus tend not to be leached out and also to concentrate in the bottom sediments of water bodies. The half-life of the most toxic is estimated to be at least ten years. Dioxins accumulate in the fatty tissues of fish and other animals.

PCDDs are often the unwanted and accidental by-products in the manufacture of certain herbicides, wood preservatives and other chemical products. The two herbicides used in Agent Orange were derived from 2,4,5-trichlorophenol (TCP). These herbicides were 2,4-dichlorophenoxyacetic acid (2,4-D) and 2,4,5-trichlorophenoxyacetic acid (2,4,5-T). The other chemical present was TCDD.

2,4-dichlorophenoxyacetic acid (2,4-D)

2,4,5-trichlorophenoxyacetic acid (2,4,5-T)

Table 13.2 *Emission sources and amounts of dioxins and furans*

	Estimated grams TEQ yr 1990	Estimated grams TEQ yr 1998
Power stations (coal and oil)	35	18
Coal combustion, domestic/industrial	38	17
Wood burning, domestic/industrial	26	26
Sinter plant	42	43
Chemical industry	12	14
Non-ferrous metal industry	27	22
Municipal waste incineration	602	14
Clinical waste incineration	51	24
Sewage sludge incineration	5	3
Chemical waste incineration	6	4
Road transport, petrol/diesel	28	11
Accidental fires and open agricultural burning	121	64
Other	86	65
Total	1,079	325

(*Notes*: Data modified from National Atmospheric Emissions Inventory)

They can also be produced by the burning of chlorine-containing materials in old, inefficient incinerators. At one time it was the burning of MSW that was the main source of both the dioxins and furans (Table 13.2). This is due to the presence of insufficient oxygen, low temperatures, and too short a residence time in the furnace area. Modern state-of-the-art incinerators do not produce dioxins in quantities likely to cause concern.

Other sources of dioxins are vehicle emissions, domestic and industrial coal combustion, and coal-fired power stations. For many years it was thought that only synthetic chemicals and activities produced these compounds. It is now known that natural forest fires and volcanic eruptions also produce a natural background concentration of these materials. Indeed, a bonfire or any wood burning fire will produce dioxins, albeit in very small quantities. Some typical values for the concentration of total PCDD/Fs in the UK are for urban air, 3.4 pg m^{-3}, urban soil, 1436 μg kg^{-1}, rural grass, 45 ng kg^{-1}, and human adipose tissue, 1.5 μg kg^{-1} of fat. Some typical values of estimated total PCDD/F emissions from known UK sources are, municipal waste incinerators 10.9 kg per year, domestic coal combustion, 5.1 kg per year, industrial coal combustion, 7.7 kg per year, and leaded petrol 0.7 kg per year.

To date (New Jersey Department of Health and Senior Services 1996), TCDD and other dioxins have been definitely proved to cause only one disease in humans, and that is chloracne. Concerns about dioxin are largely based on the very small amounts required to produce birth defects and deaths in small mammals (see Table 13.3). It is the lethal effect of TCDD on animals that has led to the belief by some that dioxins are amongst the most poisonous synthetic chemicals. Their effects on humans are still not yet proven, although they are suspected of causing miscarriages, birth defects and serious behavioural and neurological problems.

In 1994, the US Environmental Protection Agency (USEPA) reported that an unknown number of Americans had been exposed to levels of dioxins that may impair

Table 13.3 *TCDD's lethal dose in different species*

Animal	$LD_{50}/\mu g\ kg^{-1}$
Guinea pig	<1
Rat (male)	22
Rat (female)	45
Monkey	<70
Rabbit	115
Mouse	114
Dog	>300
Bullfrog	>500
Hamster	5,000

their immune system. The Agency identified the biggest source of exposure for people to be contaminated food. Dioxins are not taken up by plants but can enter the food chain by being deposited on leaves, which can be eaten by herbivores. It is suggested that dioxins cause between 1 in 1,000 and 1 in 10,000 cancers in the US. This report has led to extensive debate since the USEPA has suggested such a small upper limit (0.006 pg of dioxins per day for each kilogram of a person's body weight) that it would be exceeded by natural deposits provided by volcanoes and wood burning fires.

Pollution of the environment by benzene and other aromatic compounds

Benzene is a volatile liquid at normal room temperatures and pressure (boiling point 80 °C). It is one of the chemicals that forms part of the UK National Atmospheric Emissions Inventory. Most of the benzene that humans come into contact with is likely to have been deposited by the evaporation and combustion of petrol and oil. Benzene is a component of all petrols (about 2 per cent by volume) and therefore it is evaporated from spillages and emitted from the exhausts of petrol- and diesel-driven vehicles. In 1998 it was estimated that about two-thirds of the total emissions of benzene to the atmosphere were from road transport. Benzene is also emitted from smoke stacks, and via losses during its manufacture and usage in the chemical industry. Cigarette smokers are particularly prone to exposure to benzene present in inhaled cigarette smoke, e.g. a non-smoker living in a rural area is likely to be exposed to 120 µg of benzene a day, but a smoker living in a city may be exposed to 1,250 µg per day. Once it enters the environment the degradation of benzene takes several days. Table 13.4 shows the decline in benzene emissions between 1990 and 1998, mainly due to the introduction of the catalytic converter in 1991.

Benzene is readily absorbed into the human body via the lungs. It is soluble in fat and accumulates in fatty tissue. After two days post-exposure, about 80 per cent of this absorbed benzene is eliminated by chemical breakdown or excreted in the urine. There have been several cases, though rare, where humans have been accidentally exposed to benzene. For those acute cases where the victims have been exposed to extremely high

Table 13.4 *Sources and amounts of benzene emissions, UK, 1990 and 1998*

	1990 ktonne	1998 ktonne
Road transport combustion/evaporation	40.43	21.90
Other forms of transport	1.59	1.48
Other forms of combustion (domestic, etc.)	4.69	3.99
Production processes	6.41	4.48
Other	0.92	1.25
Total	54.04	33.07

(*Note*: Data modified from National Atmospheric Emissions Inventory)

concentrations (several million ppb) of benzene in confined areas, anaesthetic effects and death have resulted. At very high concentrations (over 5,000 ppb) severe and often fatal damage has been done to bone marrow. The chronic effects involving long-term exposure are more worrying – studies have shown that there is an increased risk of developing non-lymphatic leukaemia. Two American studies in particular have been useful in determining what levels of exposure to benzene are permissible for humans. Rinsky *et al.* (1987) in their study of 1,165 workers at a rubber plant and Wong (1995) in his study of chemical workers showed that there was an increase in risk in workers with an exposure rate estimated to be equivalent to 10 ppm every year for 20 years. Benzene acts on the genetic material of cells, i.e. it is genotoxic. It is therefore possible to say that any exposure to benzene could have serious consequences for humans. The chronic effects of benzene have also been confirmed using rats and mice – here it has been found that exposure to levels of benzene of 10,000 ppb also caused leukaemia and other malignant diseases.

It is expected that by 31 December 2003, the level of benzene in the atmosphere that should not be exceeded will be 16.25 $\mu g \, m^{-3}$ (5 ppb) at 20 °C and a pressure of 1,013 mb.

As well as being a potential carcinogen, benzene undergoes chemical reactions in the atmosphere producing products which are also dangerous. As seen earlier, ozone undergoes photolysis to produce oxygen free radicals, which then react with water vapour in the atmosphere to produce hydroxyl free radicals,

$$H_2O \; + \; O\bullet \; \longrightarrow \; 2HO\bullet$$

The hydroxyl free radicals can then attack any benzene that may be present in the atmosphere as follows,

A benzene free radical is formed, which then reacts with dioxygen to give the benzyl peroxy free radical. This latter radical can then react with nitrogen(II) oxide to yield nitrogen(IV) oxide and the oxy free radical. Any other hydrocarbon that is present, such as methane, can then react with this radical to produce phenol and the corresponding alkyl free radical. Phenol itself is both acidic and toxic.

If the benzene molecule has a side chain, then other reactions are possible, which lead to environmentally dangerous molecules. For example, methyl benzene (toluene) can undergo the following reactions,

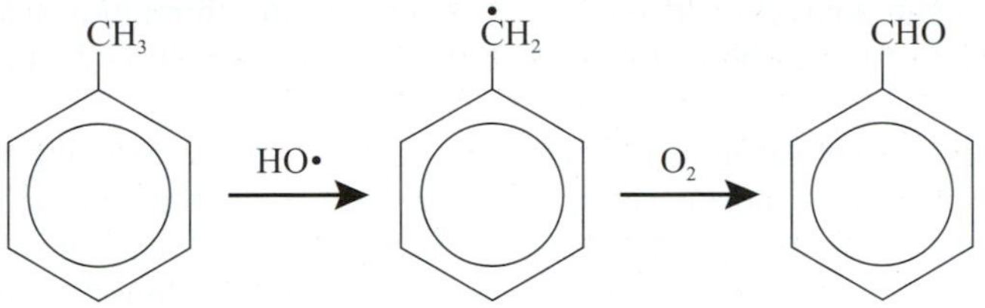

benzyl free radical   benzaldehyde

When four or more benzene rings are fused together, the resulting compounds are referred to as polycyclic aromatic hydrocarbons or PAHs. PAHs are classified by the UK IARC (International Agency for Research on Cancer) as persistent organic pollutants because they do not break down readily and have very long half-lives. Like all persistent organic pollutants, PAHs are found in trace quantities throughout the environment, all are toxic and all accumulate in plants and animals. The background level in the UK in air is $0.13\ \mu g\ m^{-3}$, which leads to the accumulation of PAHs of about $2.6\ \mu g$ per person per day. In addition, some $3.7\ \mu g$ is accumulated per person per day from food sources. There are a vast number of these compounds in existence but just 16 (Table 13.5) have been identified by the USEPA as being worthy of particular

Table 13.5 PAHs of environmental concern

PAH	USEPA	IARC	Borneff	UN/ECE
Naphthalene	Yes			
Acenaphthene	Yes			
Acenaphthylene	Yes			
Fluorene	Yes			
Anthracene	Yes			
Phenanthrene	Yes			
Fluoroanthene	Yes		Yes	
Pyrene	Yes			
Benz-a-anthracene	Yes	Yes		
Chrystene	Yes			
Benzo-b-fluoranthrene	Yes	Yes	Yes	Yes
Benzo-k-fluoranthrene	Yes	Yes	Yes	Yes
Benzo-a-pyrene	Yes	Yes	Yes	Yes
Dibenzo-ah-anthracene	Yes	Yes		
Indeno[1,2,3-cd]pyrene	Yes	Yes	Yes	Yes
Benzo-ghi-perylene	Yes		Yes	

study. Of these 16, six have been identified by the IARC as possible human carcinogens. Six referred to as the 'Borneff six PAHs' have been used in some EC emission inventories. These were selected not on toxicological grounds but on their then relative ease of detection. Some are used for the purpose of emissions inventories under the UN/ECE POP (United Nations/Economic Commission for Europe Persistent Organic Pollutants) Protocol.

In the UK, the main sources of PAHs are emissions from aluminium manufacture, and the manufacture of the carbon anodes used in aluminium smelting (termed 'anode baking').

One of the main reasons for interest in these compounds is the carcinogenic nature of some of them. The most potent carcinogens are benzo-a-anthracene, dibenzo-ah-anthracene and benzo-a-pyrene. All PAHs are semi-volatile thus making them highly mobile throughout the environment via deposition and re-volatilisation. One of the most dangerous to humans is benzo-a-pyrene or BaP shown below,

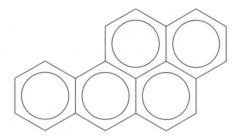

BaP or benzo-a-pyrene

BaP was the cause of skin cancers in chimney sweeps and the high mortality rates from cancer amongst gas workers prior to the use of natural gas.

PAHs are found at n g^{-1} levels in mineral oils and paraffin wax. They are formed by the burning of petrol and light diesel fuels owing to a limited supply of air. If a catalytic converter is not fitted, then BaP levels in exhaust fumes are about 50 μg per kilometre travelled. If a catalytic converter is fitted then this falls to between 0.05–0.3 μg per kilometre travelled. Aircraft engines emit approximately 10 mg of BaP during each minute of operation. Thus as leaded petrol disappears from the petrol pumps, so PAH emissions to the atmosphere will be reduced. PAHs are also formed during the combustion of fossil fuels and wood. They are to be found deposited upon the surface of particulate matter in smoke. Hence, these compounds can be easily inhaled. PAHs are also produced in the manufacture of creosote. They are also constituents of tobacco smoke – one cigarette being equivalent to 5–7 μg.

PAHs enter water bodies via fall-out and run-off from bitumen-treated roads. For example, in the River Thames at Kew, the BaP level is about 130 ng dm^{-3}. After heavy rain the level in domestic sewage can be as high as 1,800 ng dm^{-3}. BaP is found on lettuce leaves (3.1 μg kg^{-1}) and on charcoal-cooked steak (8 μg kg^{-1}).

Table 13.6 illustrates the sources of emissions in terms of the total tonnage of all 16 PAHs used in the UK as markers of air quality. This table illustrates well the move away from domestic fuel burning and the use of leaded petrol and towards abatement technologies in reducing PAH emissions in the UK since 1990.

The OSHA (Occupational Safety and Health Administration) in the US has set a limit of 0.2 mg m^{-3} of PAHs for air, and the OSHA permissible exposure limit (PEL) for

Table 13.6 *PAH emissions in the UK*

	Emissions of 16 PAHs (tonnes)		BaP emissions (tonnes)	
	1990	*1998*	*1990*	*1998*
Road transport – diesel	36	24	1.1	0.7
Road transport – petrol	229	112	8.0	3.8
Natural fires/agricultural burning	1,028	95	31.2	2.9
Creosote use	103	103	0.1	0.1
Aluminium production + anode baking	3,490	587	24.6	4.1
Coke production	104	83	1.2	1.0
Domestic wood burning	215	215	1.2	1.2
Domestic coal burning	582	261	5.1	2.3
Industrial coal burning	445	134	3.9	1.2
Other sources	26	25	0.3	0.3
Total	6,258	1,639	76.7	17.6

(*Note*: Data modified from National Atmospheric Emissions Inventory)

mineral oil mist that contains PAHs is 5 mg m^{-3} for an 8-hour exposure time. The NIOSH allows a level of 0.1 μg dm^{-3} in air at the work place over a 10-hour working day, in a 40-hour week.

Another group of persistent organic pollutants are the polychlorinated biphenyls (PCBs). Biphenyl and its numbering system are shown below.

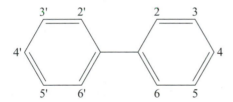

biphenyl showing the number systems
of its carbon atoms

Because of their high electrical resistance, PCBs were synthesised for use mainly as dielectric materials in electrical equipment such as transformers and capacitors. Smaller amounts have been used in lubricating oils, hydraulic fluids, etc. In 1986 the sale and new usage of PCBs was prohibited in the UK, though it is thought to be still manufactured in some other countries.

In humans, PCBs are believed to cause reduced male fertility, long-term behavioural and learning difficulties, particularly in children, and are classified as probable human carcinogens. They are insoluble in water but soluble in fats, and therefore do accumulate in the fatty tissues of animals. For example, PCBs have been detected at much greater than background levels in the blubber of whales and seals. Currently, it is likely that PCBs will continue to enter the human system via ingestion. PCBs are so persistent that the depositions on soil made many years ago can still enter the human food chain. Indeed, PCB emissions and re-deposition on soil continue.

Table 13.7 *PCB emissions in the UK for 1990 and 1998*

	1990 emissions kg	1998 emissions kg
Electrical equipment	6,228	2,193
Application of sewage sludge	70	33
Power stations	89	44
Industrial/domestic combustion	32	22
Iron and steel industry including sinter plant	529	441
Other sources	28	14
Total	6,976	2,747

(*Note*: Data from the UK National Atmospheric Emissions Inventory)

Although not manufactured in the UK for many years, PCBs still exist in old electrical equipment. It has been estimated that at least 80 per cent of emissions to the atmosphere occurring at the time of writing are due to the use of this old equipment or its disposal, primarily to landfill. PCBs are now destroyed by the use of high temperature furnaces using an excess of oxygen.

Unfortunately, it is not known with any degree of accuracy what quaantity of PCBs currently exist in electrical equipment, neither is there any accurate data on the leakage rates there from. Table 13.7 therefore shows the estimated total PCB emissions in the UK for 1990 and 1998.

Pesticides – their nature and associated problems

A pesticide is any substance or mixture of substances that kills a pest, or inhibits in some way its development. Under the UK Food and Environment Protection Act of 1985, a pesticide is defined as 'any substance, preparation or organism prepared or used, to protect plants or wood or other plant products from harmful organisms; to regulate the growth of plants; to give protection against harmful creatures; or to render such creatures harmless.' A pest is some living organism that is not required in some place because of its detrimental effects. The ideal pesticide should only attack the unwanted organism, leaving the non-targeted organism unaffected. Unfortunately, the ideal pesticide does not really exist, though some are much more selective than others. Table 13.8 shows the range of pesticides in current use.

Pollution of the environment by pesticides arises from both point sources and diffuse sources. Diffuse sources include run-off from farmland, leaching from the soil and fall-out from the atmosphere. The majority of pollution episodes though arise from point sources. These are due to spillage and washing water when equipment is cleaned on site; spillage whilst transferring pesticides from containers to applicators or whilst mixing; pesticide storage areas where the cleaning up of spillage is not correctly carried out; the improper washing out and disposal of contaminated containers; and the improper disposal of excess pesticides. Some point sources are controlled by discharge consents, e.g. from manufacturing companies.

Table 13.8 *Common pesticides and their targets*

Pesticide	Target
Acaricide	Mites and ticks
Avicide	Birds
Fungicide	Fungi
Herbicide	Plants (weeds)
Insecticide	Insects and related animals
Molluscicide	Snails and slugs
Nematicide	Nematodes (thread-worms, round-worms)
Rodenticide	Rodents (rats, mice, etc.)

Pesticides are used both indoors and out of doors. Great care must be taken in identifying areas where it may be possible to cause the unintentional injury, or indeed death, of organisms. Such areas out of doors include:

- those near surface waters
- where groundwater is near the surface or is easily accessible
- near schools, play areas, parks, hospitals, etc.
- near the habitats of endangered species and wildlife parks
- near apiaries and other animal husbandry centres
- near food or feed crops, sensitive agricultural growing areas and gardens.

Indoor areas include:

- living/working/treatment areas involving children, the sick, pregnant women and the elderly
- food preparation, processing, storage and serving areas
- areas where domestic pets are kept
- areas where indoor plants are kept.

It is possible for pesticides to migrate from where they were applied and cause problems elsewhere. When pesticides are applied in the form of liquid or solid sprays, they can **drift** or move as a result of air currents. How far they drift depends upon the particle size. Pesticides can also be carried off-site by moving water. They can be leached out of soil, and enter water bodies in the form of run-off. Run-off can cause the contamination of drainage ditches, ponds, streams, rivers and lakes. Additionally, water bodies can be polluted by the improper washing away of spillages and leaks, and by the illegal dumping of pesticides. If pesticides contaminate shoes, clothing or animal skin/fur then transference can occur to the office or the domestic situation.

The dangers with pesticides or their breakdown products (called residues) are that they can cause injury or death to a non-targeted organism.

Pesticides – their toxic nature and humans

Like most potentially harmful toxins, pesticides can enter the human body orally, dermally or by inhalation.

Oral intake is often due to careless behaviour, such as eating recently sprayed fruit and not washing the hands after using pesticides. How dangerous this proves to be depends upon the nature of the pesticide and the amount swallowed.

Dermal entry is a common occurrence and takes place usually in mixing and application processes. The pesticide may be in the form of a dry powder or wet material. Again, how serious dermal absorption is depends upon how toxic the pesticide is, how fast it is absorbed through the skin, the surface area of skin affected, the contact time and the amount of pesticide involved. What part of the body surface is affected also has a bearing on how fast a pesticide is absorbed, e.g. absorption through the scrotum is much faster than through the skin of the forearm. As long as the pesticide remains on the skin, wherever it is, then the chemical will continue to be dermally absorbed.

The results of the inhalation of a pesticide depend upon the nature of the chemical(s) involved. It occurs when pesticides in the form of vapours, dusts or sprays are mixed, poured or applied without proper protection.

All pesticides are tested to determine the type of toxicity they exhibit, and what dosage is required to produce a toxic effect. As seen in Chapter 6, testing is carefully carried out on different species of animals under stringent laboratory conditions. The results of such tests are then 'extrapolated' to human beings. Again, acute toxicity is expressed in terms of LD_{50} or LC_{50} for oral, dermal and inhalation routes. Chronic toxicity, it is to be remembered, has no standard measurement but is viewed in terms of carcinogenic, teratogenic, mutagenic and reproductive effects.

Some of the chemicals that have caused, and continue to cause, worries over their effects on the environment because of their large-scale application are herbicides and insecticides. The following sections illustrate some of the uses and dangers of *some* these compounds.

Herbicides

[EXTOXNET 2003] A herbicide is a chemical used to kill unwanted plants or inhibit their development. There are over 300 species of weeds causing problems in agriculture worldwide that are dealt with using herbicides. In addition to agricultural use, herbicides are used to destroy plants on ground selected for the erection of buildings, areas designated for industrial purposes, on road verges, railway embankments, etc. The use of herbicides has enabled the use of machines in the large-scale crop production of produce such as potatoes, sugar beet and corn. They are also much more effective and economical than using manual labour to remove weeds.

Herbicides can be classified in a number of ways (see Figure 13.1). A **selective** herbicide kills certain weeds without harming crops, but if it kills all plants it is a **non-selective** herbicide. These chemicals are applied to the foliage of weeds or to the soil according to how they act.

Herbicides are also classified according to how they enter a plant. **Contact herbicides** are those that are applied directly to plants. This type of herbicide is particularly effective against annuals. **Translocated herbicides** act by being absorbed by the roots or foliage, eventually making their way to various parts of a plant. Some translocated herbicides are effective against all weed types, whilst others are effective against perennials.

Herbicides can be classified according to the timing of application, i.e. **pre-planting** herbicides applied to an area before planting occurs, **pre-emergent** herbicides applied to an area just before the emergence of a crop or weeds, and **post-emergent** herbicides applied after the crop or weeds have emerged from underground.

Herbicides are also classified according to the *area* of application. **Band** application involves the treatment of continuous strips, e.g. along a row of crops. **Broadcast** application covers everything, including the crop; **spot** application involves a small area of weeds. **Direct spraying** is the application of herbicides to selected weeds or to the soil to avoid contact with a crop. **Over-the-top application** describes the application of a herbicide over the top of the crop and the weeds shortly after germination. This occurs when a crop is naturally tolerant to a herbicide or has been genetically engineered to be so.

Finally herbicides can be classified according to the chemical family to which they belong, e.g. amides, carbamates, bipyridliums, ureas, triazines, etc. Unfortunately, many herbicides (and insecticides) are sold under trade names and in formulations. Thus, it is sometimes difficult to know exactly what kind of herbicide is being used. There is also a wide range of organic herbicides, which have complex chemical names and are often better known by their 'commercial' names.

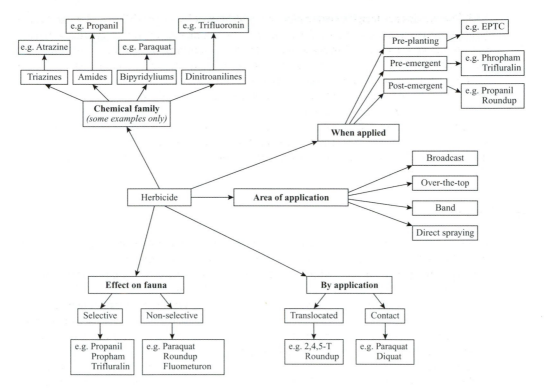

Figure 13.1 Classification of herbicides.

The solid wastes include dissolved and suspended solids. Dissolved solids are classified as those materials which will pass through a filter paper, whilst suspended solids are those that will not. Suspended solids are divided into those that will or will not settle. This is measured in terms of how many mg of the solids will settle out of one litre of waste water in one hour. All of these solids can be sub-divided into volatile or non-volatile solids, the volatile solids generally being organic material and the other solids being inorganic or mineral matter.

One of the most important measures of overall water quality is the amount of dissolved dioxygen that it contains. As indicated earlier, removal of oxygen is caused mainly by the biodegradation of organic matter.

The composition of industrial waste cannot be readily characterised by a typical range of values because its makeup depends on the type of manufacturing process involved.

The composition of infiltration depends on the nature of the groundwater that seeps into the sewers. Storm-water sewage can contain significant concentrations of bacteria, trace elements, oil and organic chemicals. Because of the unpredictability of sewage flow rates, one of the problems with the sewage system is estimating the load it is capable of taking. One estimate is to use the Dry Weather Flow (DWF). The DWF is the base flow rate through the sewers at times of low rainfall, i.e. it is the flow rate after a period of seven days in which less than 0.25 mm of rain has fallen on any one day. It should be estimated twice a year, once in the summer months and once in the winter months. The DWF is given by,

$$DWF = PQ + I + E$$

Table 13.9 *Acute toxicity data for some selected herbicides*

Type of herbicide	Example	Oral LD$_{50}$ mg kg^{-1} or ppm	Dermal LD$_{50}$ mg kg^{-1} or ppm	Inhalation
Amide	Propanil	Rats 1,080–1,217 Dogs >1,217	Rabbits 5,000	
Carbamate	Chloropropham	Rats 5,000–7,500 Rabbits 5,000		Rats 4-hour LC$_{50}$ >32 ppm
Dinitroanilines	Trifluralin	Rats >10,000 Mice >50,000 Dogs >20,000		
Phosphono amino acids	Glyphosate	Rats 5,600 Mice >10,000 Rabbits >10,000 Dogs >10,000		
Chlorophenoxy	2,4-D	Rats 375–666 Guinea pigs 320–1,000	Rats 1,500 Rabbits 1,400	
Sulphonyl ureas	Sulfometuron-methyl	Rats >5,000	Rabbits 2,000–8,000	Rats >5.3 mg dm^{-3}
Triazines	Atrazine	Rats 3,090 Mice 1,370 Rabbits 750	Rats 3,000 Rabbits 7,500	Rats 1-hour LC$_{50}$ 0.7 mg dm^{-3}
Ureas	Fluometuron	Rats 6,416–8,900	Rats >2,000 Rabbits >10,000	Rats >2 mg dm^{-3}
Bipyridyliums	Paraquat	Rats 110–150 Monkeys 50 Cows 50–70	Rabbits 236–325	Rabbits 4-hour LC$_{50}$ >20 m gdm^{-3}

where P = population served
Q = mean daily domestic waste water generated (m^3 day^{-1})
I = mean rate of infiltration
E = mean flow of industrial effluent discharge (m^3 day^{-1})

Sewage treatment works cope with DWFs from 0–3. If prolonged rain occurs, then DWFs for raw sewage in excess of 6 are achieved, and the sewage may be released into receiving tanks without further treatment because they are highly diluted.

Waste water treatment

The processes involved in municipal waste water treatment plants are usually classified as being part of primary, secondary or tertiary treatment.

Primary treatment

This is the removal of coarse and fine solids. The waste water that enters a treatment plant may contain fairly large solid items such as rags, paper and wood, that might clog or damage the pumps and machinery. Such materials are removed by screens of vertical steel bars from which they are manually or mechanically scrapped off. These can be burned or buried, but more usually are chopped up and returned upstream for reinsertion into the waste water. The waste water then passes through some kind of grinder (macerator), where leaves and other organic materials are reduced in size so that efficient treatment and removal can occur in subsequent processes. In some works, the functions

Table 13.10 *The effects of some herbicides on organisms in the environment*

Herbicide	Toxicity to birds LD_{50} mg kg^{-1} or ppm	Toxicity to fish and other aquatic life	Toxicity to the bee LD_{50}
Propanil	Mallard ducks 275	Trout 96-hour LC_{50} 2.3 mg dm^{-3} Oysters 96-hour LC_{50} 5.8 mg dm^{-3}	240 µg per bee
Chloropropham	Mallard ducks >2 000	Bass 48-hour LC_{50} 10 mg dm^{-3}	Non-toxic
Trifluralin	Ducks and pheasants >2 000	Rainbow trout 96-hour LC_{50} 0.02–0.06 mg dm^{-3} Daphnia 96-hour LC_{50} 0.5–0.6 mg dm^{-3}	Non-toxic
Glyphosate	Mallard ducks >4 500ppm	Rainbow trout 96-hour LC_{50} 86 mg dm^{-3} Shrimps 96-hour LC_{50} 280 mg dm^{-3} Daphnia 96-hour LC_{50} 934 mg dm^{-3}	>0.1 mg per bee
2,4-D	Mallard ducks 1 000 Pheasants 272	Slightly to highly toxic depending upon type of fish	0.0115 mg per bee
Sulfometuron-methyl	Mallard ducks >5 000	Rainbow trout LC_{50} >12.5 mg dm^{-3} Daphnia LC_{50} >125 mg dm^{-3}	No data available
Atrazine	Mallard ducks >20 000	Slightly toxic	Non-toxic
Fluometuron	Mallard ducks >2 974	Rainbow trout 96-hour LC_{50} 30 mg dm^{-3} Daphnia 48-hour LC_{50} 54 mg dm^{-3}	Non-toxic
Paraquat	Mallard ducks 5- to 8-day dietary LC_{50} 4 048	Rainbow trout 96-hour LC_{50} 32 mg dm^{-3} Daphnia 96-hour LC_{50} 1.2–4.0 mg dm^{-3}	Non-toxic

Glyphosate is a broad spectrum, post-emergent, moderately toxic herbicide, which inhibits nucleic acid metabolism and protein synthesis.

The formulation known as Roundup contains glyphosate as one of its components, but the acute toxicity of this formulation is thought to be linked with the other components such as the surfactant present called polyoxyethyleneamine. Roundup has caused sore throats, abdominal pain and vomiting. In more severe cases gastrointestinal bleeding has occurred. In the UK glyphosate products can be applied by any person.

The first of the **chlorophenoxy** type herbicides was 2,4-dichlorophenoyacetic acid or 2,4-D (for structure see p. 335). Other common ones are 2,4,5-trichlorophenoxyacetic acid, or 2,4,5-T, and 4-chloro-2-methylphenoxypropionic acid. These herbicides are highly selective broad-leaf weed killers, and are translocated throughout a plant. The chlorophenoxy herbicides act in a similar way to plant growth hormones – they affect phosphate and nucleic acid metabolism, and cell division processes.

Table 13.11 *The structural formulae of some selected herbicides*

Herbicide type	Herbicide type
1 3,4-dichloropropionanilide	**5** 2-(4,6-dimethylpyrimidin-2-ylcarbamoylsulfamoyl)benzoic acid

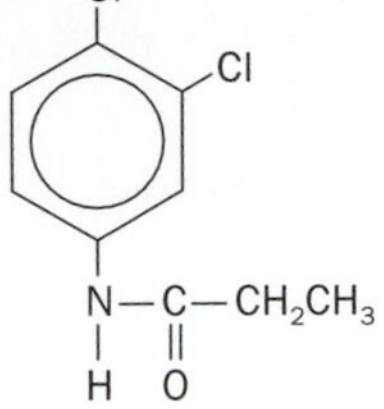

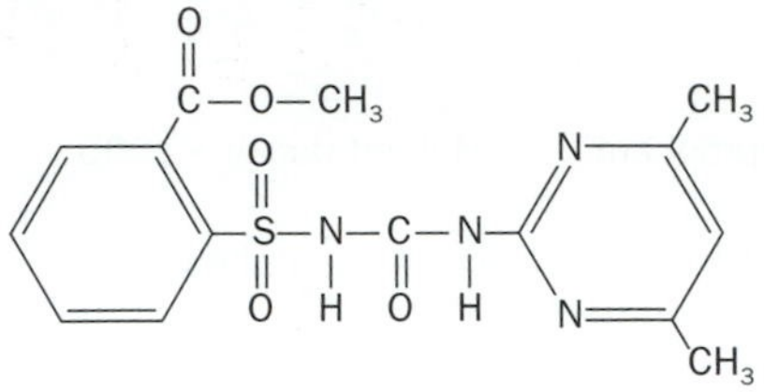

Propanil

Sulfometuron-methyl

2 isopropyl-3-chlorophenylcarbamate

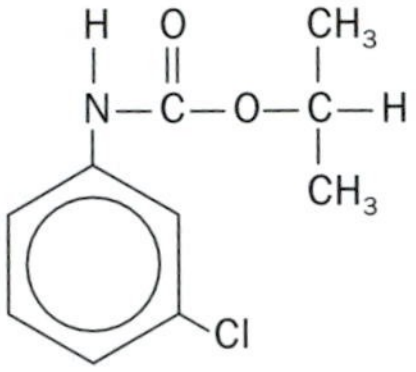

Chloropropham

6 2-chloro-4(ethlamino)-6-(isopropylamino)-s-triazine

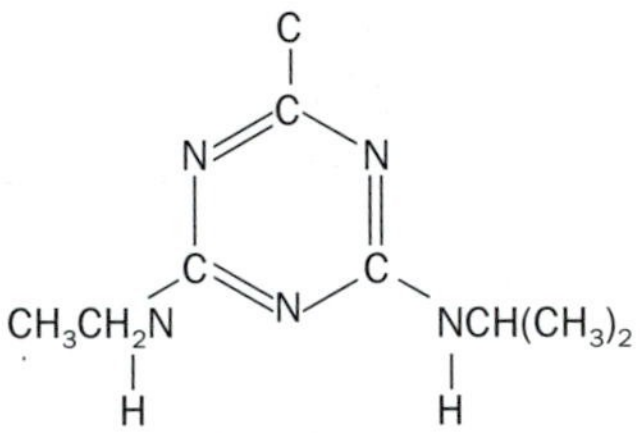

Atrazine

3 α,α,α-trifluoro-2,6-dinitro-*N,N*-dipropyl-*p*-toluidine

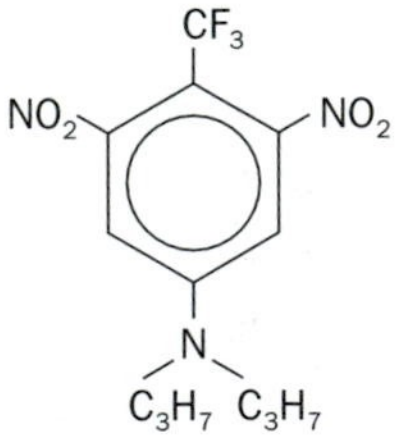

Trifluralin

7 1,1-dimethyl-3-(a,a,a-trifluoro-*m*-tolyl) urea

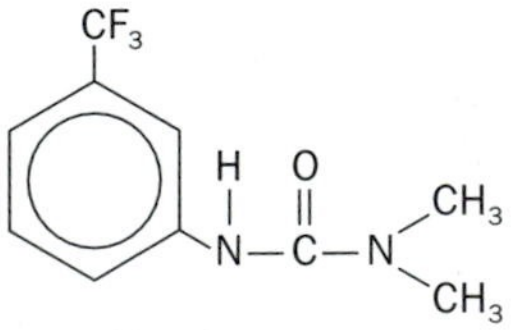

Fluometuron

4 N-(phosphonomethyl) glycine

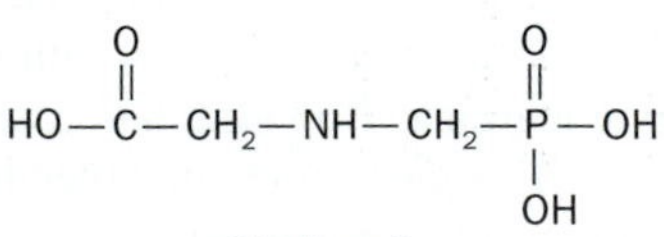

Glyphosate

8 1,1-dimethyl-4,4-bipyridylium ion

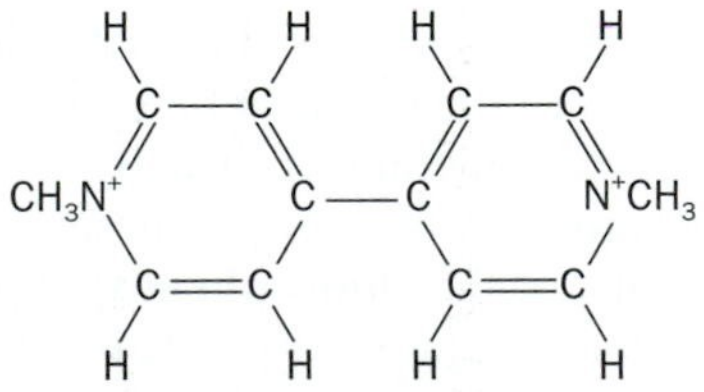

Paraquat

Table 13.12 The biological effects of some selected herbicides

Herbicide	Acute toxicity	Chronic toxicity	Reproductive effects	Teratogenic effects	Mutagenic effects	Carcinogenic effects
Propanil	Moderately toxic	Some effects on rats	None	None	None	None
Chloropropham	Slightly to non-toxic	Unknown	None	None	None	None
Trifluralin	Practically non-toxic	Inconclusive evidence	None	None	None	Unknown
Glyphosate	Virtually non-toxic	No evidence	None	None	None	None
2,4-D	Slightly to moderately toxic	Slight effects	None	None	None	Unknown
Sulfometuron-methyl	Very low toxicity	No evidence	None	None	None	None
Atrazine	Slight to moderate	Slight	None	None	None	Inconclusive
Fluometuron	Non-toxic to moderate	Some effects	None	None	None	None
Paraquat	Moderately to highly toxic	Variety of symptoms	None	None	None	Inconclusive

Table 13.13 Some environmental effects of some selected herbicides

Herbicide	Birds	Fish	Bees	Soil	Water
Propanil	Moderately toxic	Moderate to highly toxic	Non-toxic	Moderate persistence	
Chloropropham	Non-toxic	Moderately toxic	Non-toxic	Moderate persistence Strongly adsorbed on organic matter but not on soil	Leached out of soil with low organic content
Trifluralin	Non-toxic	Very highly toxic (and to other aquatic species)	Non-toxic	Moderate persistence	Low solubility Not readily leached
Glyphosate	Slightly toxic	Non-toxic	Non-toxic	Moderate persistence Strongly adsorbed	Although soluble is not readily leached
2,4-D	Slightly to moderately toxic	Slightly to highly toxic	Toxic	Low persistence	Has been detected in groundwater
Sulfometuron-methyl	Practically non-toxic	Slightly toxic to adult fish		Low persistence Not strongly adsorbed	Slightly soluble therefore potential pollutant
Atrazine	Non-toxic	Slightly	Non-toxic	Highly persistent Not strongly adsorbed	Moderately mobile in water Potential pollutant
Fluometuron	Non-toxic	Slightly	Non-toxic	Highly persistent Poorly adsorbed	Moderately soluble therefore potential pollutant
Paraquat	Moderately toxic	Slightly to moderately toxic	Non-toxic	Highly persistent Strongly adsorbed	Very soluble in water but soil adsorption prevents water contamination

In humans, prolonged breathing of 2,4-D has led to coughing, burning, dizziness and the temporary loss of muscle co-ordination, fatigue and nausea. A human given 16.3 mg in 32 days for therapeutic reasons suffered from stupor, lack of co-ordination, weak reflexes and loss of bladder control. 2,4-D has seen extensive use but has now been revoked in the UK.

The **sulphonylurea** herbicides, e.g. sulphometuron-methyl (Table 13.11, compound 5), are both soil and foliage active. They have a very specific activity and, usefully, very low application levels. Sulphonylureas act by inhibiting amino acid synthesis and thus cell division and plant growth. There are over 20 sulphonylureas currently in use.

The **triazines** are applied to the soil as a post-emergent herbicide, and are photosynthesis inhibitors. Atrazine is a well known compound of this family (Table 13.11, compound 6).

The **ureas** are relatively non-selective pre-emergent herbicides. Some are post-emergent and some are applied to foliage. These compounds are strongly adsorbed by soil. They act as inhibitors of photosynthesis. The urea selective herbicide fluometuron (Table 13.11, compound 7) can be applied both pre-emergent and post-emergent. Any person in the UK can use these herbicides.

Finally, the **bipyridyliums** are a group of non-selective contact herbicides. Plant cells and chloroplast membranes are caused to rupture within hours of application. They are not active in soil. These herbicides can only be handled and applied by weedkiller experts. Paraquat is an example of this class of compounds (Table 13.11, compound 8). Paraquat has been responsible for many cases of illness and, in some cases, death in humans – the estimated lethal dose via ingestion for humans is 35 mg kg^{-1}. In humans exposed to occupational levels of paraquat, eye, skin and nose irritation have been observed together with damage to the fingernails. The application of paraquat in the UK requires a license.

Some inorganic materials have been used as herbicides, e.g. arsenic trioxide, sodium arsenite, sodium borate, copper sulphate and sodium chlorate. However, such compounds tend to sterilise the soil, and be both non-selective and persistent. Inorganic chemicals and, indeed, organo-metallic compounds are now relatively little used in the developed countries and are being replaced by organic herbicides elsewhere.

DDT – the rise and fall of an insecticide

DDT, or dichlorodiphenyltrichloroethane, was the first synthetic organic insecticide to be manufactured and used on a very large scale. It belongs to the diphenyl aliphatics class of organochlorine compounds. First used extensively as a spray during the Second World War to combat disease-carrying mosquitoes (malaria and yellow fever) and fleas (plague and typhus), its subsequent effects on the environment caused DDT and its class of compounds to be discontinued in the UK and other countries between 1964 and 1984. The organochlorines are now mainly of historic interest, but the study of DDT well illustrates the rise and fall of a compound useful to humankind.

The choice of organochlorines as insecticides was based on their high degree of chemical stability in the environment; very low solubility in water (e.g. less than 1 mg dm^{-3} for DDT at 20 °C); high solubility in organic solvents and fats; and the high toxicity to insects and low toxicity to humans.

DDT

The World Health Organisation has stated that DDT has saved the lives of over five million people during its years of use. Unfortunately, its overuse caused a rapid rise in its concentration in the environment. Although its solubility in water is very low, DDT became lodged in sediments at the bottom of water bodies where it became adsorbed/absorbed on suspended organic particulate matter. This, coupled with its persistence, caused DDT to become more readily available to animals in the aquatic environment. Constant exposure to DDT and its metabolites led to their gradual build-up in living tissues, especially fatty tissues, to levels that caused some serious problems.

DDT is classified as slightly to moderately toxic to mammals via the oral route, and slightly to non-toxic via the dermal route (see Table 13.14).

In humans, acute toxic effects at low concentrations have resulted in sickness, diarrhoea, irritation of the eyes, nose and throat, general malaise and excitability. At higher levels of exposure tremors and convulsions have been observed.

Table 13.14 *Some oral and dermal LD$_{50}$ values for DDT*

Route	Animal	Dose (mg kg^{-1})
Oral LD$_{50}$	Rat	100–800
Oral LD$_{50}$	Rabbit	500–750
Oral LD$_{50}$	Sheep	>1,000
Dermal LD$_{50}$	Rat	2,500–3,000
Dermal LD$_{50}$	Rabbit	3,000

Chronic toxicity investigations have shown that DDT affects the nervous system, liver, kidneys and immune system in animals. Adverse effects on the liver, kidney and immune system of humans have not been observed, even in those involved in the manufacture of DDT, i.e. occupationally exposed. There is some evidence that DDT chronically affects reproduction in animals. Effects have ranged from sterility, decrease in foetal mass and a decrease in embryonic implantation. There is no evidence in humans of chronic reproductive effects. Evidence for teratogenic and mutagenic effects is inconclusive in the case of animals, and non-existent for humans. DDT causes an increase in malignant tumours in the livers and lungs of rats and mice.

DDT is slightly to practically non-toxic to birds, e.g. LD$_{50}$ mallard ducks >2,000 mg kg^{-1}. However, birds were exposed to DDT mainly through the food chain. Predators, in particular, accumulated DDT by eating fish, earthworms and other birds. This resulted in three areas of concern about the effects of DDT on birds. First, there is evidence that bird reproduction and mating have been adversely affected. Second, DDT's metabolite DDE (Figure 13.2) has caused eggshell thinning such that the nesting adult bird has caused the eggs to break. Third, DDT has caused the deaths of embryos.

DDT has proved to be highly toxic to many aquatic invertebrate species, e.g. the 96-hour LC$_{50}$ dose for daphnia is 4.7 µg dm^{-3}. It is very toxic to fish, e.g. for rainbow trout

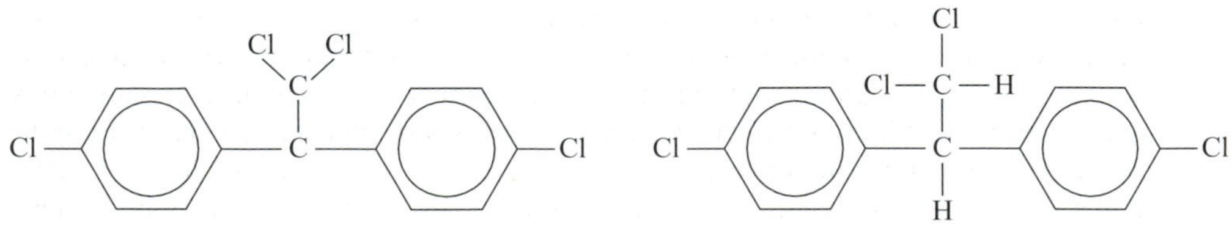

DDE
1,1-dichloro-2,2-bis(4-dichlorodiphenyl)ethene

DDD
1,1-dichloro-2,2-bis(4-dichlorodiphenyl)ethane

Figure 13.2 The metabolites of DDT – DDE and DDD.

the 96-hour LC_{50} is 8.7 $\mu g\ dm^{-3}$. The bioaccumulation factor for DDT is very high in aquatic species. DDT is however non-toxic to the honeybee (LD_{50} is 7 μg per bee).

DDT has a half-life in most soils of between 2 and 15 years and is therefore highly persistent. It is removed very slowly by various processes: run-off, volatilisation, photolysis and aerobic/anaerobic degradation. Its breakdown products DDE and DDD (Figure 13.2) are both very persistent and are similar to DDT in both chemical and physical properties. DDT enters surface waters via run-off, deposition from the atmosphere or by direct application. Its half-life is 56 days in lake water and about 28 days in river water. Aquatic animals readily take up and store DDT and its metabolites. It is not though taken up and stored by plants.

Insecticides – some types and their modes of action

[EXTOXNET 2003] Insecticides are classified according to the biological process interfered with, i.e. cuticle production, energy production and water balance, or system attacked, i.e. the nervous system and endocrine system.

Insecticides that affect the nervous system are the **pyrethroids**, **organophosphides** and **carbamates**. The synthetic pyrethroids are derivatives of naturally occurring insecticides called pyrethrins. The pyrethrins are found in the flowers of plants belonging to the chrysanthemum family. The original natural compounds were not used on a large scale because of their expense and the ease with which they were decomposed by sunlight. The pyrethroids are very stable in sunlight and are effective against a range of insects at a very low level of application. These compounds have undergone such development that the current set of compounds are referred to as the fourth generation. The first major agricultural insecticides in this family were the third-generation ones which included permethrin (Table 13.15, compound 1).

Just as in the case of herbicides, insecticides are investigated to determine:

- acute and chronic toxic effects
- mutagenic, teratogenic, reproductive and carcinogenic effects
- effects on organisms in the environment, including the possibility of bioaccumulation
- persistence in soil
- persistence in water.

Some of the properties of permethrin and other insecticides are shown in Tables 13.16 and 13.17.

The fourth-generation pyrethroids are particularly useful because they do not undergo photochemical decomposition and have very low volatilities. These properties, coupled with the even lower amounts that need to be applied for effectiveness, mean cheapness and relatively longer periods of time in which they can stay active. The pyrethroids act by combining with a protein which affects the sodium channel activity in the axon of a nerve. Normally, this channel is able to open and close, thus controlling nerve impulses. The pyrethroids cause the channel to remain open, thereby causing continuous nerve stimulation. The result is that the insect suffers tremors followed by paralysis and death. Any person in the UK can use permethrin.

The organophosphides are much more toxic to vertebrates than other insecticides, and are more chemically unstable and non-persistent. These compounds are classified

Table 13.15 *The structure of some insecticides*

Insecticide type	Insecticide type
1 3-phenoxybenzyl(1RS)-*cis,trans-*3-(1,2-dichlorovinyl)-2,2-dimethylycyclopropanecarboxylate	**4** 1-napthylmethylcarbamate

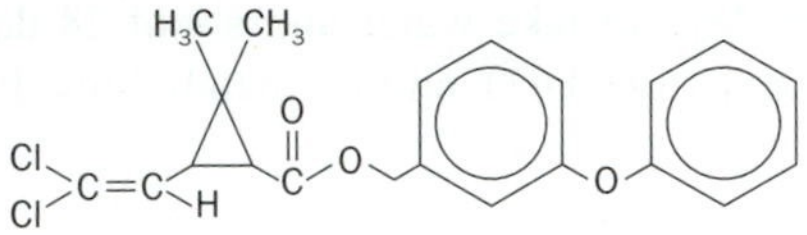

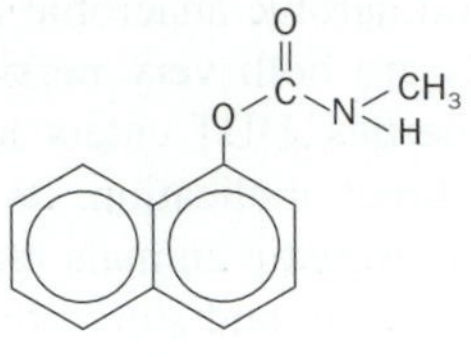

Permethrin

Carbaryl

2 diethyl(dimethoxythiophosphorylthio)-succinate	**5** 1-(4chlorophenyl)-3-(2,6-difluorbenzoyl)urea

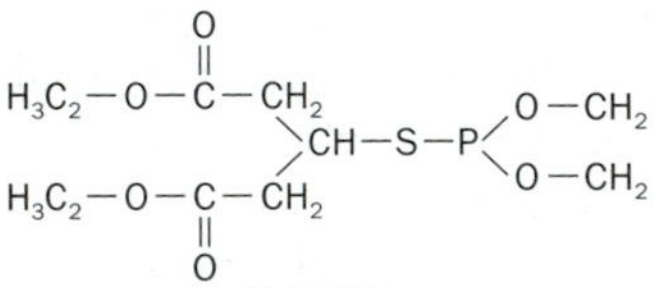

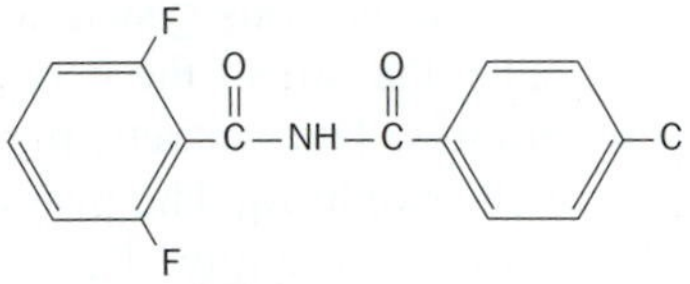

Malathion

Diflubenzuron

3 *O,O*-diethyl-*O*-4-nitrophenyl phosphorothionate	**6** *O,O*-diethyl-*O*-2-isopropyl-6-methyl (pyrimidine-4-yl)phosphorothionate

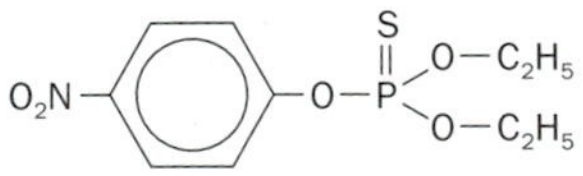

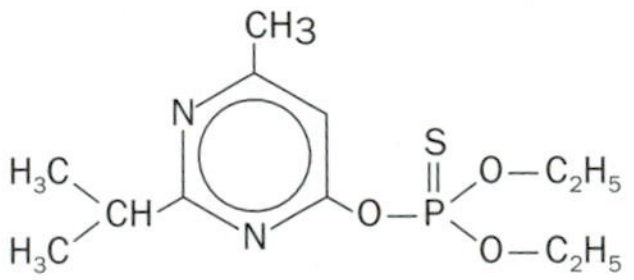

Parathion

Diazinon

according to their 'central' chemical structure. The aliphatic organophosphides contain carbon chains, e.g. malathion (Table 13.15, compound 2).

Many people working with pesticides and small children accidentally exposed have suffered from the acute effects of malathion poisoning. The symptoms have included numbness, tingling sensations, lack of co-ordination, headache, dizziness, tremor, sickness, abdominal cramps, sweating, blurred vision, difficulty in breathing and slow heart beat.

Females are more prone to poisoning than males. However, when given at very low dosage to volunteers for a short period of time malathion has showed no chronic toxic effects in humans. Again, this is an insecticide that can be applied in the UK by any person.

The phenyl derivatives are much more persistent than the aliphatic-type compounds, e.g. parathion (Table 13.15, compound 3). Parathion is highly toxic to agricultural workers and has been known to cause human fatalities. The lowest dosage that produces

Table 13.16 *Acute toxicity data for some selected insecticides*

Type of insecticide	Example	Oral LD$_{50}$ mg kg^{-1} or ppm	Dermal LD$_{50}$ mg kg^{-1} or ppm	Inhalation
Pyrethroids	Permethrin	Rats 430–4,000	Rats >4,000	4-hour LC$_{50}$ 23.4 mg dm^{-3}
Organophosphides	Malathion	Rats >4,000 Mice >4,000	Rats >4,000	
Organophosphides	Parathion	Rats 2–30 Mice 5–25 Rabbits 10 Dogs 5	Rats 6–50 Mice 19 Rabbits 15	4-hour LC$_{50}$ rats 84 mg dm^{-3}
Organophosphides	Diazinon	Rats 300–400	Rabbits 3,600	4-hour LC$_{50}$ 3.5 mg dm^{-3}
Carbamates	Carbaryl	Rats 250–850 Mice 100–650	Rabbits >2,000	4-hour LC$_{50}$ 200 mg dm^{-3}
Benzoylphenyureas	Diflubezuron	Rats >4,640 Mice >4,640	Rats >10,000 Rabbits >4,000	
Carbamates	Fenoxycarb	Rats >10,000	Rats >2,000	Rats LC$_{50}$ >0.45 mg dm^{-3}
Trifloromethyl aminohydrazones	Hydramethylon	Rats 1,100–1,300	Rabbits >5,000	Rats 4-hour LC$_{50}$ >5 mg dm^{-3}

Table 13.17 *The effects of some insecticides on organisms in the environment*

Insecticide	Toxicity to birds LD$_{50}$	Toxicity to fish and other aquatic life	Toxicity to bees
Permethrin	Mallard ducks >9,900	Rainbow trout 48-hour LC$_{50}$ 0.005 mg dm^{-3}	Extremely toxic
Malathion	Mallard ducks 1,485 Starlings >100	Trout LC$_{50}$ 0.1 mg dm^{-3} Goldfish LC$_{50}$ 10.7 mg dm^{-3}	Highly toxic
Parathion	Mallard ducks 2.1 Pigeons 3.0	Fish 96-hour LC$_{50}$ 1.4 mg dm^{-3}	
Diazinon	Birds 2.75–40.8	Rainbow trout LC$_{50}$ 2.6–3.2 mg dm^{-3}	Highly toxic
Carbaryl	Mallard ducks >2,000	Rainbow trout LC$_{50}$ 1.3 mg dm^{-3}	Highly toxic
Diflubezuron	Mallard ducks dietary LC$_{50}$ >4,640	Rainbow trout LC$_{50}$ 240 mg dm^{-3}	Non-toxic
Fenoxycarb	Mallard ducks >3 000	Rainbow trout LC$_{50}$ 1.6–10.3 mg dm^{-3} Daphnia LC$_{50}$ 1.6 ng dm^{-3}	Non-toxic
Hydramethylon	Mallard ducks >2,510	Rainbow trout 96-hour LC$_{50}$ 160 μg dm^{-3} Daphnia 48-hour LC$_{50}$ 1.14 mg dm^{-3}	

Table 13.18 Some biological effects of selected insecticides

Insecticide	Acute toxicity	Chronic toxicity	Reproductive effects	Teratogenic effects	Mutagenic effects	Carcinogenic effects
Permethrin	Oral – moderate to non-toxic Dermal – slightly toxic Inhalation – non toxic	Negligible	None	None	None	Inconclusive
Malathion	Oral – slightly toxic Dermal – slightly toxic	None at low dosages	None	None	Inconclusive	Inconclusive
Parathion	Acutely toxic via all routes	Chronic toxic effects via all routes	Some evidence	None	None	Maybe
Diazinon	Slight to moderate	Some symptoms	No data available	Inconclusive	Inconclusive	None
Carbaryl	Moderate to very toxic	No details available	None	None	Unlikely	Unlikely
Fenoxycarb	Oral – non-toxic Dermal slightly toxic Inhalation – moderately toxic	None	None	None	No data available	No data available
Diflubenzuron	No signs	Some	None	None	None	None
Hydramethylon	Slight	Some	None	None	None	None

Table 13.19 Some environmental effects of some selected insecticides

Insecticide	Birds	Fish	Bees	Soil	Water
Permethrin	Non-toxic	Highly toxic	Extremely toxic	Readily adsorbed by soil. Moderate persistence	Very low solubility therefore no leaching. Not found in groundwater
Malathion	Moderately toxic	Slight to very toxic according to species	Highly toxic	Not readily adsorbed by soil. Low persistence	Soluble and undergoes chemical reaction in water. Danger of leaching
Parathion	Highly toxic	Moderately toxic		Strongly adsorbed on soil. Moderate persistence	Undetectable in water within one week due to adsorption on sediments
Diazinon	Moderately toxic	Highly toxic		Low persistence	Slight solubility therefore pollutes groundwater. pH affects rate of decomposition
Carbaryl	Non-toxic	Moderately toxic	Highly toxic	Low persistence	Detected in run-off and groundwater. Stability pH dependent
Diflubenzuron	Non-toxic	Practically non-toxic		Low soil mobility. Low persistence	
Fenoxycarb	Practically non-toxic	Non-toxic		Strongly adsorbed on soil. Low persistence	Not leached. The higher the pH the faster the decomposition
Hydramethylon	Practically non-toxic	Slightly toxic		Strongly adsorbed by soil. Low soil mobility. Low soil persistence	Slightly soluble. Not very water mobile. Readily hydrolysed at high pH

toxic effects in humans is 240 μg kg^{-1}. It therefore can only be applied by licensed personnel, and to crops other than fruit, nuts or vegetables. It is likely to be withdrawn both in the UK and US in the near future.

The compound diazinon (Table 13.15, compound 6) was the first member of the heterocyclic family of these types to be used as an insecticide. Diazinon's acute toxic effects on humans include weakness, headache, sweating, nausea and diarrhoea. Death has occurred at very high levels of oral and dermal exposure. This insecticide can only be applied by licensed personnel. Chronic toxicity experiments have shown it to cause enzyme inhibition in some animals.

Of the carbamates, carbaryl (Table 13.15, compound 4) was the first to be used worldwide. The toxicity of this insecticide depends upon what it is mixed with, i.e. its formulation. Symptoms in humans via oral and dermal routes are typical of the carbamates, i.e. nausea, stomach cramps, diarrhoea and excess salivation. At higher dosage sweating, lack of co-ordination and convulsions result.

The organophosphides and carbamates also attack the insect nervous system, in particular the synapse. The synapse is the junction between two nerves where signals are transferred from one to the other. These materials bind with the acetylchlorinesterase, an enzyme found in the synapse. The role of this enzyme is to stop a nerve impulse after it has crossed the synapse. These insecticides prevent this enzyme action. Again, continuous stimulation of the nerves occurs leading to tremors, paralysis and death of the insect.

Chemicals that cause the inhibition of cuticle production are known as chitin synthesis inhibitors. The **benzoylphenyl ureas**, e.g. diflubenzuron (Table 13.15, compound 5), are used against such insects as fleas, caterpillars, beetle and fly larvae. These insecticides prevent the insect from making chitin which is the major component of the exoskeleton of an insect, so preventing them moulting successfully to the next stage of their life cycle. Diflubenzuron is rapidly degraded by microbes present in the soil. It has very low soil mobility and degrades rapidly in water.

Insecticides that are designed to attack insect endocrine systems are basically growth inhibitors. They work by mimicking the insect's juvenile hormone, keeping the insect in the immature state. Hence, they are unable to undergo metamorphism to the adult stage and therefore cannot reproduce. Such compounds, e.g. fenoxycarb, are often insect specific and act on their endocrine or hormone system. The advantage of such insecticides is their very low mammalian toxicity and non-persistence in the environment.

Currently, very few chemicals are used to inhibit energy production in insects. Hydramethylon, a **trifluoromethyl aminohydrazine**, is one such chemical and works by binding to the protein cytochrome in the electron transport system of the mitochondrion. This prevents the production of ATP and thus energy for cellular metabolism. Another chemical called sulphuramide works by being broken down by enzymes in the insect's body to products that are toxic to that insect.

Hydramethylon is known to cause irritation of the eyes and mucous membranes of the upper respiratory system in humans. Chronic toxicity investigations have shown liver and other organ mass increases in dogs and rats.

Insects have a thin coating of wax on their bodies, which helps to regulate water loss. Materials like diatomaceous earth and specially developed dusts are very effective at absorbing oils. Thus, when they are applied to insects they remove the coating of oil and therefore the insect loses too much water. The insect dies of desiccation. Borates have also been used to upset the water balance in insects, but their mode of action is not yet fully understood. Tables 13.18 and 13.19 show some of the biological and environmental effects of some selected insecticides.

Toxic organic waste disposal

All organic materials can be virtually destroyed by incineration. The main products will be CO_2 and H_2O plus minor components such as SO_2, NO_x, etc. depending upon the elements present. However, unless the incineration is properly carried out some organic compounds such as the PCBs will not be destroyed or unwanted materials such as dioxins will be formed. These would then become part of the emissions from the incinerator which could pollute the environment. Hazardous organic waste is normally destroyed by incinerators that operate at a **destruction** and **removal efficiency** (DRE) of better than 99.9999 per cent thus creating very little potential for, or real, pollution of the atmosphere. The DRE is calculated for difficult-to-burn substances such as PCBs by determining the amounts that are fed to the incinerator and present in the flu gases.

Hazardous waste incinerators are specifically designed to deal with wastes that can be pumped into the furnace or can be fed to it via drums. Some are able to accept bulky items such as electrical transformers. The conditions necessary for incineration are the correct temperature, degree of turbulence and amount of time of combustion. Turbulence ensures good mixing of air/oxygen, fuel and the organic chemicals within the combustion chamber. Turbulence also ensures that no 'cold spots' will be present in the furnace. Figure 13.3 is a schematic diagram of the incineration process used to destroy hazardous chemicals.

The temperature is rigidly controlled so as to be high enough to overcome the activation energy of combustion and ensure complete combustion of the organic materials. For example, to destroy PCBs the temperature must be at least 1,100 °C and for non-chlorinated organics at least 850 °C. The residence time in the furnace should be at least two seconds to prevent the formation of dioxins. If turbulent burning is not possible, e.g. in the case of electrical transformers, then the burning time must be extended for several hours. Excess air is added to ensure that the amount of oxygen present is sufficient to cause complete oxidation. Thus, in the exhaust gases there will be some unused oxygen. Some incineration plants that specialise in the destruction of organic compounds use pure oxygen.

Gases which leave the furnace are rapidly cooled using water sprays, which bring the gas temperature down to about 80 °C in a few seconds, to prevent the formation of dioxins from residual wastes. This water removes some of the particulate matter and some gases. Complete removal is done by using sodium hydroxide scrubbers to remove acid gases, such as SO_2 and HCl, and electrostatic precipitators to remove particulate matter. The resulting clean, cool, wet gas can then be reheated using hot air and vented to the atmosphere through a tall stack. Any solid residue (ash or slag depending upon temperature of burning) from the incinerator, dust from the precipitators and the filter cake from the slurry produced in the scrubbers are sent to landfill.

There are a number of other devices that can thermally destroy organic waste. These include the use of a plasma torch, infrared heating, pyrolysis (heating in the absence of air/oxygen), supercritical water oxidation, catalytic incineration, microwave heating and the use of concentrated solar energy.

Other chemical techniques are mainly based on either de-chlorination or oxidation. De-chlorination can be performed, for example, by reacting sodium, potassium or calcium metals with PCBs, resulting in the metal chloride salt and hydrocarbons. A second way is to react the PCBs with high temperature hydrogen and steam, which produces hydrocarbons and hydrogen chloride. Other methods being examined are the use of hot pressurised sodium hydroxide, methane or hydrogen with a catalyst, and potassium hydroxide in the presence of polyethylene glycols.

The UK Atomic Energy Authority developed a technique for the removal of organic materials from radioactive waste called the WINWOX process. Here, hydrogen peroxide

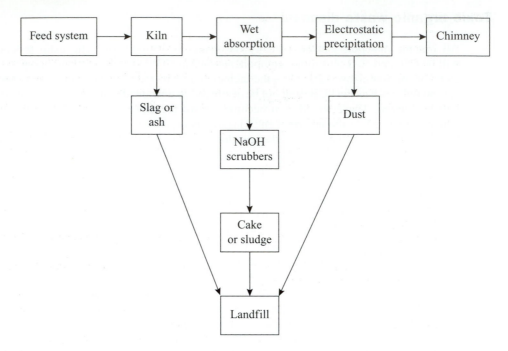

Figure 13.3 Schematic diagram of incineration process for hazardous organic chemicals.

is used to oxidise the organic materials. The hydrogen peroxide is mixed with the liquid or sludge containing the organics at 80 °C in a batch process. Once the exothermic reactions start, there is no need to continue heating the mixture. Carbon dioxide, steam and a little dioxygen are the gaseous by-products of reaction, and any chlorine or sulphur present is converted to chloride ion and sulphate(VI) ion in solution. This process has also been used in other industries.

Electrochemical incineration is sometimes used for the destruction of pesticides and some pharmaceuticals. The electrolyte is dilute nitric acid containing Ag^+ ions. At the anode of the electrochemical cell silver(I) ions are oxidised to silver(II), Ag^{2+}. This latter ion then rapidly oxidises any organic material that is present. Silver(II) ions are regenerated during the passage of electric current through the cell. Nitric(V) acid is reduced to nitric(III) acid which needs to be re-oxidised occasionally.

Some organic materials are toxic to bacteria or have structures that are so incompatible to bacteria that they are unable to metabolise them. In sewage treatment, the activated sludge process is capable of destroying some hazardous wastes, providing they are not too concentrated. This process involves the oxidation of carbon compounds to mainly carbon dioxide and water. Anaerobic processes have also been used to change organic materials into useful materials like methane. Bioremediation of land previously contaminated with organic chemicals involves the use of natural bacteria already present in that contaminated land. By increasing the amounts of nutrients, e.g. nitrogen, oxygen, for the bacteria to feed on their microbial action is increased. Very often groundwater on a site is pumped out, fed the nutrients, and then allowed to soak back into the ground where it eventually meets the organic waste. Bacteria then use these nutrients in the biodegradation of the waste. Bacteria find it very difficult to break down chlorine containing compounds since they are so rare in nature. However, the use of genetically engineered bacteria to destroy waste is being investigated.

Summary

- Dioxins and furans are persistent organic pollutants (POP).
- TCDD is one of the most toxic and is the standard to which all other dioxins and furans are referred. It is given an arbitrary Toxic Equivalent Factor (TEF) of 1.0. TEF values can be used to determine the risk associated with mixture of dioxins and furans.
- The chemical and therefore the biological stability of furans and dioxins increase with increase in number of chlorine atoms.
- Anthropogenic sources of dioxins and furans include herbicide production, incineration, combustion of petrol and oil, and the combustion of coal. Natural emissions are caused by volcanic activity and natural fires.
- Benzene is a carcinogen and enters the environment via evaporation and combustion of petrol and oil.
- Benzene can undergo chemical reactions in the atmosphere which produce toxic products. In particular it can form polycyclic aromatic hydrocarbons (PAHs).
- PAHs are formed by the burning of petrol and diesel, and during the combustion of wood and fossil fuels.
- The use of polychlorinated biphenyls (PCBs) is banned in the UK, but they can still be found as dielectric materials in old electrical equipment.
- PCBs are fat soluble and therefore can bio-accumulate. They are probable carcinogens.
- Pesticides can cause point or diffuse source pollution, primarily the former. Injury and death can be caused to non-target organisms either by the pesticide itself or its residues.
- There are a wide variety of herbicides which can be classified in a number of ways.
- Herbicide chemical families include the amides, carbamates, dinitroanilines, phosphono amino acids, chlorophenoxys, sulphonylureas, triazines, ureas and bipyridyliums.
- Insecticides are classified according to how they attack or interfere with an insect's biological system, i.e. nervous system, cuticle production, endocrine system, energy production and water balance.
- Toxic organic waste disposal is largely via oxidation in highly efficient incinerators.
- Chemical techniques can be used to dispose of organic materials. Other methods such as electrolysis and the use of bacteria have also been investigated.

Questions

1 Fungicides are used in large quantities in Western Europe. Find out what main types are used and identify any toxic or environmental effects they have had on non-targeted organisms.

2 Compare and contrast the toxicities of the furans and dioxins with those of the PAHs.

3 Using appropriate insecticides, carefully explain what is meant by each of the terms 'biomagnification' and 'bioconcentration'.

4 Toxaphene is no longer in use. What kind of chemical is it, and why is it still being deposited in the Great Lakes of Canada?

5 Identify the functional groups in TCP, 2,4-D, DDT, DDE, DDD, glyphosate, malathion and trifluralin.

6 Both synthetic insecticides and herbicides are more of a threat to humans and the environment than they are worth. Discuss this statement.

References

Buckingham, W.A. (2000) *Operation Ranch Hand: Herbicides in SE Asia 1961–71*. Available at: http://cpug.org/user/billb/ranchhand/ranchhand.html.

EXTOXNET (2003) *Pesticide Information Profiles*. Available at: http//ace.ace.orst.edu/info/extoxnet/pips/ghindex.html.

Profiles used in this chapter were those of propanil, chloropropham, trifluralin, gylphosphate, 2,4-D, sulphometuron-methyl, atrazine, fluometron, paraquat, DDT, permethrin, malathion, parathion, diazinon, carbaryl and diflubezuron.

Institute of Medicine (1998) *Executive Summary: Veterans and Agent Orange – Update 1998*. National Academy Press, Washington, DC.

National Atmospheric Emissions Inventory (NAEI) (1970–98) *Annual Report, UK Emissions of Air Pollutants Section 6.2: Persistent Organic Pollutants*. Available at: http://www.aeat.co.uk/netcen/airqual/naei/annreport/annrep98/naei98.html.

NATO/CCMS (1988) *International Toxicity Equivalence Factor (I-TEF) Method of Risk Assessment for Complex Mixtures of Dioxins and Related Compounds*. Pilot Study, Report No. 176 NATO, Committee on the Challenges of Modern Society.

New Jersey Department of Health and Senior Services (1996) *Hazardous Substances Fact Sheet TCCD*. Trenton, NJ.

Rinsky, R.A., Smith, A.B. and Hornung, R. (1987) Benzene and leukaemia: an epidemiological risk assessment. *New England Journal of Medicine*, **316**, 1044–50.

Van der Berg, M., Birnbaum, L., Bosveld, B.T.C. *et al.* (WHO) (1998) Toxic equivalency factors (TEFs) for PCBs, PCDDs, PCDFs for human and wildlife. *Environmental Health Perspectives*, **106**, 775–91.

Westing, A.H. and Pfeiffer, E.W. (1995) Dioxins in Vietnam. *Science*, **270**, 5234.

Wong, O. (1995) Risk of acute myeloid leukaemia and multiple myeloma in workers exposed to benzene. *Occupational and Environmental Medicine* **52**(6), 33,380–384.

Further reading

The Course Team Open University (1993) S237 Environmental Control and Public Health, Unit 10: *Hazardous Wastes Management*. Open University, Milton Keynes.

Although quite an old book, it contains much that is still relevant. It deals with how all types of hazardous wastes are generated, managed and disposed of, and what problems such wastes cause. Very readable, very interesting with self-assessment questions and answers.

EXTOXNET 2003 *Pesticide Information Profiles*. Available at: http//ace.ace.orst.edu/info/extoxnet/pips/ghindex.html.

This Internet site offers extensive information about the effects of pesticides on living organisms. The data is linked to an extensive set of references. It is maintained by several US universities.

National Atmospheric Emissions Inventory (NAEI) (1970–98) *Annual Report, UK Emissions of Air Pollutants*. Available at: http://www.aeat.co.uk/netcen/airqual/naei/annreport/annrep98/naei98.html.

Everything you wanted to know about UK atmospheric emissions.

14 Energy production: coal, oil and nuclear power

- Sources of energy
- Fuel energy terms and quantities
- UK production and use of fossil fuels
- Formation and nature of oil, coal and natural gas
- Pollution effects of fossil fuel combustion. Combustion of coal. Combustion of natural gas
- Four-stroke petrol engine – engine knock and leaded petrol. Pollution and the petrol engine. The three-way catalytic converter
- Fractional distillation of crude oil
- Origin of nuclear energy
- Chain reaction and criticality
- The components and structure of a thermal reactor
- Radiation and risk assessment. Biological effects of radiation
- Nuclear accidents
- The disposal of spent fuel

Sources of energy

Since energy can neither be created nor destroyed, the total energy in the environment is essentially constant – there can never be an energy crisis. What is required is energy in a variety of useful forms together with a range of different energy converters. A source of useful energy is called a **fuel**. It is both the amounts of fuels that are available, and the efficiency of conversion of these fuels, that has led to the term 'energy crisis' – it should more accurately be described as a 'fuel crisis'.

When an electric light bulb is switched on, its primary function is to convert electrical energy to light energy. Unfortunately, much electrical energy is also converted to unwanted 'heat' energy and a little to sound energy. An electric light bulb is not an efficient energy converter. In addition, the electrical energy it uses may have been produced from stored chemical energy in coal at a coal-fired power station. Hence, before the electrical energy's end-use consumption there have been several inefficient energy conversions *after* the coal was placed in the power station.

Coal is a **primary source** of energy. Other primary sources include (i) solar energy, (ii) fossil fuels, (iii) nuclear energy, and (iv) 'other' sources of energy.

Solar energy can be used **directly** or **indirectly** as a source of useful energy. Direct methods include:

- using the Sun's rays to heat air or water, e.g. for use in central heating;
- focusing of solar energy on water to produce steam that will drive a turbine and therefore produce electricity, i.e. solar thermal electrical energy;

- use of photovoltaic cells to convert solar energy to electricity;
- heating and lighting of domestic, commercial and industrial areas by passive (no moving parts) solar energy;
- crop drying and other drying processes.

Indirect methods include:

- the burning of wood and other organic matter (biomass) produced by photosynthesis;
- the conversion of biomass to other fuels, e.g. ethanol is produced by fermentation of starches or sugars; methane from landfill sites; methanol from wood, and bio-diesel from rapeseed oil and waste cooking oils;
- the use of wind energy to drive turbines – the Sun is the direct cause of winds;
- wave energy – winds cause ocean waves which can be harnessed to drive turbines.

Solar energy is a renewable form of energy but currently only provides about 6 per cent of world demand for energy.

The fossil fuels coal, crude oil and natural gas are derived from carbohydrates that were made using solar energy many millions of years ago. They provide some 90 per cent of the world's current energy needs. Coal is still the main fuel used throughout the world to produce steam in both power stations and industrial boilers. Refined oil is used in vast quantities for transport. It is also used as a domestic boiler fuel and for space heating. Natural gas is the cleanest of the fossil fuels and is used by the developed nations mainly for space heating. It is also used for cooking and for some chemical industry processes requiring methane as a feedstock. Natural gas is being increasingly used to generate electricity. Fossil fuels are non-renewable forms of energy.

Nuclear energy is derived from nuclear fission and provides about 20 per cent of the world demand for electricity via steam-driven power stations. In this respect it is second only to coal. Nuclear energy can be regarded as a non-renewable form of energy.

'Other' sources of energy include geothermal energy and tidal energy. Geothermal energy is energy that can be extracted because of the temperature difference between the interior of the Earth and the surface. The interior is hot because of the energy that was transferred to it during accretion, and the radioactive decay of the long-lived isotopes thorium-232, uranium-238 and potassium-40. Heat transfer to the surface occurs

Table 14.1 *Fuel energy – some units in common use*

Unit	Example use	Equivalent number of joules
British Thermal Unit (Btu)	Gas bills	1 Btu = 1.055×10^3 J
Kilowatt-hour (kWh)	Electricity bills	1 kWh = 3.6×10^6 J = 3.412×10^3 Btu
Calorie (C)	Information about diet	1 C = 4.184×10^3 J
Electron-volt	Nuclear power	1 eV = 1.602×10^{-19} J
Erg	Some scientific literature	1 erg = 1×10^{-7} J
Quadrillion Btu (Quad) – used in the US	Annual energy consumption	1 quad = 1×10^{15} J
Exajoule (EJ)	Large amounts of energy	1 EJ = 1×10^{18} J
Terawatt year (tWyr)	World energy budgets	1 tWyr = 10^9 kW yr = 31.54 EJ = 29.89 quad

Table 14.2 *Some non-SI units for fossil fuel*

Resource	Example use	Energy content/J
Coal	tonne	24×10^9
Oil	barrel	5.7×10^9
	tonne	42×10^9
Gas	therm	105×10^6
	m^3	38×10^6
Electricity	kWh	3.6×10^6

principally by convection currents in hot, deformed solids, and at the surface, where the crust is stiffer, by conduction. Tides and therefore tidal energy are caused primarily by gravitational interactions between the Earth and the Moon. Both of these forms of energy are renewable.

The units of energy used in dealing with fuels and energy demand are wide and varied (Tables 14.1 and 14.2). In the UK, energy calculations, fuel production and fuel usage figures are often expressed in millions of tonnes of oil equivalents (mtoe).

Figure 14.1 shows the total UK production of primary fuels (mtoe) for 1970 and 1999 – it can be seen that the total production of primary fuels has risen by 169 per cent since 1970 mainly because of the increases in production of petroleum and natural gas.

In 1970 UK consumption of coal, gas, oil and primary electricity (produced mainly by nuclear power stations) was some 212 mtoe, and in 1999, 228 mtoe. About 31 per cent of this energy was lost in energy conversion and distribution, and used by the energy-producing industry. Only 69 per cent was available to the end-user. Figure 14.2 shows the origins of the *total* energy consumption, i.e. including the energy unavailable to an end-user, in the UK in 1970 and 1999.

Since 1970 there has been a large increase in the use of natural gas and a large decrease in the use of coal. The amount of electricity used from nuclear reactors also significantly increased.

Between 1970 and 1999, there was a substantial decline in energy use in industry, and a corresponding increase in the use of energy in transport (Figure 14.3).

In 1970, the UK continental shelf production of oil and gas was about 10 mtoe, and in 1999, 249 mtoe. It was estimated in 1999 that the total reserves in known oil and gas deposits were approximately 4100×10^6 tonnes of oil and 3200×10^9 m^3 of gas.

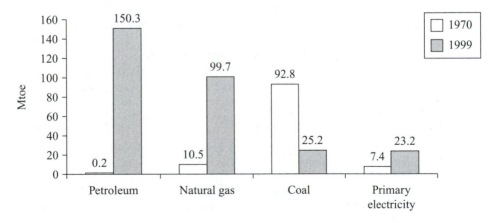

Figure 14.1 Production of primary fuels in the UK in 1970 and 1999.

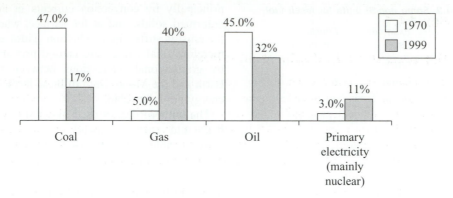

Figure 14.2 Total primary fuel consumption for the UK in 1970 and 1999.

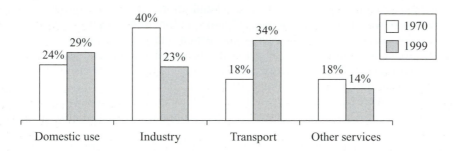

Figure 14.3 Percentage end use of energy in the UK in 1970 and 1999.

The formation and nature of oil, coal and natural gas

The generic term 'petroleum' includes all naturally occurring solid, liquid and gaseous hydrocarbons that are stored in the Earth (Table 14.3), whilst 'oil' usually means crude oil or natural gas liquids.

Crude oil is a mixture of over 1,200 hydrocarbons, and is described according to its density, i.e. light crude oil, medium crude oil, heavy oil and extra heavy oil. It is believed to have originated from the decay of small marine organisms such as plankton over many millions of years. During the formation of crude oil (and natural gas), complex organic molecules were broken down into much simpler ones. Marine organisms that settled to the bottom of the oceans were gradually buried by sediments. As the depth of burial increased, the temperature and pressure increased, causing the organic material to undergo a series of changes. When the temperatures and pressures were relatively low and bacterial action was possible, a complex material **kerogen** and **biogenic methane** gas were produced. Kerogen is composed of a wide variety of organic compounds containing sheets of aromatic hydrocarbon rings and heterocyclic compounds with attached alkane chains. Biogenic methane constitutes about 20 per cent of the

Table 14.3 *Petroleum hydrocarbon mixtures*

Solids	Liquids	Gases
Asphalt	Crude oil	Natural gas
Bitumen	Liquefied natural gas	
Tar		

Box 14.1

The causes and effects of oil slicks

Oil slicks at sea cause many problems, ranging from the deaths of marine organisms to the spoiling of beach and other amenities. About one-third of the oil that is released into the environment is actually caused by shipping accidents, though these tend to be more dramatic when they do occur. The other two-thirds is caused by deliberate discharges from ships and industrial facilities; urban run-off; leaks from pipelines, oilrigs and during the exploration and production of oil; deposition from the atmosphere; and by natural sources.

Oil slicks are dealt with in three main ways. First, booms may be used to skim off the oil from the water's surface. This method is environmentally friendly but is slow. In addition, booms usually only manage to retrieve a relatively small amount of oil, which requires immediate storage facilities. A second method is the application of appropriate detergents, which will disperse the oil. This is quick and large areas can be covered. However, detergents are largely ineffective against thick oil slicks and can cause environmental problems. Burning, i.e. during calm seas it is set alight and allowed to burn off, can also eliminate slicks. Unfortunately, this causes large amounts of pollutants to be emitted to the atmosphere.

The effects of oil slicks on wildlife can be catastrophic. The lighter molecules found in crude oil have anaesthetic properties. The heavier aromatic molecules are carcinogens and can become concentrated in the food chain. Volatile compounds can also burn the eyes and skin, and irritate or damage other sensitive membranes. In addition, crude oil coatings can cause suffocation. If pollution is caused by refined oil, then toxicity becomes more problematical. Refined oil products do tend to degrade more rapidly in the environment than the less toxic crude oil.

What happens to oil over a long period of time? Fortunately, there are a number of ways that oil can be degraded and dispersed, i.e. by weathering, evaporation, emulsification, oxidation and biodegradation. It may though take a number of years before all traces of an oil spillage finally disappear.

world's methane. As the temperature and pressure increased, kerogen was converted into crude oil and natural gas. At the highest temperatures crude oil could not form, only natural gas with a high methane content.

Crude oil is termed **conventional oil** when it is found mainly in sandstone sedimentary deposits. It is composed of 82–7 per cent carbon and 12–15 per cent hydrogen by mass, plus small amounts of sulphur, oxygen and nitrogen. The sulphur and nitrogen content has caused the formation of the pollutants SO_2 and NO_x when crude oil has undergone combustion.

Unconventional oil is derived from oil shale, tar sands and heavy oil. Oil shale is sedimentary rock containing oil with deposits of kerogen. Tar sands and heavy oil are formed when oil migrates to the surface where evaporation of the more volatile components occurs, leaving behind the heavier components.

Natural gas is primarily methane plus other saturated hydrocarbons, e.g. US natural gas contains 88–96 per cent methane by volume. Types of natural gas are shown in Table 14.4.

Table 14.4 *Types of natural gas*

Name of natural gas	Origin/composition
Biogenic gas	Produced by action of relatively low pressures and temperatures plus bacterial action on marine organisms
Thermogenic gas	Formed at high temperatures. Found with oil and coal
Abiogenic gas	Natural gas not formed from organic matter
Associated gas	Methane found with oil
Non-associated gas	Biogenic and gas associated with coal
Dry gas	Almost pure methane, e.g. coal gas, biogenic gas
Wet gas	Contains hydrocarbons other than the alkanes greater than 4% by volume
Gas liquids	Natural gas that contains hydrocarbons other than alkanes

Coal is also believed to have formed by the compression of organic plant matter by the deposition of sediments. The plant matter was first converted to peat (about 60 per cent C by mass). As the pressures and temperatures increased, peat was transformed to lignite (about 73 per cent C), then to bituminous coal (about 84 per cent C) and finally to anthracite (about 93 per cent C).

Some hydrocarbon fuels are described as **primordial fossil fuels**, which were trapped at depths greater than 300 km as the Earth formed by accretion. This is supported by two pieces of evidence. First, $^4_2$He is normally produced by radioactive decay but some gases have been found enriched in $^3_2$He, which suggests a primordial origin. Second, methane and carbon dioxide have been found trapped in diamonds, which were formed at great depth.

The combustion of coal

How coal is burnt depends on its particle size. Particles greater than about 5 mm in diameter can be burned on grates, usually in small to medium sized coal burning furnaces. When the particle size is about 3–5 mm in diameter the coal is burned in a fluidised bed furnace operating at 800–900 °C. In this case, coal is mixed in a bed of finely divided limestone or dolomite, which is highly agitated by blowing air through the bed from below. At the temperatures involved, the limestone/dolomite chemically reacts with any SO_2 formed from the burning coal to produce calcium sulphate, which is periodically removed. The amount of thermally produced NO is very low because of the relatively low temperatures involved. Any ash that is produced remains mainly in the bed. The nitrogen in the coal is converted to N_2O, but if the bed temperature exceeds 900 °C, then NO_x is produced.

Coal undergoes combustion more often in pulverised coal burning furnaces. Powdered coal containing particles of about 70 μm in diameter is mixed with and carried by streams of air to the combustion chamber.

Coal can also undergo combustion by gasification, the resulting gases being burned in gas turbine engines.

When coal is burned, it first decomposes to gases and volatile liquids, the quantities of which depend on the type of coal. Gases and vapours released are composed of H_2, CO, hydrocarbons such as CH_4, C_2H_6, C_2H_4, etc., oxygenated hydrocarbons of medium relative molecular mass, tar (containing high relative molecular mass hydrocarbons),

CO_2 and H_2O. These compounds then burn, leaving small unburned carbon particles and ash,

$$C_nH_{2n+2} \longrightarrow C_nH_{2n} + H_2$$

$$C_nH_{2n} \longrightarrow CO + H_2 + H_2O$$

$$H_2 + 1/2\,O_2 \longrightarrow H_2O + heat$$

$$CO + 1/2\,O_2 \longrightarrow CO_2 + heat$$

The carbon particles then start to oxidise,

$$C + 1/2\,O_2 \longrightarrow CO$$

$$C + O_2 \longrightarrow CO_2$$

As burning continues, some of the mineral content of the coal melts and then solidifies to form larger ash particles. Meanwhile, other components volatilise and form very small solid particles. These latter particles can carry relatively large amounts of toxic heavy metals, e.g. Hg, Ni, Pb and As. UK coal also contains chlorine, so HCl can be emitted. Radioactive isotopes are also present in emissions from coal burning furnaces.

Coal production in the UK fell from 147×10^6 tonnes in 1970 to 37×10^6 tonnes in 1999. In 1999, power stations in the UK used 72 per cent of that year's coal production, compared with 43 per cent in 1970. Figure 14.4 compares the use of coal in 1970 and 1999, the other most significant difference being the big reduction in the domestic use of coal.

The 1999 proved reserves of coal in the UK amounted to about $1,500 \times 10^6$ tonnes, and in the US approximately $250,000 \times 10^6$ tonnes. At the present rates of production these reserves should last about 40 years and 250 years, respectively. Generally, the current rate of extraction of coal from known resources throughout the world would suggest that there is no immediate shortage of coal since it is estimated it will last about 230 years.

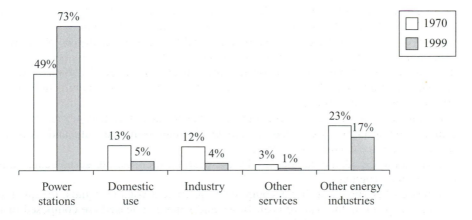

Figure 14.4 The use of coal in the UK in 1970 and 1999.

Natural gas

Any gaseous fuel has to be first mixed with air before it is set alight. Once burning, the exothermic chemical reactions that occur generate enough thermal energy to produce highly reactive free radicals, e.g. $O^\bullet$, $CH_3^\bullet$ and $^\bullet OH$. When more fuel and air are injected into a combustion chamber, the heat and free radicals present cause it to ignite and the reactions proceed via free radical chain reactions.

In the case of methane and air, the overall reaction is,

$$CH_4(g) + O_2(g) = CO(g) + H_2(g) + H_2O(g) \quad \Delta H^0_{298\,K} = -277.5 \text{ kJ mol}^{-1}$$

This is followed by the reaction,

$$H_2(g) + 1/2\, O_2(g) = H_2O(g) \qquad\qquad\qquad \Delta H^0_{298\,K} = -241.8 \text{ kJ mol}^{-1}$$

and then,

$$CO(g) + 1/2\, O_2(g) = CO_2(g) \qquad\qquad\qquad \Delta H^0_{298\,K} = -283.0 \text{ kJ mol}^{-1}$$

The latter reaction is kinetically slower than the first two.

Within a combustion chamber there are flame zones of different temperatures. In the hottest region, free radicals are readily formed and the first two reactions in the above sequence go to completion. Because the third reaction is slower, CO continues to be oxidised in other parts of the flame until what is left of it is finally exhausted from the combustion chamber.

In the case of higher alkanes, e.g. propane, the first step in the combustion process is the formation of an alkene(s) which then undergoes reactions of the kind described above, e.g.

$$C_3H_8(g) = C_3H_6(g) + H_2(g)$$

Then, $C_3H_6(g) + 2O_2(g) = 3CO(g) + 2H_2(g) + H_2O(g)$

$$H_2(g) + 1/2\, O_2(g) = H_2O(g)$$

$$CO(g) + 1/2\, O_2(g) = CO_2(g)$$

If methane, ethane, propane, etc., do not enter a flame region that is hot enough to ensure their oxidation, or they are not mixed properly with the correct amount of air, then unburned hydrocarbons together with products of partial oxidation, such as alkenes, and aldehydes will be emitted.

When CO is detected in exhaust gases it means that it hasn't experienced the temperatures necessary for its conversion to CO_2 – this can be the result of the CO not spending enough time inside a high temperature region or the mixture not containing enough dioxygen.

Nitrogen(II) oxide can be formed in the hottest part of the flame by the reaction between dioxygen and dinitrogen, and since the free radical concentrations here are so much higher than elsewhere, its formation is much faster. However, the residence times of dioxygen and dinitrogen are short. NO is also formed in other parts of the flame, where it is referred to as **thermal** NO. Some NO is oxidised to NO_2 inside the combustion chamber. Hence, both NO and NO_2, i.e. NO_x, are emitted from burning hydrocarbon

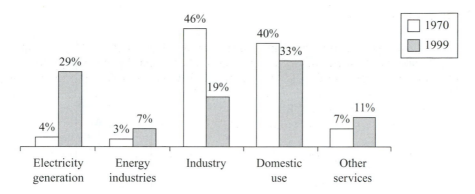

Figure 14.5 The use of natural gas in the UK in 1970 and 1999.

fuels. Since the formation of NO requires such a high temperature, in order to limit the amount formed the temperature of a flame must be reduced or the residence time of a fuel mixture inside the hot regions reduced. Unfortunately, this would increase the CO emissions. NO_x removal is usually carried out by catalytic reduction, e.g. if NH_3 is added to the exhaust gases and then passed over a suitable catalyst NO_x will be reduced to N_2, H_2O and O_2.

Figure 14.5 shows how natural gas was used in 1970 and 1999. It should be noted that the total end-usage for natural gas in the UK in 1970 was some 46 TWh and in 1999 about 1,065 TWh, a 22-fold increase. The use of natural gas in the domestic situation alone rose by nearly 200 per cent.

The four-stroke petrol engine

The petrol engine is simply a simple reaction chamber (the cylinder) fitted with a piston, sparking plug and valves, where highly exothermic reactions occur between hydrocarbons and dioxygen. A typical reaction is,

$$C_8H_{18}(l) + 25/2O_2(g) = 8CO_2(g) + 9H_2O(g) \quad \Delta H^0_{comb.} = -5\,512.5 \text{ kJ mol}^{-1}$$

Thus a great deal of high-pressure, hot gas is produced.

The petrol engine consists of a cycle of four events (Figure 14.6). The first is the **induction stroke** – as the piston moves down, the inlet valve opens and the outlet valve closes. A mixture of air and petrol vapour is sucked into the cylinder. The second step is the **compression stroke**. Here, the piston rises when both inlet and outlet valves are shut and so the reaction mixture becomes greatly compressed. Just before maximum compression the spark is generated. The gaseous mixture explodes, resulting in the production of much heat (temperatures in excess of 1,500 °C are achieved) and gas. This causes a powerful expansion, forcing the piston down. This is called the **ignition stroke** and produces the power necessary to make the vehicle move. The final step is the **exhaust stroke**. The inlet valve remains closed, the outlet valve opens and the piston rises, pushing out the product gases at about 900–1,000 °C to the exhaust system. The exhaust fumes are vented to the atmosphere at about 200 °C when the engine is at its working temperature. Thus the petrol engine is described as a four-stroke internal combustion engine. The cycle is repeated many times per minute in a row of four or six cylinders.

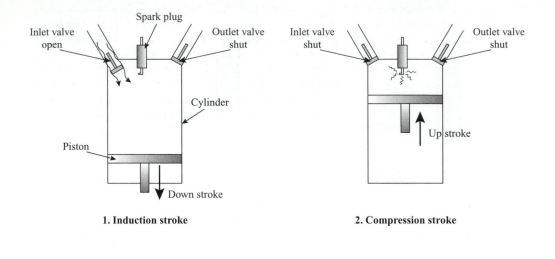

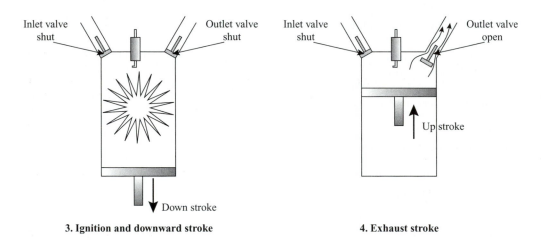

Figure 14.6 The four strokes of a petrol engine.

This engine appears to exhaust to the atmosphere, via the car exhaust pipe, just the 'greenhouse' gases CO_2 and H_2O. So why was it necessary to use leaded petrol as opposed to unleaded? Why is the motorcar accused of causing NO_x pollution? Why are modern cars fitted with catalytic converters?

Engine knock and leaded petrol

The ratio of the volume V_1 of the reactant gases contained within the cylinder at the maximum point of the induction stroke to the volume V_2 that those gases occupy just prior to ignition is called the compression ratio of an engine, i.e.

Compression ratio $= V_1/V_2$

The higher the compression ratio, the more powerful is the engine. Compression ratios of 10–14 are common in modern vehicles. The higher the compression ratio, the more the reactant gas mixture is compressed, so that when ignition occurs the more violent is the explosion. Ideally, the fuel mixture should burn with a smooth flame front that expands from the point of ignition throughout the cylinder. However, the advancing flame front causes the unexploded fuel pressure to increase and its temperature to go up. This results in the fuel in the 'corners' of the cylinder prematurely igniting before the flame front reaches it. This is called **knocking** and if not corrected gives rise to loss in engine power and engine wear.

Engine knock and the research octane number

In searching for an answer to engine knock, it was found that hydrocarbons with branched chains gave little or no knocking, whilst those without long-branched chains were the worst. In order to determine a scale of knocking, 2,2,4-trimethyl pentane (TMP, otherwise known as iso-octane) was chosen as the standard branched chain compound, and heptane the long chain hydrocarbon. TMP gave very little engine knock and was found to be an excellent car engine fuel. Heptane readily causes engine knock.

Since TMP contains eight carbon atoms it gave rise to the term **octane number**. The **research octane number** (RON) is a way of expressing a material's potential as a fuel. The RON value of TMP is arbitrarily set at 100 and that of heptane at 0. Fuels are burned in a test engine in which the compression ratio can be changed. The compression ratio is changed until the test fuel produces knocking. Then mixtures of TMP and heptane are introduced in the engine until the appropriate mixture is found that gives the same degree of knocking for the now fixed compression ratio. The RON of the test fuel is then expressed as the percentage of TMP by volume in the TMP/heptane mixture. For example, if 1 dm^3 of heptane and 4 dm^3 of TMP gave the same knocking, then the RON value would be given by (4 dm^3/5 dm^3 × 100) or 80 per cent. Some fuels are worse than heptane in producing knocking and they are allocated negative RON values, e.g. octane has a RON of −19. Fuels better than TMP are given values greater than 100, e.g. benzene RON is 106.

In general, fuels that are straight chain hydrocarbons have very low RON values (e.g. hexane RON 25), branched chain ones are much higher (e.g. 2,3-dimethylbutane), whilst cycloalkanes (e.g. cyclohexane RON 83) and aromatic hydrocarbons are higher still. Between 1960 and 1970, petrols were manufactured by cracking and reforming which gave between 85–90 RON. The final rise to 97 RON for four-star petrol was attained by adding tetraethyl lead to the petrol. How this compound works in stopping knocking is still not fully understood. Unfortunately, it is the use of leaded petrol in the motor car that has caused lead pollution in the environment.

Tetraethyl lead is a colourless liquid with a boiling point of 200 °C, and readily dissolves in hydrocarbons. The structure of tetraethyl lead is,

CH_3CH_2
|
Pb ⋯⋯ CH_3CH_2
╱ |
CH_3CH_2 CH_3CH_2

This compound undergoes combustion in the cylinder with the petrol/air mixture and is oxidised to lead(II) oxide, PbO. Lead(II) oxide is deposited on the spark plugs and on the cylinder walls, where its abrasive action shortens engine life. To prevent this, compounds like 1,2-dibromoethane were added to the petrol, so that during combustion lead(II) bromide, $PbBr_2$, was formed instead of the oxide. This compound is a much more volatile material and leaves the cylinder via the exhaust stroke, where it condenses in the exhaust system and is evacuated to the atmosphere. Hydrogen bromide gas is also produced. This latter compound dissolves in water to produce hydrobromic acid which corrodes the exhaust system.

Because of the toxic nature of lead, alternatives to tetraethyl lead had to be found. Without it, the octane number of petrols would have had to be reduced and therefore the compression ratio of engines. This would mean less efficient engines and therefore more CO and unburned hydrocarbons. The answer lay in more intensive refining and an increase in the amount of aromatic compounds in petrol. Unfortunately, this leads to more PAHs being formed and is therefore not only more costly but perhaps even more dangerous. Another way of increasing the RON values of petrols is to add oxygenates, i.e. compounds with oxygen atoms in their structures. For example, the RON value of ethanol is 111, whilst that of 2-methoxy-2-methylpropane is 117. These burn readily and do not produce unpleasant pollutants. However, large volumes of these compounds need to be added to petrol, so the use of oxygenates would be expensive.

The fractional distillation of crude oil and the making of petrol

Since crude oil is a complex mixture of alkanes, cycloalkanes and aromatic hydrocarbons it must be first be separated into useful components. Although these compounds have a wide range of boiling points, only a partial separation is needed, and so fractional distillation is used. Crude oil is heated at the bottom of a tall tower fitted with many horizontal trays and bubble caps (Figure 14.7(a)). The more volatile components start to boil off first and rise up the column. As the vapours pass through the bubble caps they condense on the trays and transfer their heat to the trays themselves. As the oil in the boiler gets hotter and hotter, some of the less volatile components start to boil and their vapours start to pass up the column. As these cool down on the trays, they pass their heat on to the condensed, more volatile components which are reheated and start to vaporise again and rise higher up the column. This process is repeated until there is a temperature gradient along the column. The components of crude oil are distributed along this column according to their temperature ranges. The most volatile, lighter components appear at the top of the column where it is cooler, whilst the heaviest components stay in the boiler. Thus the crude oil is separated into fractions with different boiling point ranges. A typical column temperature distribution and corresponding fractions of the crude oil are shown in Figure 14.7(b).

The light and heavy gasoline fractions are combined in one stage of making petrol, i.e. the range 40–200 °C. These combined fractions make up about 20 per cent of the total volume of crude oil and contain mainly straight chain hydrocarbons composed of 5 to 10 carbon atoms. The RON value of this combined fraction is about 60–70. Petrols produced from this are called **straight run petrols**. A typical composition by volume of a straight run petrol is 60 per cent alkanes, 25 per cent cycloalkanes and 15 per cent aromatic compounds. To increase the RON value of straight run petrols, catalytic reforming of the straight chain compounds is undertaken. The conditions for

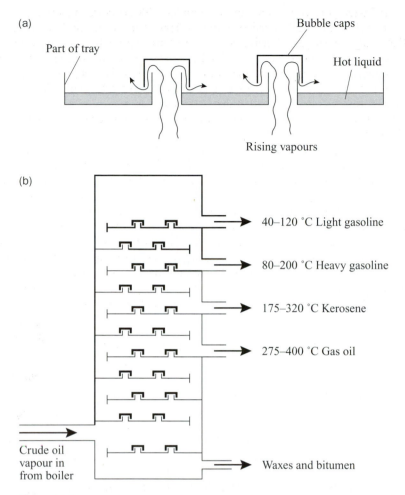

Figure 14.7 (a) Bubble tray detail of fractional distillation of crude oil. (b) A schematic diagram of the primary distillation of crude oil.

catalytic reforming are pressures between 4 and 10 atmospheres, a temperature of 500 °C and a platinum/rhenium catalyst. Figure 14.8 illustrates the catalytic reforming of hexane.

The other fraction used for the manufacture of petrols is the gas oil fraction, which contains straight chain alkanes with 12 to 18 carbon atoms. These molecules are broken down or catalytically cracked to smaller molecules. This takes place at 525 °C over zeolite catalysts,

e.g. $CH_3(CH_2)_{10}CH_3(g) \longrightarrow CH_3(CH_2)_6CH_3(g) + CH_2{=}CHCH_2CH_3(g)$

The products of the two processes are then combined to produce petrols, thus considerably increasing the volumes available from crude oil (Figure 14.9).

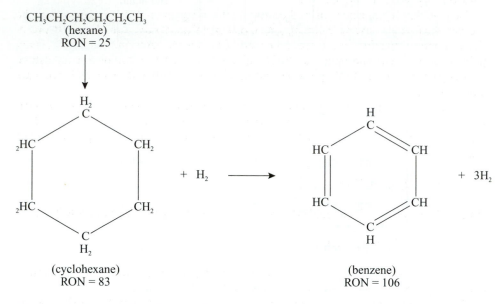

Figure 14.8 The catalytic reforming of hexane.

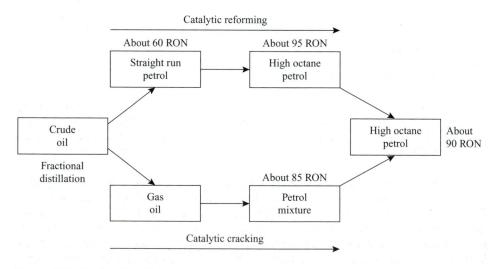

Figure 14.9 Crude oil to high octane petrol.

Pollutants and the petrol engine

The petrol engine has been responsible for substantial pollution of the atmosphere and consequently soil and surface waters.

For complete combustion of hydrocarbons, the right amount of dioxygen must be present, the fuel/air mixture thoroughly mixed, and sufficient time available for the reaction to go to completion before cooling occurs. The exhaust gases contain three primary pollutants, NO_x, CO and unburned hydrocarbons 'HC' due to incomplete combustion. This is particularly so at idling speeds and deceleration. Carbon dioxide and steam are also emitted.

The dinitrogen in air would normally only react with dioxygen at high temperatures. Such temperatures exist in the internal combustion engine. Nitrogen(II) oxide is thus formed in the high pressure, high temperature region behind the expanding flame front, e.g.

$$N_2(g) + O_2(g) = 2NO(g)$$

When NO is exhausted and cools down it reacts with dioxygen to produce $NO_2(g)$,

$$2NO(g) + O_2(g) = 2NO_2(g)$$

As the piston moves down the cylinder during the expansion/power stroke, the temperature and pressure start to fall, which stops the formation of NO.

There is also nitrogen present in some of the compounds that make up the fuel, so NO_x will also be formed by the oxidation of that nitrogen. In addition, cyanide and ammonia will be formed.

The correct air/fuel ratio (A/F ratio), i.e. the mass of air used to mass of fuel used, is vital if the formation of toxic CO is to be minimised. Because air contains about 20 per cent dioxygen, in order to ensure there is enough dioxygen to stoichiometrically oxidise the fuel then the air:fuel ratio has to be about 14.7:1. If the air/fuel ratio is rich in fuel, then CO as well as CO_2 will form at high temperatures. In addition, CO is formed by the thermal dissociation of CO_2, i.e.

$$CO_2(g) \rightleftharpoons CO(g) + 1/2\ O_2(g)$$

Hence, CO is always present in about 1–2 per cent by volume inside the cylinder. As the temperature falls, the equilibrium in the above reaction should move to the left and CO_2 should reform. However, the temperature becomes so low that the reverse reaction is kinetically too slow. Carbon monoxide is therefore always found as part of the exhaust gases.

Unburned hydrocarbons consist of those found in the fuel and others formed by reactions occurring in the cylinder. Any fuel that is burned at high temperatures should be converted to CO at least. If CO is present in exhaust gases, then parts of the fuel cannot have been in contact with the hot flame, i.e. some of the hydrocarbons must be 'hidden' in the corners of the cylinder and trapped in the piston rings and other places, particularly if the engine is worn. As the pressure drops after ignition and the power expansion, then these trapped gasses become available for oxidation, but it is by then too late. Hence, unburned hydrocarbons will be exhausted to the atmosphere.

Diesel engines are more efficient and powerful than petrol engines because of their higher compression ratios and lean air to fuel ratio. Diesel and air are not pre-mixed before being sprayed into the cylinders, neither is a spark plug used. Air is first compressed in the cylinder and becomes very hot. Diesel fuel is then sprayed into the air causing an explosion to take place. Because of the nature of these explosions, noise emission is very high. However, because of the overall fuel-lean A/F ratio, CO and 'HC' emissions are low. The high temperatures and pressures involved ensure that NO emissions are high. Sooty particles are also a serious problem particularly under high engine loads. This soot can also carry adsorbed unburned hydrocarbons and those that are formed during combustion, e.g. PAHs.

In order to control emissions to the atmosphere, the three-way catalytic converter is now fitted to cars, and the sale of leaded petrol in the UK banned.

The three-way catalytic converter

The three-way catalytic converter (CAT) reduces the amounts of the three classes of primary pollutants exhausted to the atmosphere by 90 per cent or more. The converter contains mainly rhodium(Rh) and platinum (Pt), and to a lesser extent palladium (Pd), as the catalysts, together with ceria (CeO_2), alumina (Al_2O_3) plus other metal oxides.

Typically, a converter contains a ceramic monolith (also known as the substrate or brick) of cordierite ($2 Mg.2Al_2O_3.5SiO_2$) which is constructed in a honeycomb. It has porous walls and long parallel channels (about 60 cm long). The design ensures that a high rate of flow of exhaust gases can pass through the monolith. Cordierite is used because of its resistance to both the high temperatures necessary for reactions and the rates of thermal expansion that can occur. To increase the surface area of the mono-lith honeycomb, its internal surfaces are treated with a rough coating of alumina containing oxides such as barium oxide (BaO) and germanium oxide (GeO_2). The former oxide acts as a **structural promoter** and the latter as a **chemical promoter**. The surface area available for the catalysed reactions to take place on is very large. Platinum, palladium and rhodium amount to 1–2 per cent by mass of the alumina coating applied and are widely dispersed through the coating. Currently, in the UK, converters containing between 1 and 2 g of Pt and Rh are mainly used. Unfortunately, before the catalytic converter starts to operate effectively, the catalysts must reach a temperature of between 250 and 300 °C. Thus a few miles will be travelled from cold starting before the catalysts start to work, and consequently there will be some tail-pipe emissions. All CATs are therefore fitted as close to the engine manifold as is possible to cut down their warm-up time.

A catalytic converter first removes any CO that is produced as a result of fuel-rich conditions with insufficient dioxygen. Platinum and rhodium are the catalysts used to accelerate the oxidation of CO, and ceria the chemical promoter used in this reaction,

$$CO(g) + H_2O(g) = CO_2(g) + H_2(g)$$

Then,

$$2H_2(g) + O_2(g) = 2H_2O(g)$$

Second, it removes unburned hydrocarbons 'HC' using Pt and Rh as the catalysts,

$$'HC' + O_2(g) = CO_2(g) + H_2O(g)$$

Third, a converter removes NO using rhodium as the catalyst via the reaction,

$$2NO(g) + 2CO(g) = N_2(g) + 2CO_2(g)$$

The three-way catalytic converter can be easily damaged by overheating or can be poisoned by lead (in the case of Pd) and by phosphorus or sulphur (in the case of platinum).

The origin of nuclear energy

The nucleus of a deuterium atom is composed of one proton and one neutron.

The sum of the rest mass of the proton and the neutron is 3.34755×10^{-27} kg. The rest mass of the deuterium nucleus is 3.34358×10^{-27} kg. Thus there is a mass loss of 0.00397×10^{-27} kg when the deuterium nucleus is made by adding together its nucleons,

$$\underbrace{{}_{0}^{1}R + {}_{0}^{1}P}_{3.34755 \times 10^{-27} \text{ kg}} \longrightarrow \underset{3.34358 \times 10^{-27} \text{ kg}}{{}_{1}^{2}H}$$

This mass loss has been converted to energy and reflects an increase in stability of the formed nucleus. The Einstein equation relating mass (m) and energy (ε) is given by,

$$\varepsilon = mc^2$$

where c is the velocity of light ($c = 3.0 \times 10^8$ m s^{-1}). In the case of the above mass,

$$\begin{aligned}
\varepsilon &= (0.00397 \times 10^{-27} \text{ kg}) \times (3.0 \times 10^8 \text{ m s}^{-1})^2 \\
&= 3.57 \times 10^{-13} \text{ J} \\
&= 2.23 \times 10^6 \text{ eV} \\
&= 2.23 \text{ MeV}
\end{aligned}$$

The total energy lost is called the **binding energy**, and is the total energy released as a consequence of a small loss in mass when the required number of nucleons come together to form a nucleus. Alternatively, it is the total energy required to separate the nucleons when the atom is caused to disintegrate. Because different atoms have different numbers of protons and neutrons, it is more convenient to determine the **average binding energy** per nucleon by dividing the binding energy by the total number of nucleons. Figure 14.10 shows the average binding energy per nucleon plotted against mass number. From this graph, the average binding energy rises steeply to a maximum at mass number 56 (Fe) and declines gradually. It might therefore be expected that the more nucleons that can be fused together will cause a rise in nuclear stability up to $A=56$, and for nuclei with $A>56$ their fission might lead to nuclear stability.

Because of the problems with the technology of building a working efficient nuclear fusion power station, nuclear fusion is still in its 'scientific investigation/experimentation' phase.

Nuclear fission is the process from which nuclear energy is currently extracted. A heavy radioactive nucleus is caused to absorb a neutron which makes that nucleus even more unstable and thereby liable to undergo nuclear fission. The isotope ${}_{92}^{235}$U is the one in major use. Here a neutron from a suitable source is absorbed by the uranium-235 which then undergoes fission to produce **fission products**, e.g.

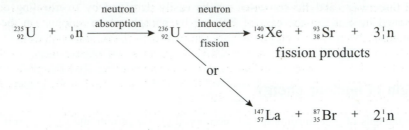

The uranium-235 is called a **fissile isotope** because the process of neutron absorption occurs irrespective of that neutron's kinetic energy. Uranium-236 can break down or undergo fission to several different combinations of isotopes. In the above sequence just two possible disintegrations are shown.

When the unstable isotope of uranium-236 breaks down, binding energy is released. The amount of binding energy produced is approximately 200 MeV per fission, though it does depend upon the relative stabilities of the resulting fission products. The average number of neutrons produced per fission is approximately 2.5. The fission products themselves are radioactive and continue to decay primarily by the emission of β-radiation, γ-rays and other sub-atomic particles.

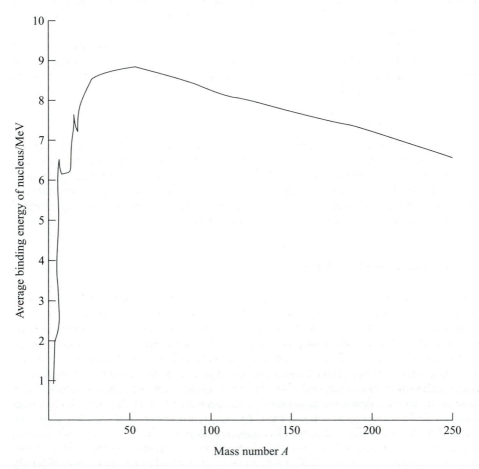

Figure 14.10 A schematic diagram of the average binding energy of nuclei vs their mass number.

Uranium-235 is a fissile isotope, but there are those nuclei that require a neutron to have a minimal amount of kinetic energy, the **threshold energy**. Thorium-232 and uranium-238 are such isotopes and are referred to as **fissionable isotopes**.

Of the 200 MeV released, about 170 MeV appear as the kinetic energy of the fission products. Kinetic energy is reflected in the temperature of a body and so the fission products become very hot. It is this source of heat that is used in the nuclear reactor to boil water, produce steam and then electricity.

About 25 MeV of the energy is associated with the emission of the β-radiation, γ-rays and other sub-atomic particles. The remaining 5 MeV is the kinetic energy of the neutrons produced, i.e. about 2 MeV per neutron.

Natural uranium is composed of 99.28 per cent uranium-238 and 0.72 per cent uranium-235. How each isotope interacts with neutrons depends upon the kinetic energy of those neutrons. If their kinetic energy is of the order of 1 MeV, then they are described as being **fast neutrons**. Neutrons with this energy are much more likely to be scattered by uranium-238 atoms than be absorbed by uranium-238, thus making them unavailable to the uranium-235. Hence, in order to ensure absorption the neutrons must either be slowed down or the amount of uranium-235 available increased. The latter is referred to as **enrichment**. Neutrons with kinetic energy of the order of 0.025 eV are termed **thermal neutrons** and neutrons as slow as this have a greater chance of being absorbed by a uranium-235 nucleus. Thus, slowing down the neutrons and enriching the uranium to between 2 and 3 per cent of uranium-235 will give conditions conducive to neutron capture and fission.

How does the energy produced in the burning of carbon in a chemical reaction compare with the binding energy released when an atom undergoes fission? In the reaction below, where one atom of carbon is reacted with one molecule of dioxygen, 6.54×10^{-19} J of energy is liberated,

$$\text{i.e. } C(s) + O_2(g) = CO_2(g) - 6.54 \times 10^{-9} \text{ J}$$

In nuclear reactions, energy is measured in **electronvolts (eV)** and is the energy acquired by an electron when it is accelerated through a potential difference of one volt. One electronvolt is equivalent to 1.602×10^{-19} J. Hence, in terms of eV the energy liberated in burning one atom of carbon is equivalent to 4.08 eV. Nuclear energy is measured in **mega** electrovolts or millions of eV. Thus, on a per atom basis, nuclear energy obtained from an atom undergoing fission is very much greater.

Chain reaction and criticality

Once the fission of uranium-235 is started, then the two or three neutrons that form part of the fission products will be used to sustain a chain reaction, i.e. these neutrons are used to create new fissions which then carry on to cause the next and so on (Figure 14.11). If 10^{19} fission chains take place in a reactor at any time, then this would produce 1,000 MW of electricity. Fuel in a reactor is contained in fuel elements. The neutrons will leave a fuel element, enter another element and even leave the nuclear reactor altogether unless they are controlled. The fuel in a reactor must first be surrounded by a material that slows down the neutrons that are produced within the fuel. Such a material is called a **moderator**. When a small ball (A) hits another small ball (B) of similar size and mass, both balls will move after collision. Ball A will have less kinetic energy, and therefore velocity, whilst ball B will have gained kinetic energy. What is one ball's loss is another's gain! If ball B was massive in size and mass, ball A would hardly lose any energy upon collision and would continue almost at its original velocity. This

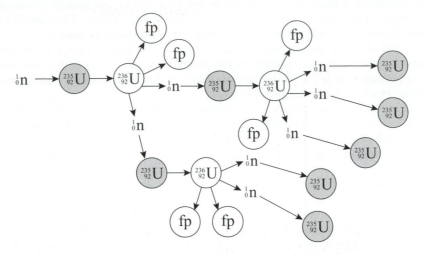

Figure 14.11 A schematic diagram showing the start of a chain reaction. (*Notes*: Two or three neutrons are produced per fission. A fission is represented by fp. The formation of fission products is accompanied by emission of γ-rays.)

principle applies to the collisions undergone by neutrons. Thus, a moderator, if it is to slow down a neutron, must be composed of light atoms. Moderators therefore contain hydrogen, deuterium and carbon (graphite) atoms.

In order to have a successfully operating nuclear reactor, the amount of fissile material must be carefully selected; the diameter of the fuel rods must be of the right length, the size of the spacing between fuel rods correct, and the right moderator chosen. Reactors are designed with all these factors in mind. A chain reaction, which is self-sustaining and produces a steady, controllable amount of energy, is called a **critical system**. Neutrons can leak out of a nuclear reactor if it is too small and the chain reaction will be broken – such a system is termed a **sub-critical system**. If the reactor is too large, few neutrons can escape and therefore more are available for fission. If too many are present, then the system may become **super-critical** and the chain reaction may become uncontrollable. Most reactors are operated under critical conditions or just slightly super-critical. Boron is a good **neutron absorber** and is used in **control rods** to reduce the number of neutrons. These rods can make a reactor become sub-critical.

A reactor that uses neutrons with very low kinetic energy is termed a **thermal reactor** and has a moderator present, whilst reactors that use fast neutrons are called **fast reactors** and have no moderator present.

The components of a thermal nuclear reactor

Figure 14.12 shows the structure of a thermal nuclear reactor. The **reactor core** contains the nuclear fuel. This fuel contains the fissile isotope ^{235}U (sometimes ^{239}Pu) mixed with ^{238}U in the form of solid UO_2. The reactor core consists of cylindrical rods or pins (or stacks of pellets) of this fuel, encapsulated in a **cladding** designed to contain the fission products. The cladding is a thin-walled metal container often made of a zirconium alloy because it must be resistant to chemical attack. The fuel rods are between 5 and 25 mm in diameter and a metre in length. A **fuel element** consists of a number of fuel pins, and a

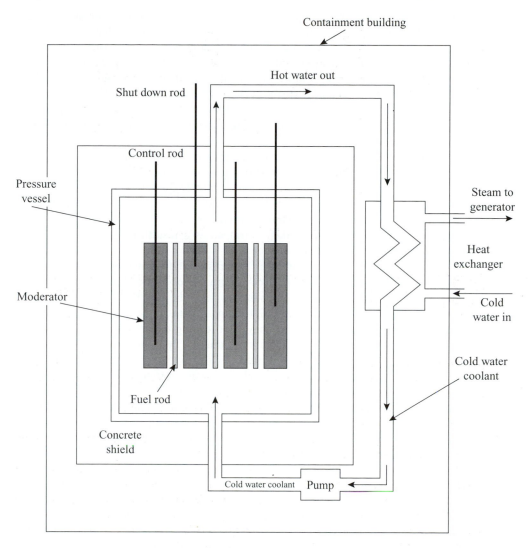

Containment building

Hot water out

Shut down rod

Control rod

Pressure
vessel

Steam to
generator

Heat
exchanger

Moderator

Cold
water in

Cold water
coolant

Fuel rod

Concrete
shield

Cold water coolant   Pump

Figure 14.12 A schematic diagram showing the components of a thermal reactor.

reactor core can contain several hundred fuel elements. The fuel elements will be used for 3–6 years before they are replaced. The total mass of fuel used to produce 1,000 MW of electricity is some 100 tonnes and the reactor operates with an efficiency of about 32 per cent. At the end of their productive lives, the fuel elements contain **spent fuel** which is equivalent to the ash produced in a conventional power station – 1.1 tonne of fissile material will produce 1 tonne of waste per year.

The heat produced by fission is carried away by a **coolant**. The coolant is in contact with the cladding and flows through the fuel elements. The coolant can be either a gas or a liquid. The majority of gas-cooled reactors use carbon dioxide(CO_2) or helium(He), whilst liquid-cooled ones use water(H_2O) or heavy water(D_2O). Light water will absorb neutrons and therefore the fuel needs to be enriched. This is unnecessary if heavy water is used, e.g. in Canada's CANDU reactors.

In gas-cooled reactors, the hot gas is pumped directly to the steam generator. In light water-cooled reactors the water may be allowed to boil and the steam fed directly to the turbines. In the latter case it is termed a **direct steam cycle**. Where such a system is used, there are often coolant leaks and therefore heavy water is too expensive to use. If the water is not boiled but is allowed to heat a *separate water* supply, via a heat exchanger, then this is an **indirect steam cycle**. In this case, D_2O is usually used. After heat is extracted from the coolant it is pumped back to the reactor via the **primary cooling circuit**. This circuit has several backup circuits as an insurance against pump failure.

The higher the temperature of the steam entering the turbine, the greater is the efficiency of electricity generation. Steam at 100 °C has a very low efficiency, whilst steam at 350 °C has an acceptable efficiency of approximately 30 per cent. Hence, the pressure on the water in a water-cooled reactor is raised from about 1 atmosphere to about 218 atmospheres so that its boiling point is raised from 100 °C to 374 °C. The problem with this is that, if the pressure suddenly dropped as a result of a leak, then this hot liquid water would boil with great violence.

The heat produced in a thermal reactor is a result of fission product heating. It is always produced once the radioactive decay begins, even if the reactor is shut down. The amount of heat produced depends upon how long the fuel has been in the reactor – towards the end of its life it is very high compared to start-up. Thus heat must be removed at all times otherwise the fuel may melt leading to **meltdown**.

In thermal reactors, the fuel elements are surrounded by a **moderator**. The purpose of the moderator is to slow down the neutrons to low kinetic energies, when the probability of fission in the fissile isotope is highest. Common moderators are light water, heavy water and graphite. If cheap light water is used as a coolant, it also acts as the moderator because it absorbs neutrons. In such a case, the uranium fuel must be enriched to between 1.5 and 4.0 per cent to give a critical system.

The power produced in a nuclear reactor is regulated using **control rods**, and the chain reaction is stopped using **shutdown rods**. The former are raised and lowered in the core, and regulated minute by minute to ensure the sustainability and control of the chain reaction. Upon demand, all of the shutdown rods can be fully inserted into or out of the core, and can thus stop the chain reaction from occurring. Both sets of rods are made from boron, a good absorber of neutrons.

A **pressure vessel** made of steel surrounds the reactor core. This is then surrounded by a concrete **biological shield**, typically 2–3 metres thick. The latter protects the operating staff from any neutrons and gamma rays that may escape from the core. The **reactor containment building** surrounds the reactor and provides a barrier that prevents the escape of fission products to the outside world.

During normal reactor operations, one hazard may be the accidental release of gaseous radioactive isotopes. For example, tritium is formed as a fission product or via interaction of neutrons with hydrogen in the coolant/moderator water and, because it is a light gas, is particularly difficult to contain. Krypton-85 is also a fission product that can escape through pinholes which may develop in the fuel cladding.

Tritium atoms may also be exchanged for hydrogen atoms in the water molecule and thus escape during water losses. In addition, because some 0.037 per cent of oxygen atoms in water are the isotope ^{17}O, these can interact with neutrons to produce helium and carbon-14. This latter isotope may find its way out of a reactor in the form of CO_2 gas or, if it reacts with water, as simple hydrocarbons.

Fortunately, most of these sources of radiation contribute very little to the collective dose of workers and are not hazardous to the general public.

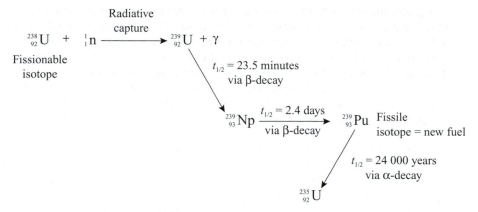

Figure 14.13 The production of a fissile isotope.

The production of fissile isotopes

If uranium-238 is subjected to fast neutrons, then a process called **radiative capture** can occur. The new isotope of uranium then undergoes a series of radioactive decays until the fissile isotope plutonium-239 is produced (Figure 14.13). If the reactor produces more new fissile isotopes than the original isotope used to produce them, then the reactor is referred to as a **fast breeder reactor**. Isotopes such as uranium-238 that can be transformed into fissile material are termed **fertile isotopes.**

Radiation and risk assessment – Hiroshima and Nagasaki/ ankylosing spondylitis

On 6 August 1945 at 8.16 a.m. the first atomic bomb was dropped on the Japanese city of Hiroshima. A second bomb was dropped on Nagasaki on 9 August at 7.55 a.m. These bombs finally ended the Second World War in September 1945. In Hiroshima between 75,000 and 100,000 are believed to have died in the fireball and accompanying blast with some 68,000 injured. At Nagasaki, the Japanese estimated that 87,000 people were killed. Not only were the people of these two cities exposed to fire and blast but also to an intense burst of radiation (mainly γ-rays). Within five years of the dropping of the bombs about half of the population of both cities had died of the injuries sustained and disease caused from the nuclear explosions.

Cancer is one of the most important delayed effects of exposure to radiation. One of the estimates of the risk associated with cancers caused by radiation is based on the survivors of these atomic bombs. Between 1950 and 1982 at Hiroshima and Nagasaki, over 100,000 people were monitored. These included some 54,000 who received a significantly large dose of radiation. During this time 3,832 of the 54,000 people died of cancer. In a normal population of Japanese people of the same size, 3,601 cancer-related deaths were reported. Thus, there were 231 excess deaths in the 'bomb' group. Cancer is a common disease and approximately 7 per cent mortality rate would be expected without the extra exposure to radiation. The excess deaths are attributed to radiation from the bombs. There is no evidence, as yet, that suggests life expectancy has been shortened in the Hiroshima/Nagasaki victims other than by deaths caused by cancer. There has also been no statistically significant increase in genetic effects observed amongst the children born to the atomic bomb survivors.

Ankylosing spondylitis is a non-cancerous disease of the spine and was treated by irradiation from X-rays for many years. The treatment was partially successful in slowing down the progression of the disease, but a number of patients developed cancer. Between 1935 and 1954, 14,106 patients were treated and subsequently their health monitored to establish the level of risk associated with cancer. It was found that 18 patients developed leukaemia and 70 others cancers of the heavily radiated parts of their bodies. This total of 88 cancers was more than would be expected for a normal group. However, when taken over along period of time, the number of extra cancers diminishes until it is no different from a normal population. This indicates that early cancers did develop as a result of the irradiation but there was no overall increase in cancers with time.

Evidence such as the kind described above would indicate a degree of uncertainty concerning the development of cancer and exposure to radiation. It is also interesting to note that people who are in the nuclear power industry are amongst the healthiest in the UK. This is because they are so closely monitored that it results in early diagnosis of not just cancer but of other diseases.

The biological effects of radiation and its units

All types of radiation produced in a nuclear reactor (α-particles, β-particles, γ-rays and neutrons) have the ability to damage biological systems because they cause ionisation within living cells. Humans are composed of eukaryotic cells, which have nuclei containing chromosomes made of DNA. Sections of DNA are the genes that carry inherited information. When ionisation occurs, it is possible that some of the chemical bonds in DNA, or some other important molecule, may be broken and lead to unwanted chemical reactions. Cells are composed mostly of water, which can be ionised to form highly reactive free radicals,

$$H_2O = H_2O_+^{\cdot} + e^-$$

$$H_2O_+^{\cdot} = H^+ + \cdot OH$$

$$H^+ + e^- = \cdot H$$

$$\cdot OH + \cdot OH = H_2O_2$$

Notice in the last equation, that the strong oxidising agent hydrogen peroxide is produced.

When biological cells reproduce they do so in one of two ways. **Mitosis** is simple asexual cell division which produces genetically identical cells. Such cells are termed somatic cells. One cell produces two cells containing the same number of chromosomes of the same type. Such cell division enables the repair of damaged tissues and growth of a living organism.

In **meiosis**, specialised sex cells called gametes are involved (the ova and sperm cells). Each of these cells contains one of a pair of 23 chromosomes When an ovum and a sperm cell fuse to form a zygote in the process of fertilisation, the chromosomes pair to give a total of 23 pairs or 46 chromosomes. These chromosomes carry all the genetic information that dictates what the new person becomes, i.e. colour of eyes, sex, longevity, etc.

Radiation causes the worst damage to the nuclei of cells. In the case of somatic cells, the damage can prevent the cells from dividing and thus stop repair of tissues. In addition, DNA mutation can result in the uncontrolled division of somatic cells, leading to the formation of cancers. If gametes are damaged, this can lead to somatic cell damage in children, which can lead to birth deformities and other serious problems.

How much damage to humans is caused by radiation depends upon how high the dosage received, how frequent the exposure and the type of radiation. The most ionising radiation and therefore the most damaging are α-radiation and neutrons. The former positively charged particle can transfer its energy to a cell within a very short distance. Neutrons are similar in this respect because they produce charged particles within a cell by interaction with the nuclei of the atoms making up that cell. These particles are said to have a high linear energy transfer (LET) and produce large numbers of ionised particles.

Negatively charged β-particles and the electrons produced by γ-rays (and X-rays) have longer ranges in tissues than γ-particles. However, they have a much lower LET and therefore any damage caused to cells can be more easily repaired.

One major difficulty in interpreting how much radiation a person might receive lies in the wide variety of terms used to describe **dosage**. Indeed, dosage and **activity** are often misrepresented in the media when reporting anything to do with nuclear energy or weapons. The term dosage refers to the amount of radiation received – the term activity refers to how much radiation and what type is being emitted by a radioactive source.

The unit of dose is the grey (Gy). One grey is equivalent to 1 joule of energy being absorbed by 1 kg of tissue. Thus if 1 kg of tissue receives 100 J of energy then it has received an **absorbed dose** of 100 Gy.

Different types of radiation have differing abilities to cause cellular damage and this must be taken into account. The type of radiation is reflected in the **Relative Biological Effectiveness (RBE)**. RBE is a ratio in which a particular radiation's effects on cells is compared with a reference radiation, i.e. it is defined by,

$$\text{RBE} = \frac{D_0 \text{ for the reference radiation}}{D_0 \text{ for the radiation being investigated}}$$

where D_0 is the absorbed dose that kills 63 per cent of the cells under test.

For example, X-rays of 100 keV are used as a reference radiation which results in $D_0 = 2.8$ Gy and are given an arbitrary RBE value of 1. Thus for α-particles of 5 MeV and $D_0 = 0.2$ Gy, then the RBE of these latter particles is given by,

$$\text{RBE} = 2.8 \text{ Gy}/0.2 \text{ Gy} = 14.0$$

The radiation necessary to kill 63 per cent of the test cells depends on the type of cell. To take this into account a **Quality Factor (QF)** is introduced for each type of radiation (Table 14.5). This allows the biological dangers associated with a particular type of radiation to be expressed as **dose equivalent**, i.e.

$$\text{dose equivalent} = \text{absorbed dose} \times \text{QF}$$

Dose equivalent is measured in **sieverts (Sv)**.

Thus if tissues receive an absorbed dose of 50 Gy from fast neutrons, then the dose equivalent will be given by,

$$\text{dose equivalent} = 50 \text{ Gy} \times 10 \text{ Sv Gy}^{-1}$$
$$= 100 \text{ Sv}$$

Not only has the type of radiation to be taken into account but also the fact that different organs respond differently to radiation. The **Effective Dose Equivalent** is used in this context. A quality factor is used, which weighs

Table 14.5 *Average quality factors for different types of radiation*

Radiation type	QF/Sv Gy^{-1}
β-rays, γ-rays, X-rays	1
Thermal neutrons	2–3
Fast neutrons	10
α-particles	20

the dose equivalent received by each organ and therefore takes into account organ sensitivity. The effective dose equivalent is also measured in sieverts, and is often simply referred to as the dose.

When dealing with a group of people, the **collective dose equivalent** is often used. This is the sum of the effective dose equivalents of the individual members of the group, and is measured in **man-sieverts (manSv)**. If the collective dose is over a period of time, then this is also specified, e.g. man Sv yr^{-1}.

Other dose units in use are the **roentgen**, the **rep** and the **rad**. The roentgen is defined in terms of the ionization produced in air by X-rays and γ-rays only. One roentgen (1 R) produces 3×10^9 coulombs of electrical charge in 1 cm^3 of dry air at one atmosphere pressure and a temperature of 0 °C. This unit is used to specify the amount of radiation to which a sample is exposed. A dose of 1 R corresponds to (i) 83.8×10^{-7} J per gram of air, at nearly all X-ray and γ-ray energies, or (ii) about 93–5×10^{-7} J per gram of living tissue. The unit rep stands for **roentgen equivalent physical**. If 1 kg of soft tissue absorbs 9.3×10^{-3} J, then this is equivalent to 1 rep. It is approximately equal to 1 R of about 200 keV X-rays in soft tissue.

The **radiation absorbed dose** (rad) is a unit that expresses the amount of energy dissipated in a unit mass of matter. It is applicable to all types of radiation. An absorbed dose of 1 rad dissipates 1×10^{-2} J kg^{-1} of tissue. One rad thus equals 0.01 Sv; or 1 Gy $= 100$ rad $= 1$ J kg^{-1}.

The degree of activity a radioactive source has is expressed in terms of the **Becquerel (Bq)**, which is one disintegration per second.

Radiation and risk estimates

There are two levels of radiation dose (effective dose equivalent) that are distinguished. At 2 Sv or greater, the outcomes associated with radiation are fairly predictable and are referred to as **non-stochastic effects**. Below this dose, outcomes are unpredictable and are called **stochastic effects**. Within the latter there are two groups. The first are somatic effects, which involve genetic mutation leading to possible cancer. The second are **genetic effects**, where the results appear in the offspring. What is an acceptable safe level of radiation is linked to what is an acceptable risk.

The risks associated with radiation must be placed in the context of what humans receive from their environment (Table 14.6). The average total radiation exposure is about 2,150 μS or 2.2 mSv.

An instantaneous dose can lead to observable non-stochastic effects and a level of risk associated with stochastic effects. A dose greater than 20 Sv delivered to the whole or a substantial part of the body for a few minutes will cause irreversible damage to the central nervous system and radiation sickness. Almost invariably such a dose will cause death. A dose of 10 Sv will also result in unlikely survival. At about 5 Sv the gastrointestinal form of radiation sickness manifests itself and the lenses of the eyes may be damaged. If untreated, the victim has a 50–50 chance of survival. The damage caused by the same dose given over a year may well be repaired by the body's immune system. A dose of 2 Sv or more will cause skin burns and the bone marrow form of radiation sickness. Between 0.1 and 1 Sv there is a temporary low blood count and temporary sterility. The chance of early death from a single exposure to radiation of the whole body rises from zero at 1 Sv to 100 per cent at 10 Sv and above.

In the case of stochastic results, cancer is the main delayed effect. The International Commission on Radiological Protection has concluded that, at an additional dose of 0.1 Sv above the background radiation level, the risk of a fatal cancer is 1 in 800 which falls to 1 in 80,000 at a dose of 0.001 Sv (1 mSv). The latter figure means that, if a

Table 14.6 *Sources of radiation for the average individual in the UK*

Source	Dose µS	Percentage of total
Inhaled, i.e. radon and thoron	800	37
Rocks and soil	400	19
Food and drink, e.g. potassium-40	370	17
Cosmic rays	300	14
Medical, i.e. X-rays and radioisotopes	250	12
Fallout (nuclear testing residues)	10	0.4
Occupational, e.g. dentistry, nuclear power industry, medicine, etc.	8	0.4
Discharges from nuclear industry	1	<0.1
Other, e.g. cosmic radiation in aeroplanes, coal burning	10	0.4

population of one million people were exposed to an extra 1 mSv the incidence of cancer deaths would rise from 200,000 to 200,012, which would be within the normal variation in cancer frequency. Indeed, the mortality rates of workers in the radiation industry are indistinguishable from those of other groups of workers. In addition, no radiation-induced hereditary damage has ever been observed in humans. Where studies have been carried out on populations exposed to chronic low levels of radiation above normal background levels, e.g. Denver, Colorado, background above 10 mSv per year, they have shown no adverse biological effects.

Living organisms on Earth have been subject to radiation from the beginning. It could be argued that this natural radiation has contributed to the development of life on this planet. Certainly, living organisms have the ability to readily recover from irradiation.

Disposal of spent fuel

The biggest potential for environmental impact originates in the spent fuel which contains uranium-238, unused uranium-235, fission products and plutonium. The spent fuel is very much more radioactive than new fuel, mainly due to the fission products. In the UK spent fuel may undergo reprocessing. This process separates uranium and plutonium which can be recycled back into nuclear fuel production. Throughout the reprocessing sequence, liquid, gaseous and solid low-level radioactive wastes are produced. The gases are discharged into the atmosphere and the solids are buried in concrete-lined trenches. The liquid wastes are pumped into the Irish Sea after much of the radioactive elements present have been removed by chemical treatment. The remaining waste is then stored as solid waste.

High-level radioactive solid waste has to be stored in a secure environment. This waste requires hundreds to thousands of years to decay to a safe level and must be cooled for decades. It is suggested that the best method of long-term storage/disposal is by burying it deep underground in specially selected geological areas. The solid waste would be vitrified, i.e. entombed in glass, and placed at depths of between 300 and 1,000 m. However, the difficulty in selecting a suitable site together with public opposition ('not in my backyard') has prevented final decision taking from occurring. Wherever the waste is stored it must not be capable of being moved by flowing water;

the area must be geologically stable, e.g. not subject to movement; the rocks must be thermally stable; any pathways that enable water to move to the surface must be long; and ground water flow must be very slow.

Nuclear accidents

The main opposition to nuclear power is the fear of nuclear accidents and what to do with the radioactive wastes that are left over after a fuel rod is spent or a power station decommissioned. The fear arises from the perceived and real dangerous effects of nuclear radiation. Although there have been many nuclear accidents, perhaps the most infamous has been the Chernobyl event in the Ukraine on 26 April 1986.

The reactor involved was a Russian type called the RBMK-reactor. There were four such reactors at Chernobyl, each producing some 1,000 kW of electricity. There are currently 14 of these reactors still in use in the regions of the former USSR. This reactor used enriched UO_2 fuel rods encased in a zirconium alloy, with a graphite moderator. The coolant is light water passed through the reactor core in zirconium alloy pressure tubes. The pressurised water enters the reactor at the bottom of the core at 270 °C and leaves as a mixture of water and steam at 285 °C at the top of the core. The steam is then used directly to drive a turbine. Water is recycled through the core.

It is the combination of graphite and the light water coolant that maintains this kind of reactor at a critical level. Water is a better coolant and neutron absorber than is steam. Thus, if a bubble of steam is formed near a fuel rod, a void occurs and the number of neutrons that can be absorbed by the coolant decreases. If the reactor is operating under critical conditions, it is possible that the occurrence of such steam voids can cause it to become super-critical. Whether or not this happens depends also upon the amount of fuel present, whether it is enriched and how the fuel rods are arranged within the core. For example, if the amount of moderator decreases, then more neutrons could be captured by the uranium-238 present, thus counteracting the effects of decreased neutron absorption by the light water. If the effect of a void in a reactor is to increase the rate of fission, then it is referred to as having a positive void coefficient. This is the case in RMBK reactors.

Some fission products decay to form other products that can be a threat to the reactor itself. For example, tellurium-135 is a fission product that decays to form xenon-135, which readily absorbs neutrons to produce the stable isotope xenon-136. How much xenon-135 is produced depends upon the power level that the reactor is working at – the more power produced, then the more xenon-135 is produced! It must also be remembered that, when a reactor is shut down, the fission products continue to decay. If the power level is reduced, the amount of xenon-135 present actually rises initially to a maximum value and then gradually decreases. This is because there are fewer neutrons in the core and therefore less chance of a xenon-135 nucleus becoming xenon-136. The rise occurs about 12 hours after power decrease and is accompanied by a decrease in the number of neutrons causing sub-criticality (termed xenon poisoning). To make the reactor critical again, the control rods must be raised.

So what caused the accident?

Reactor Four was to be shut down for routine maintenance on 25 April 1986. It was therefore decided that a test would be run to see, in the event of a shutdown, if enough electricity would still be continued to be produced to operate the emergency equipment and core cooling pumps until the diesel-powered generators came on line. The test also required the core cooling system to be switched off. It was known that the RMBK reactor had a large positive void coefficient if operated at less than 20 per cent full power. Hence the test was to be performed at 25 per cent full power. As the power was reduced, the xenon-135 level increased and therefore the control rods were withdrawn further. An order was received to continue to supply electricity at 50 per cent full power. The xenon-135 level continued to rise and therefore the control rods were withdrawn more. When the testing was restarted the control rods were lowered and the power fell uncontrollably to about 7 per cent of full power. Therefore positive voids and even more xenon-135 were formed. At this point automatic shutdown should have occurred, but was prevented by operator intervention. The steam supply to the turbine was also shut off. There were only seven control rods in the core when there should have been a minimum of 30. Indeed, the seven inserted rods were themselves too far out of the core – a violation of operating instructions.

The overall result was an increase in the amount of steam generated and therefore a positive void coefficient leading to super-criticality. The control rods could not be inserted fast enough to restore criticality and therefore the power level rose to 100 times the maximum design value of the RMBK reactor. Most of the core therefore melted and explosion resulted.

The explosion was not a nuclear one, but was caused by chemical reaction between the molten fuel and cooling water and also between the zirconium alloy cladding and coolant to produce H_2 gas. The graphite also caught fire, producing CO and CO_2. Reactions of the following kinds thus occurred,

$$Zr(l) + 2H_2O(g) = ZrO_2(s) + 2H_2(g)$$

$$2H_2(g) + O_2(g) = 2H_2O(g)$$

$$C(s) + O_2(g) = CO_2(g)$$

$$2C(s) + O_2(g) = 2CO(g)$$

Unfortunately, the reactor building was not of the containment type and blew apart, releasing into the atmosphere radioactive substances present in the reactor core.

The immediate effect of the accident was that over 30 people were immediately killed. Over 135,000 persons were evacuated from within a 20-mile radius of the reactor site. Lethal dosage was received by a number of small mammals within a 10 km radius of the site, and coniferous trees also perished. By 1989 the local environment had started to recover.

There is evidence that there has been an increase in thyroid gland cancers in children from the Ukraine and an increase in birth deformities. In the period 1981–5, prior to the accident, the average number of such cancers was 4–6 per million. From 1986–97 this had risen to 45 per million. Many people have suffered from anxiety, stress and depression as a result of the accident, which has had a number of social consequences such as a reduced birth rate.

More recently, on 30 September 1999, the worst nuclear accident to occur in Japan happened in a nuclear processing plant at Tokaimura in the Ibaraki Prefecture, some 87 miles north-west of Tokyo. In the plant, on 29 September, three workers started to prepare enriched (18.8 per cent) uranyl nitrate(V), $UO_2(NO_3)_2$, solution by dissolving uranium powder in nitric(V) acid. Instead of performing the reaction in the mixing tank provided, these workers used a 10 dm^3 stainless steel bucket. They then appear to have fed seven batches of uranyl nitrate(V) solution (equivalent to 16.6 kg of uranium) into the precipitation tank using a 5 dm^3 stainless steel bucket and a funnel. The tank was designed to limit the mass to one batch containing 2.4 kg of uranium. Unfortunately, a critical amount of uranium was consequently present and resulted in a flash of blue light and an initial nuclear fission chain reaction. The initial fast fission reaction gradually slowed down over a period of 20 hours.

The result of this accident was that a number of workers were exposed to a high dose of radiation. At least three workers received radiation doses that can lead to death. In addition, 21 workers received lower doses. People working within a 350-metre radius of the accident site were evacuated, whilst people living/working within a 10-kilometre radius were advised to remain within doors for approximately 24 hours. What the ultimate outcomes of this accident will be remain to be seen.

Summary

- Combustion of fossil fuels contributes to the greenhouse effect, acid rain, photochemical smog, chemical toxicity and the deposition of particulate matter.
- Particulate matter emitted during coal burning can contain toxic levels of heavy metals and radioactive isotopes.
- Non-uniform burning of natural gas caused by variations in flame temperatures and incorrect mixing can result in incomplete combustion and the liberation of unburned hydrocarbons, alkenes, aldehydes and carbon monoxide to the atmosphere.
- Tetraethyl lead was added to petrols to reduce engine wear by preventing premature ignition.
- Organic materials were added to petrol to enable the lead additive to be converted to more volatile lead(II) compounds, thus causing pollution of the environment.
- More extensive refining of petroleum and the addition of oxygenates have removed the need for the use of lead additives.
- Exhaust gases from petrol engines can contain three primary pollutants, NO_x, CO and unburned hydrocarbons 'HC'.
- Diesel combustion products are low in CO and 'HC' but NO emissions are high. Sooty particles and PAHs are also a problem.
- The three-way catalytic converter contains the metals Rh, Pt and Pd, which are used as catalysts in the oxidation of CO to CO_2, the combustion of 'HC' and the conversion of NO to N_2.
- Uranium-235 undergoes fission when it absorbs a neutron with a very low kinetic energy, and forms the basis of the fuel in a thermal reactor.
- A fissionable isotope is one that requires a neutron to have a minimum amount of kinetic energy.
- A chain reaction is a series of controlled fissions.
- A critical system is one in which a chain reaction is self-sustaining and produces a steady, controllable amount of energy.
- Radiation causes most damage to the nucleus of a cell.
- Radiation dosage is measured in greys.
- Relative Biological Effectiveness is an indication of the different abilities of different radiations to cause cellular damage. It is measured in sieverts.
- The Effective Dose Equivalent takes into account how different organs respond to different radiations and is measured in sieverts.
- Collective Dose Equivalent applies to groups of receivers and is the sum of the effective dose equivalents of the individual members of a group. This is measured in man-sieverts.
- Radioactive activity refers to the amount of radiation emitted by a source, and is expressed in becquerels.
- Risk associated with radiation is estimated in terms of stochastic and non-stochastic effects. In stochastic effects genetic mutation leads to the formation of cancers, whilst genetic mutations are found in the offspring.

Questions

1. (a) Calculate the energy supplied in 365 solar days by a coal-fired power station delivering 25 GW of electricity to a national grid.

    (b) If the power station converts energy stored in coal with an efficiency of 30 per cent, how much stored energy was there in the original coal?

    (c) If 1 tonne of coal supplies 24×10^9 J of energy, how many tonnes of coal would need to be supplied to the power station per year?

    What would this be equivalent to in terms of barrels of oil?

    (d) Explain *why* a coal-fired power station has such a low energy conversion efficiency.

2. Find out what kinds of energy storage devices have been, or are being developed, and give their respective uses.

    What effects have these devices had on the environment?

3. In no more than 400 words describe the effects that the internal combustion engine has had and continues to have on the environment.

4. Calculate the energy in joules, electron volts and kilowatt-hours released when 1.0 kg of uranium-235 undergoes the following fission process,

    $$^{235}_{92}U + ^{1}_{0}n = ^{135}_{52}Te + ^{100}_{40}Zr + ^{1}_{0}n$$

    Given

atomic mass of uranium-235	3.90283×10^{-25} kg
atomic mass of tellurium-52	2.24030×10^{-25} kg
atomic mass of zirconium-40	1.65920×10^{-25} kg
mass of neutron	1.67495×10^{-27} kg
velocity of light	$c = 3.0 \times 10^8$ m s^{-1}
Avogadro number	$N_A = 6.022 \times 10^{23}$ mol^{-1}

5. You are to prepare a briefing paper on the RBMK nuclear reactor for a group of people whose level of scientific understanding and knowledge corresponds to that of the dual award GCSE in science. Write a suitable description of the RBMK reactor that includes an *explanation* of all terms used, e.g. moderator, coolant, etc. The description should not exceed 500 words.

6. Describe how a thermal nuclear reactor works, and explain the real and potential environmental effects of this type of reactor.

7. The improvement of old technologies and the introduction of new ones have lead to energy conservation, and thus have made a major contribution to environmental and economic sustainability. To what extent do you agree with this statement?

Further reading

Blunden, J. and Reddish, A. (eds) (1996) *Energy, Resources and Environment*. Hodder & Stoughton, London.
A readable book covering a wide variety of topics. The first three chapters deal with the principles of energy, the effects of present policies and energy sustainability.

DTI (2002) *Digest of UK Energy Statistic*. Her Majesty's Stationery Office, London.
A very valuable and comprehensive source of energy statistics. There is also a booklet available, *UK Energy in Brief*, which highlights all the main points. The data is also available at: www2.dti.gov:uk/epa/esi.

Howes, R. and Fainberg, A. (1991) *The Energy Source Book*. American Institute of Physics, New York.

Although a little old now, it still contains some very interesting material. The chapters are written by experts in their own fields and range from fossil fuels, fusion technology, photo-voltaic cells, geothermal energy, etc.

The Open University (1994) S280 Science Matters, Book 3: *Nuclear Power*. Open University, Milton Keynes.

An excellent book that covers the scientific principles very clearly. It describes the production of nuclear energy, the risks associated with it, the disposal of waste, economic principles and even nuclear weapons.

Answers to questions

Chapter 1

1   Element atomic number 32 is Germanium, Ge, of electronic configuration
$$1s^2 2s^2 2p^6 3s^2 3p^6 4s^2 3d^{10} 4p^6$$
Ge is in Group 14 of the Periodic Table.
Element atomic number 56 is Barium, Ba, of electronic configuration
$$1s^2 2s^2 2p^6 3s^2 3p^6 4s^2 3d^{10} 4p^6 5s^2 4d^{10} 5p^6 6s^2$$
Ba is in Group 2 of the Periodic Table.

3   (a) 4.949×10^{-3} mol of I_2 containing 2.981×10^{21} molecules

     (b) 0.48 mol of S_8 containing 2.893×10^{23} molecules

     (c) 0.139 mol of Hg containing 8.377×10^{22} atoms

4   Assuming beer has a density of 1.00 g cm^{-3}, then 0.576 dm^3 has a mass of 576 g.
One pint of beer contains 0.55 mol of ethanol.

5   (a) 5.844 g

     (b) 13.23 g

     (c) 1.403 g (4 significant figures)

     (d) 25.48 g (2 decimal places)

6   (a) 2.5×10^{-2} g in 1 dm^3

     (b) 1.9×10^{-7} g

     (c) 2.2×10^3 μg dm^{-3}

7   pH = 3.40 is equivalent to 3.98×10^{-4} mol dm^{-3} of hydronium ions.
pH = 3.60 is equivalent to 2.51×10^{-4} mol dm^{-3} of hydronium ions.

8   (a) 2.91

     (b) 3.98×10^{-6} mol dm^{-3}

     (c) 3.16×10^{-10} mol dm^{-3}

9   (a) (i) III (ii)V (iii) IV (iv) VI (v) VI (vi) IV

     (b) (i) Fe^{3+} reduced to Fe^{2+}

              S^{2-} oxidised to S

        (ii)   Zn oxidised to Zn^{2+}

              H^+ reduced to H_2

        (iii) MnO_2 reduced to Mn^{2+}

              Cl^- oxidised to Cl_2

        (iv) MnO_4^- reduced to Mn^{2+}

              I^- oxidised to I_2

Chapter 2

1 (a) $+6.95$ kJ mol^{-1} (b) $+107.3$ kJ mol^{-1}.

 The reaction requires a catalyst.

 The formation of ammonia is disfavoured by a rise in temperature.

 An increase in pressure would favour the formation of product.

2 $K^{\ominus} = 0.114$

3 $K_p = 1.78 \times 10^{15}$ and is unitless.

4 Units of the rate constant are dm^3 mol^{-1}

5 Second order in A and zero order with respect to B.

 Overall, it is a second-order reaction.

6 $Cl_2(g) + CHCl_3(g) = HCl(g) + CCl_4(g)$

 Reaction intermediates are $Cl(g)$, $CCl_3(g)$

 $J = k_r[CHCl_3][Cl_2]^{1/2}$

7 $E_a = 103$ kJ mol^{-1}

Chapter 3

1 (a) (i) and (ii); (vii) and (viii); (xi) and (xii)

 (b) (i) and (ii);

 (c) (v) and (vi); (ix) and (x)

 (d) (vii) and (viii)

 (e) (iii) and (iv)

 (f) none

2 Three possible structural formulae may be drawn:

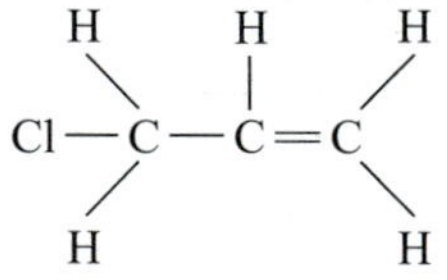

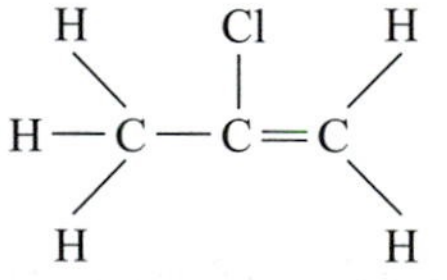

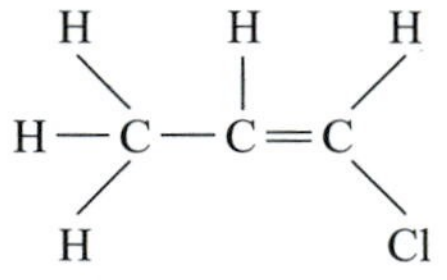

```
      H       H      H
       \      |     /
   Cl — C — C = C
       /            \
      H              H
```

```
      H      Cl      H
       \      |     /
   H — C — C = C
       /            \
      H              H
```

```
      H       H      H
       \      |     /
   H — C — C = C
       /            \
      H              Cl
```

3 The last structure in Question 2 can exhibit geometric (*cis/trans*) isomerism.

4   (a)   Add bromine water which would decolourise.

(b)   X = $C_4H_6O_2$ and contains a $\begin{array}{c}\diagdown \quad \diagup \\ C=C \\ \diagup \quad \diagdown\end{array}$ group and a $\begin{array}{c}-C-O-C \\ \parallel \\ O\end{array}$ group.

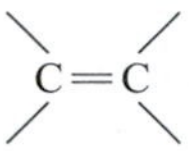

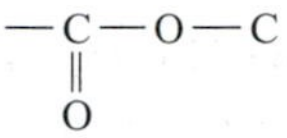

Thus the five possible structures are

$HCOOCH_2CH=CH_2$ **A**

$H_2C=CH\ COOCH_3$ **B**

$HCOOCH=CHCH_3$ **C**

$\underset{\underset{CH_3}{|}}{HCOOH=CH_2}$ **D**

$CH_3\ COOCH=CH_3$ **E**

Compound C above exists as a pair of geometric isomers.

(c)   Compound A gives $HOCH_2CH=CH_2$ **F**

Compound B gives $HOCH_3$ **G**

Compound C gives $HOCH=CHCH_3$ **H**

Compound D gives $\underset{\underset{CH_3}{|}}{HOC=CH_2}$ **I**

Compound E gives $HOCH=CH_2$ **J**

Compounds F and H show positional isomerism.

5   (c) $CH_3CH_2OH + CH_3COOH \longrightarrow CH_3COOCH_2CH_3 + H_2O$

(d)   The carboxylic acid is $CH_3CH_2C\equiv CCH_2CH_2CH_2COOH$

The alcohol is CH_3CH_2OH

6   (a)   The new product B is the *trans* isomer and C the *cis* isomer.

(b) $CH_3CH_2CH_2CHBrCH_2CH_2CH_2COOCH_2CH_3$ **K**

$CH_3CH_2CHBrCH_2CH_2CH_2CH_2COOCH_2CH_3$ **L**

(c)   The chiral centres on compounds K and L are highlighted thus,

$CH_3CH_2CH_2\mathbf{CH}BrCH_2CH_2CH_2COOCH_2CH_3$

$CH_3CH_2\mathbf{CH}BrCH_2CH_2CH_2CH_2COOCH_2CH_3$

(d)   Yes!

Chapter 4

2   (a) $^{58}_{26}Fe + ^{1}_{0}n \longrightarrow ^{59}_{26}Fe$

$^{59}_{26}Fe \longrightarrow ^{59}_{27}Co + ^{\ 0}_{-1}\beta$

(b) $^{241}_{95}Am + ^{4}_{2}He \longrightarrow ^{243}_{97}Bk + 2^{1}_{0}n$

The half-lives of Americium and Berkelium are very short and therefore any that may have been formed during the Earth's accretion and development would have long gone.

5   765 Ma or 7.65×10^8 years.

Chapter 7

3   The lattice energy of $CdBr_2$ is $-2\,405.5$ kJ mol^{-1} determined using a Born–Haber cycle. The calculated value is $-2\,537$ kJ mol^{-1}. There is good agreement between the experimentally determined value and calculated value, suggesting the bonding in $CdBr_2$ is ionic.

4   $\Delta H^0_{sol}\{CdBr_2(s)\} = -71.3\,kJ\,mol^{-1}$

5   (a) Aluminium is protected by a passive layer of aluminium oxide, which prevents reaction with water.

From the given data, using $\Delta G^0 = -nE^0F$, the Gibb's Free Energy change for the reaction shown below is $-240.2\,kJ\,mol^{-1}$.

$$3H_2O(l) + Al(s) \rightleftharpoons Al^{3+}(aq) + 3OH^-(aq) + 3/2H_2(g)$$

The reaction is thus thermodynamically favourable.

(b) One representation of the cell is $Ni(s)\,|\,Ni(aq)\,\|\,Sn^{2+}(aq)\,|\,Sn(s)$

For which $E_{cell} = -0.14\,V - (-0.23\,V) = +0.09\,V$.

The EMF of the cell is thus 0.09 V and the right-hand electrode is the negative pole of the cell.

Chapter 8

3

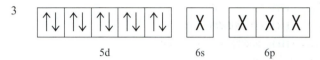

       5d   6s   6p

The mercury(II) ion becomes sp^3-hybridised and a tetrahedral structure results where X = a lone pair of donated electrons from four ammonia molecules.

4   (a) hexaamminecobalt(II) chloride

     (b) potassium hexacyanoferrate(II)

     (c) tetraaquacopper(II) sulphate dihydrate

5   (a) $[Ni(NH_3)_4(H_2O)_2]Cl_2$

     (b) $[Co(NH_3)_5NO_2]Br_2$

     (c) $K[PtCl_4]$

     (d) $K_6[Rh(ox)_3][RhClOH(ox)_2].8H_2O$

Chapter 9

1   Number of moles = 1.373

Number of particles = 8.26×10^{23}

In 1 cm^3 of air there are 8.26×10^{17} particles. This number is called the number density.

2   12 cylinders

3   Concentration of pure methane = $712\,g\,m^{-3}$

Concentration in air sample = $2\,g\,m^{-3}$

4   $13.2\,mg\,m^{-3}$

5   Molar mass of unknown gas = 121

Chapter 10

2　(a)　Number of particles $= 2.331 \times 10^9$

　　(b)　Total surface area $= 1.87 \text{ m}^2$

　　(c)　New mean radius $= 3.713 \times 10^{-6} \text{ m}$

　　　　New total surface area $= 4.04 \text{ m}^2$

　　　　The smaller the particles in a fixed mass of collected material, the greater the surface area.

3　(a)　$3.55 \times 10^{-19} \text{ J}$

　　(b)　240 nm

4　$1.12 \times 10^{-6} \text{ m}$

5　$2.49 \times 10^{-7} \text{ m}$

　　$7.15 \times 10^{-6} \text{ m}$

6　(a)　Before $[H_3O^+(aq)] = 1.0 \times 10^{-4} \text{ mol dm}^{-3}$

　　　　After $[H_3O^+(aq)] = 1.0 \times 10^{-6} \text{ mol dm}^{-3}$

　　　　Amount of hydrogen ions in lake before $= 5 \times 10^6 \text{ mol}$

　　　　Amount of hydrogen ions in lake after $= 5 \times 10^4 \text{ mol}$

　　　　Amount to be neutralised is $4.95 \times 10^6 \text{ mol}$

　　(b)　Number of moles of limestone $= 2.48 \times 10^6$ and the mass $= 2.48 \times 10^5 \text{ kg}$

Chapter 11

3　(a)　$4.18 \times 10^8 \text{ J}$　　(b)　$2.26 \times 10^{10} \text{ J}$　　(c)　$3.96 \times 10^4 \text{ J}$

4　(a)　$4.61 \times 10^4 \text{ Pa}$　　(b)　$23.3 \times 10^2 \text{ Pa}$　　(c)　$250 \times 10^2 \text{ Pa}$

5　(a)　pH $= 5.04$　　(b)　pH $= 8.92$

6　pH $= 6.0$

7　(a)　$7.36 \times 10^{-11} \text{ mol dm}^{-3}$

　　(b)　$K_{sp} = 8.5 \times 10^{-40} \text{ mol}^3 \text{ dm}^{-9}$ and $pK_{sp} = 44.1$

Chapter 12

3　$15\,000 \text{ m}^3 \text{ d}^{-1}$

5　Day 2 0.287 g N s^{-1}　　　Day 2 0.403 g N s^{-1}

6　(a)　$3.22 \times 10^9 \text{ m}^3$　　(b)　$6\,708 \text{ dm}^3$

Chapter 14

1　(a)　$7.88 \times 10^{17} \text{ J}$　　(b)　$2.63 \times 10^{18} \text{ J}$　　(c)　1.1×10^8 tonne or 4.61×10^8 barrels

4　$7.679 \times 10^{13} \text{ J}$

　　$4.798 \times 10^{32} \text{ eV}$

　　$2.133 \times 10^7 \text{ kWh}$

 # Glossary

acid

An acid is a substance that can donate one or more protons (hydrogen ions H^+).

acid rain

Acid rain is rainfall with a pH of less than 5.6. It is caused by natural events as well as the burning of fossil fuels. The main culprits are SO_2 and NO_x.

adsorption

Not to be confused with absorption. Adsorption is the interaction between gases and liquids and the surfaces of solids or liquids, i.e. there is a concentration and adherence of one material to the surface of another. Desorption is the reversal of adsorption. Absorption involves the penetration of one material by another and is often accompanied by adsorption.

aerobic

Conditions that involve dioxygen.

algae

A large group of essentially aquatic plants found in both fresh and salt water. They range in size from microscopic plants that form green scums on ponds to the huge brown kelps more than 45 m long.

alveoli

Sac-like structures one cell thick that protrude from the end of a bronchiole. Each contains a network of capillary blood vessels that permit the exchange of gases between the blood and the air.

amino acids

Organic compounds containing carbon, hydrogen, oxygen and nitrogen. They contain at least one carboxyl group (COOH) and at least one amino group (NH_2). These acids are of great biological significance because they link together to produce proteins.

anaerobic

Conditions in which dioxygen is absent.

anion

Negatively charged species, e.g. Cl^-, SO_4^{2-}. If these species are able to move under the influence of an electrical potential difference, i.e. are in solution or molten, then they will move towards the positive electrode or anode.

aquifer

Sandstone or limestone rock that, because of its porosity and permeability, can store water in recoverable amounts.

aromatic compounds

Aromatic compounds contain one or more benzene rings. The benzene ring has a particular stability, enabling the cyclic hexagonal structure of carbon atoms to be maintained during chemical reactions.

autotrophic

Denotes a living organism that can produce organic molecules from inorganic ones. Most do this via photosynthesis using the energy provided by sunlight. Some can obtain the required chemical energy from inorganic reactions – these are termed chemoautotrophs.

Avogadro's constant/number

The number of atoms in exactly 12.000 g of the isotope carbon-12. It has the value 6.023×10^{23}. The number is also called a mole of carbon atoms.

bacteria

Simple unicellular microscopic organisms, usually classified in the plant kingdom. They lack a clearly defined nucleus and most do not contain chlorophyll. There are three typical forms, rod-shaped (bacillus), round (coccus) and spiral (spirillium). They are a major source of disease in humans but many are harmless and beneficial.

bar

A bar is the air pressure which will support a column of mercury 75.007 cm high. Normal atmospheric pressure is about 1.01325 bars or 1013.25 millibars.

base

A base is a substance that can accept one or more protons.

becquerel

The degree of activity of a radioactive source. It is one disintegration per second.

biodegradable

Capable of being broken down by the action of bacteria.

biomagnification

This is the increase in the concentration of a chemical that can occur in a living organism when the rate of accumulation is greater than that of excretion. The amount of a particular chemical in an organism can thus be greater than that of its environment. If the organism forms part of a food chain then the chemical can be concentrated further up that chain.

biosphere

The name given to that part of the Earth's crust and surrounding atmosphere that includes all living organisms, both animal and plant.

biota

All living matter.

bronchi

Large air passages in a lung. Each lung has one main bronchus which originates from the end of the windpipe (trachea). This bronchus divides into smaller branches which then subdivides into bronchioles.

buffers

Solutions that undergo only small changes in pH upon the addition of small quantities of acids and bases.

carbohydrate
An organic compound composed of carbon, hydrogen and oxygen. Carbohydrates are a component of many different kinds of food. Examples of carbohydrates are the sugars such as glucose and fructose.

carcinogen
Any agent that causes cancer in humans, e.g. tobacco smoke, high-energy electromagnetic radiation or asbestos fibres. Chemicals form the largest group of cancer-forming agents. Polycyclic aromatic hydrocarbons (PAHs), which occur in tobacco, smoke, soot, pitch and tar, are amongst the most dangerous.

catalyst
A substance that can speed up or slow down a chemical reaction without itself being consumed. It enables an alternative pathway for a reaction to proceed which involves a lower activation energy barrier. The amount of product is not increased but the time taken to achieve equilibrium is shortened or lengthened.

cations
Positively charged species, e.g. Na^+, Cu^{2+}. If these species are able to move under the influence of an electrical potential difference, i.e. are in solution or molten, then they will move towards the negative electrode or cathode.

cation exchange capacity
The number of exchangeable adsorbed cations expressed as milliequivalents per 100 g of oven-dried soil at a specified pH. The milliequivalent is 1 millimole of H^+ ions or the amount of other cations that will displace it. It is a good indicator of a soil's fertility. The adsorption sites are found mainly on clays and organic matter.

cilia
Protruding hair-like filaments which beat in unison to create currents of liquid over the surface of cells.

complex ion/molecule
A central metal ion or atom that is surrounded by ions or molecules capable of donating a lone pair of electrons. The bonds formed are called co-ordinate or dative covalent bonds. The donating species are called ligands. Molecules or ions capable of donating more than one pair of electrons are chelating agents.

covalent bond
A type of chemical bond that is formed by the sharing of electrons between atoms. Each single covalent bond contains two electrons. Because the sharing is directional, this gives rise to different geometries/shapes of molecules. More than one covalent bond can be formed between certain elements.

crystal lattice
A regular three-dimensional array of ions, atoms or molecules. These particles are often represented as points. There is a repeating unit cell from which the whole crystalline structure can be built, e.g. face-centred cubic lattice.

desorption *see* **absorption**

electronegativity
The power of an atom in a molecule to attract electrons to itself. Metals have a low electronegativity whilst non-metals have high electronegativity. Water is a liquid at room temperature primarily because of the difference in electronegativity between the oxygen atom and the hydrogen atom.

electronic configuration

The arrangement of electrons in atoms, ions or molecules. In particular, it is this arrangement that helps to explain the bonding types between atoms.

endocrine system

In humans the system of ductless glands that secrete hormones directly into the blood stream. These glands are not structurally connected to each other. An example is the thyroid gland located in the throat, which secretes thyroxin, a growth hormone.

endothermic

Denotes the absorption of heat. The sign associated with an endothermic change is positive.

enthalpy change, ΔH

The change in heat at constant pressure.

entropy, S

A measure of the 'unavailable' energy in a system or the likelihood of a particular state of a system existing. In any changing, isolated system the associated change of entropy is always positive. The energy used in achieving this is not available to do useful work.

enzymes

Proteins which act as specific biological catalysts. They are suffixed by the ending -*ase*. Thus enzymes that catalyse hydrolytic reactions are termed *hydrolases*.

epidemiology

The study of disease, such as cholera, plague, influenza, in a population. The population is carefully counted and each person is defined in terms of age, sex, occupation, etc. The incidence (number of cases in a given time) and prevalence (number of people who are suffering at any given time) of the diseases are determined.

equilibrium

The point in any change at which no further change is seen to take place. In chemical reactions equilibrium is achieved when the concentrations of remaining reactants and the products remain unchanged. An equilibrium constant can be measured which shows the relationships between these concentrations at a particular temperature. The larger this constant, the more products are present at equilibrium than reactants.

evaporation

The process by which a liquid turns into a vapour. It occurs from the surface of the liquid at all temperatures up to and including the boiling point.

fats

Organic compounds containing carbon, hydrogen and oxygen, but the proportion of oxygen present is very low when compared with carbohydrates. They are oily, greasy, waxy substances which when pure are normally tasteless, colourless and odourless.

fatty acids

Fatty acids are straight chains of carbon atoms with some side chains. They contain a single carboxyl group. They can be produced by reacting water with fats under alkaline conditions. Acetic acid, found in vinegar, is one of the simplest fatty acids.

food chain

A feeding order. At the bottom are the green plants (producers) followed by the herbivores (primary consumers) followed by the carnivores (secondary consumers). At every stage, including the end of the chain, bacteria and enzymes act to break down waste and dead matter into forms that can be absorbed by plants, thus perpetuating the chain.

free radical

A chemical species that contains an unpaired electron. These species are very reactive.

fungi

A wide variety of plants that cannot make their own food by photosynthesis. They include mushrooms, moulds, truffles, smuts and yeast.

gall bladder

A muscular sac found in the abdomen which stores and concentrates bile which it receives from the liver. Bile is used in both digestion and excretion processes.

Gibbs' Free Energy change, ΔG

The Gibbs' Free Energy change is a measure of the energy that is available to do work once any entropy changes have been accounted for. A large negative value indicates a spontaneous thermodynamically favourable reaction with a large equilibrium constant. However, a reaction may be spontaneous in terms of thermodynamics, but there may be an energy (kinetic) barrier that would prevent reaction.

halons

Organic compounds containing one or two carbon atoms together with fluorine, chlorine, bromine or iodine. They are gases and were widely used in fire extinguishers until banned in 1994. One example is halon-1211 or bromochlorodifluoromethane, $CBrClF_2$.

heavy metals

The general term used for metals having a density greater than 6 g cm^{-3}. The common ones often associated with environmental problems are cadmium, chromium, copper, mercury, nickel, lead and zinc.

heterotroph

Living organisms that cannot manufacture their own energy sources, but need to feed and obtain their energy from other organisms. Heterotrophs that feed on other living organisms are called *consumers*, whilst those that feed on dead organisms are called *decomposers*.

hydrocarbon

Organic compounds composed of the elements carbon and hydrogen only. When burned in a plentiful supply of air such compounds provide a lot of energy and produce carbon dioxide and water vapour. Incomplete combustion can yield toxic carbon monoxide and unburned hydrocarbons, and soot.

igneous rocks

Rocks that have been formed by the cooling of molten magma found deep within the Earth.

immune system

A collection of cells and proteins that work to protect the body from potentially harmful infections from micro-organisms such as bacteria and viruses. The system also plays an important role in the control of cancer and is responsible for allergic reactions and rejection problems in surgical transplant operations.

inorganic materials

Materials which do not include carbon in their composition. The exceptions to this rule are the carbonates, carbides and oxides of carbon.

isotopes

Atoms with the same number of protons but different numbers of neutrons in their nuclei. Their chemical properties are the same but there are some differences in rates of reaction and physical properties.

leaching

The removal of soluble substances from soils by percolating water.

metamorphic rocks

Rocks that have had their physical nature changed by intense heat and/or pressure.

mole

The Avogadro number of any material contains 6.024×10^{23} stated particles of that material. This number is known as a mole.

molecule

A combination of atoms in a fixed ratio joined by chemical bonds. A molecule does not carry an electrical charge and is therefore neutral. An element composed of molecules contains the atoms of the same atomic number. A molecule that contains atoms of different atomic number is a compound.

monomer

The single unit from which a polymer is built, e.g. amino acid units link together to form proteins.

non-polar

Denotes covalent molecules which show no uneven distribution of electrons that may give rise to partial charges at different ends of the molecule. The solvent water is a polar solvent because it contains polar molecules, whereas hexane is a non-polar solvent.

oesophagus

The muscular tube which carries swallowed food from the throat to the stomach. In human adults it is about 30 cm long.

ore

A mineral or combination of minerals from which metals or non-metals can be profitably extracted.

organic compounds

Compounds that contain carbon, usually hydrogen, and other elements such as nitrogen, phosphorus, chlorine and oxygen.

oxidation

This is part of an overall chemical reaction called a redox reaction. It is the removal of one or more electrons, or the addition of oxygen.

ozone

A very close relative of oxygen. It has the formula O_3 and can decompose readily to form O_2 and free oxygen atoms. It is the latter which can make ozone dangerous to living organisms. It is a vital component of the upper atmosphere, where it protects the Earth's life forms from the effects of harmful solar radiation.

pancreas

An elongated soft gland, which lies slightly to the left behind the stomach. Its function is to provide enzymes, which help in digestion, and the hormone insulin, which helps to control blood sugar levels.

pathogen

A micro-organism or substance that causes disease.

pH scale

A logarithmic scale devised to describe how acidic a substance is. Acidic substances have a pH below 7.00, alkaline substances have a pH above 7.00, and neutral substances have a pH of 7.00.

phase

A homogeneous region in a system. One way of classifying phases is solids, liquids and gases.

photosynthesis

The chemical processes occurring in green plants and other organisms, where carbon dioxide, water, sunlight and chlorophyll produce oxygen and glucose.

precipitation (atmospheric)

Moisture falling on to the Earth's surface from clouds, i.e. rain, hail or snow.

precipitation (chemical)

The formation of an insoluble solid by a chemical reaction between two or more solutions.

proteins

Organic compounds containing carbon, hydrogen and nitrogen. They consist of hundreds of amino acid 'building blocks'. About 20 different amino acids can occur in proteins.

protozoa

Unicellular organisms found extensively in marine and fresh water, either free living or as parasites. They have the ability to move and some contain chlorophyll.

redox reaction

An overall chemical reaction that involves simultaneous oxidation and reduction reactions.

reduction

The addition of one or more electrons, or the addition of hydrogen.

saline

Usually taken to mean the presence of the salts of potassium, sodium and magnesium.

sedimentary rocks

Rocks that have been formed by the compression and cementing together of minerals and organic particles. These particles were formed by erosion at one or more places and deposited by wind, water and glacial ice or precipitated from solution at another place.

sievert

The SI unit of the effective dose equivalent of radiation absorbed by living tissues.

stereoisomerism

Two or more forms of a compound of the same molecular formulae, but whose atoms are arranged differently in three-dimensional space. A different spatial arrangement about a plane will result in geometrical isomerism, whilst that about an asymmetric centre optical isomerism.

synergic

It is possible that the effects of two or more chemicals on a living organism can be greater than their individual effects. That is, chemicals can mutually enhance their effects. Such an effect is called a synergic effect.

trachea

Windpipe. It begins immediately below the voice box (larynx) and runs down the centre of the front of the neck to end behind the upper part of the breastbone (sternum). Here it divides to form the two main bronchi.

volatile

Changes readily to a vapour.

Index

ESSENTIAL READING

Environmental Physics

Clare Smith

Hb: 0–415–20190–X
Pb: 0–415–20191–8   **Routledge**

Using Statistics to Understand the Environment

Penny A. Cook and C. Phillip Wheater

Hb: 0–415–19887–9
Pb: 0–415–19888–7   **Routledge**

Environmental Biology

Allan Jones

Hb: 0–415–13620–2
Pb: 0–415–13621–0   **Routledge**

Natural Environmental Change

A. M. Mannion

Hb: 0–415–13932–5
Pb: 0–415–13933–3   **Routledge**

Biodiversity and Conservation

Mike Jeffries

Hb: 0–415–14904–5
Pb: 0–415–14905–3   **Routledge**

Information and ordering details
For price, availability and ordering visit our website www.tandf.co.uk
Subject Web Address: www.geographyarena.com
Alternatively our books are available from all good bookshops.